SpringerWienNewYork

Sponsored by the
European Association of Neurosurgical Societies

Advances and Technical Standards in Neurosurgery

Vol. 30

Edited by

J. D. Pickard, Cambridge (Editor-in-Chief),
N. Akalan, Ankara, C. Di Rocco, Roma,
V. V. Dolenc, Ljubljana, R. Fahlbusch, Erlangen,
J. Lobo Antunes, Lisbon, M. Sindou, Lyon,
N. de Tribolet, Lausanne, C. A. F. Tulleken, Utrecht

SpringerWienNewYork

With 40 partly coloured Figures

Printed in Austria

SpringerWienNewYork is a part of Springer Science Business Media
springeronline.com

Printed on acid-free and chlorine-free bleached paper

ISSN 0095-4829

ISBN 978-3-211-99874-8 e-ISBN 978-3-211-27208-4

Preface

As an addition to the European postgraduate training system for young neurosurgeons, we began to publish in 1974 this series of *Advances and Technical Standards in Neurosurgery* which was later sponsored by the European Association of Neurosurgical Societies.

This series was first discussed in 1972 at a combined meeting of the Italian and German Neurosurgical Societies in Taormina, the founding fathers of the series being Jean Brihaye, Bernard Pertuiset, Fritz Loew and Hugo Krayenbuhl. Thus were established the principles of European cooperation which have been born from the European spirit, flourished in the European Association, and have been associated throughout with this series.

The fact that the English language is now the international medium for communication at European scientific conferences is a great asset in terms of mutual understanding. Therefore we have decided to publish all contributions in English, regardless of the native language of the authors.

All contributions are submitted to the entire editorial board before publication of any volume for scrutiny and suggestions for revision.

Our series is not intended to compete with the publications of original scientific papers in other neurosurgical journals. Our intention is, rather, to present fields of neurosurgery and related areas in which important recent advances have been made. The contributions are written by specialists in the given fields and constitute the first part of each volume.

In the second part of each volume, we publish detailed descriptions of standard operative procedures and in depth reviews of established knowledge in all aspects of neurosurgery, furnished by experienced clinicians. This part is intended primarily to assist young neurosurgeons in their postgraduate training. However, we are convinced that it will also be useful to experienced, fully trained neurosurgeons.

We hope therefore that surgeons not only in Europe, but also throughout the world, will profit by this series of *Advances and Technical Standards in Neurosurgery*.

The Editors

Contents

Advances

Cranial Venous Outflow Obstruction and Pseudotumor Cerebri Syndrome. B. K. Owler[1,2], G. Parker[3], G. M. Halmagyi[1], I. H. Johnston[1], M. Besser[1,2], J. D. Pickard[4], and J. N. Higgins[5], [1] T. Y. Nelson Departments of Neurosurgery and Neurology, Institute of Clinical Neurosciences, Royal Prince Alfred Hospital, Sydney, Australia, [2] Department of Surgery, University of Sydney, Sydney, Australia, [3] Department of Radiology, Royal Prince Alfred Hospital, Sydney, Australia, [4] Department of Neurosurgery, Academic Neurosurgical Unit, University of Cambridge, Addenbrookes Hospital, Cambridge, UK, [5] Department of Neuroradiology, Addenbrookes Hospital, Cambridge, UK,

Technical Standards

Sacral Neuromodulation in Lower Urinary Tract Dysfunction. J. R. Vignes[1], M. De Seze[2], E. Dobremez[3], P. A. Joseph[2], and J. Guérin[1], [1] Department of Neurosurgery, Medical School Hospital, Hôpital Pellegrin, Bordeaux, France, [2] Neuro-rehabilitation Unit, Medical School Hospital, Hôpital Pellegrin, Bordeaux, France, [3] Department of Pediatric Urology, Medical School Hospital, Hôpital Pellegrin, Bordeaux, France

List of Contributors

Besser, M., T. Y. Nelson Departments of Neurosurgery and Neurology, Institute of Clinical Neurosciences, Royal Prince Alfred Hospital, Sydney, Australia

Chiaretti, A., Paediatric Intensive Care Unit, Catholic University Medical School, Rome, Italy

Dardis, R., Section of Neurosurgery, Department of Clinical Neurosciences, King's College, London, UK

De Seze, M., Neurorehabilitation Unit, Medical School Hospital, Hôpital Pellegrin, Bordeaux Cedex, France

Dobremez, E., Department of Pediatric Urology, Medical School Hospital, Hôpital Pellegrin, Bordeaux Cedex, France

Guérin, J., Department of Neurosurgery, Medical School Hospital, Hôpital Pellegrin, Bordeaux Cedex, France

Halmagyi, G. M., T. Y. Nelson Departments of Neurosurgery and Neurology, Institute of Clinical Neurosciences, Royal Prince Alfred Hospital, Sydney, Australia

Higgins, J. N., Department of Neuroradiology, Addenbrookes Hospital, Cambridge, UK

Hummel, T., Smell and Taste Clinic, Department of Otolaryngology, University of Dresden Medical School, Dresden, Germany

Johnston, I. H., T. Y. Nelson Departments of Neurosurgery and Neurology, Institute of Clinical Neurosciences, Royal Prince Alfred Hospital, Sydney, Australia

Joseph, P. A., Neurorehabilitation Unit, Medical School Hospital, Hôpital Pellegrin, Bordeaux Cedex, France

Lacroix, J.-S., Unité de Rhinologie-Olfactologie, Service d' Oto-Rhinologie-Laryngologie, Hopitaux Universitaires de Genève, Genève, Switzerland

Landis, B. N., Smell and Taste Clinic, Department of Otolaryngology, University of Dresden Medical School, Dresden, Germany

Langer, A., Paediatric Intensive Care Unit, Catholic University Medical School, Rome, Italy

Lopes da Silva, F. H., Section Neurobiology, Swammerdam Institute for Life Sciences, University of Amsterdam, Amsterdam, The Netherlands

Owler, B. K., Department of Surgery, University of Sydney, Sydney, Australia

Parker, G., Department of Radiology, Royal Prince Alfred Hospital, Sydney, Australia

Pickard, J. D., Department of Neurosurgery, Academic Neurosurgical Unit, University of Cambridge, Addenbrookes Hospital, Cambridge, UK

Strong, A. J., Section of Neurosurgery, Department of Clinical Neurosciences, King's College, London, UK

Vignes, J. R., Department of Neurosurgery, Medical School Hospital, Hôpital Pellegrin, Bordeaux Cedex, France

Advances

Offprint from
Advances and Technical Standards in Neurosurgery, Vol. 30
Edited by J.D. Pickard

Depolarisation Phenomena in Traumatic and Ischaemic Brain Injury

A. J. STRONG and R. DARDIS

Section of Neurosurgery, Department of Clinical Neurosciences, King's College, London, UK

With 11 Figures

Contents

Abbreviation List

ADC	Apparent diffusion coefficient
ATP	Adenosine triphosphate
Ca^{2+}	Calcium ion
CA1	The CA1 region of the hippocampus
Cl^-	Chloride ion
Cl^-_e	Extracellular chloride ion
CSD	Cortical spreading depression

CBF	Cerebral blood flow
DC	Direct current
ECoG	Electrocorticography
ECS	Extracellular space
Hb(O)	Haemoglobin (oxidised form)
HSP	Heat shock protein
HSD	Hypoxic spreading depression – like depolarisation
IEG	Immediate early gene
IL	Interleukin
IP3	Inositol trisphosphate
K^+	Potassium ion
K_e	Extracellular potassium ion
MCAO	Middle cerebral artery occlusion
mM	Millimoles per litre
mRNA	Messenger ribonucleic acid
mV	Millivolts
Na^+	Sodium ion
PID	Peri-infarct depolarisation
pO_2	The partial pressure of oxygen
NO	Nitric oxide
Na-K ATPase	Sodium-potassium ATPase
NIRS	Near infrared spectroscopy
nm	Nanometres
NAD(H)	Nicotinamide adenine dinucleotide (reduced form = NADH)
Vm	Neuronal membrane potential
BDNF	Brain derived neurotrophic factor
NF-κB	Nuclear factor kappa-B
NMDA	N-methyl-D-aspartate
TPA	Tissue plasminogen activator
TBI	Traumatic brain injury

History, Definitions and Introduction

In 1944 a young Brazilian physiologist, Aristides Leão, was studying for his doctorate in Harvard University. According to Somjen [1], he was attempting to study propagation of epileptic activity in the cerebral cortex, and he approached the problem by applying electrical stimulation to the frontal convexity cortex of anaesthetised rabbits, and recording from an array of corticography electrodes posterior to this (Fig. 1). Instead of seeing propagating epileptic activity, he observed a period of electrical silence, which was first seen adjacent to the stimulating electrodes, and did indeed propagate from the site of stimulation backwards along the cere-

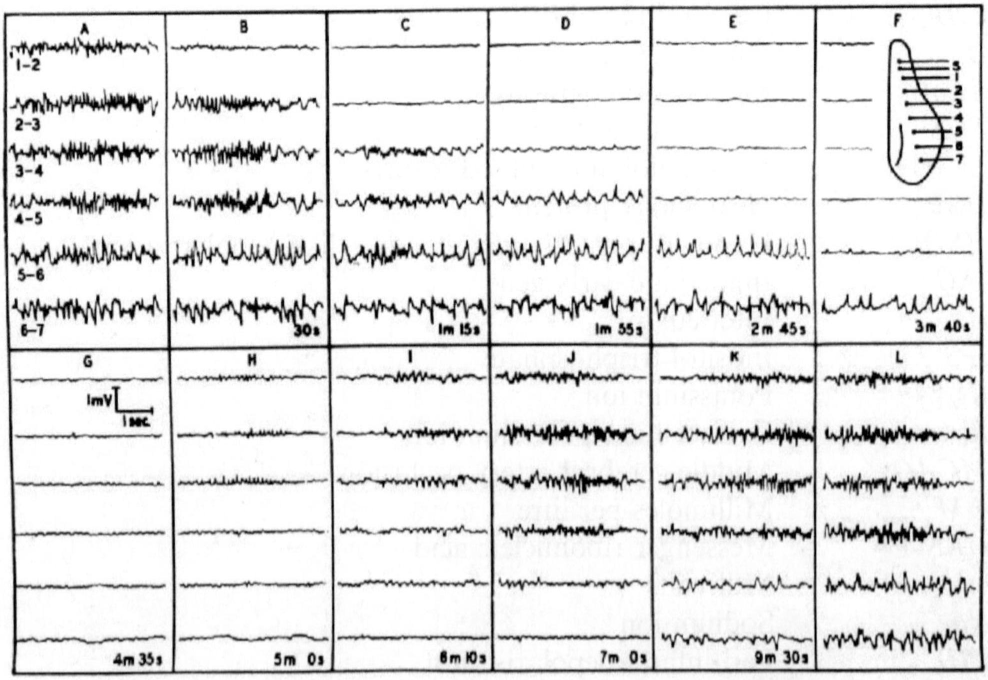

Leão, 1944

Fig. 1. Leão's original demonstration of cortical spreading depression, demonstrating a time sequence of twelve separate recordings spanning some 10–11 minutes, from a linear array of seven electro-corticographic (*ECoG*) electrodes extending antero-posteriorly over the right hemisphere of a rabbit anaesthetised with barbiturate. A pair of bipolar electrical stimulating electrodes are placed at the front of the hemisphere, and following stimulation, a wave of electrical silence is seen to propagate backwards from the site of stimulation, followed after approximately 7–9 minutes by spontaneous recovery at each site. (Reproduced with permission from Leão [2])

bral hemisphere – at a rate of some 3 millimetres per minute. The phenomenon resolved after 5–15 minutes, with – apparently – full resumption of cortical electrical activity. He reported his findings in a landmark paper entitled "Spreading depression of activity in the cerebral cortex" [2]. The event which he described became known as "spreading depression" or "cortical spreading depression" [of Leão] (CSD), and has remained a subject of intense interest to neurophysiologists. Although the electrophysiological and haemodynamic features have become very well characterised, with mass focal depolarisation of neurones and glia as the defining event, its most enigmatic challenges have remained its uncertain physiological role in grey matter, and its relevance – if any – to human disease states.

Since 1977–1978, stroke research laboratories have become aware of a feature of cerebral cortex in the ischaemic penumbra which shares certain characteristics with CSD, but also differs from it in critical aspects. "Peri-infarct depolarisations" (PIDs) arise spontaneously in cortex at the edge of the core ischaemic territory and propagate in the penumbra, but unlike CSD, they are harmful in that they cause progressive recruitment of the penumbra into the core territory, thus enlarging the infarct [3]. Somjen refers to such events as *hypoxic spreading depression-like depolarisations (HSD)* [1]. The evolution of this concept, and increasing awareness among some clinicians of its existence, has prompted increasing speculation as to whether CSD or PIDs occur in the injured human brain. Demonstrations of CSD-like events in models of traumatic brain injury, the imaging in the laboratory of propagation of PIDs across the cerebral cortex in models of focal cerebral ischaemia, the knowledge that not only cerebral cortex but also deep nuclei and the hippocampus may be subject to CSD, and particularly the recent confirmation that such events do indeed occur in patients with serious head injury [4], seem likely to open a fresh chapter in clinical brain injury research. This is an area of research to which neurosurgeons are uniquely placed to contribute.

The features of cortical spreading depression as it is observed in the experimental laboratory have been the subject of a number of authoritative reviews extending over many years, and the reader seeking the most detailed information is directed to them [1, 5–7]. We have relied extensively on these reviews as well as on the original sources. In this review, we shall draw together the principal physiological, chemical and haemodynamic features of CSD and PIDs, and consider their possible functions and effects in the context of acute ischaemic and traumatic injuries to the human brain. We shall also explore methods for detection of depolarisations in the injured human brain, and the actual and potential impact of this information on our understanding of the pathophysiology of the injured human brain and on our clinical management of traumatic and ischaemic brain injury. The broader term "depolarisation" will be used where neither CSD nor PID is specifically under discussion.

Cortical Spreading Depression

The "Onset" Phase of CSD

Initiation of CSD

Leão's observations were made in rabbits under barbiturate anaesthesia, and the stimulus to the cortex was bipolar electrical current delivered from an induction coil, but several other stimuli are also effective. Dialysis through an implanted microcatheter or superfusion of the exposed cortex with potassium chloride (KCl) at 130 mM or more is effective in the rat brain [8], as is local application of KCl with a wick. Neurosurgeons should also be aware that needling of the cortex is effective, and it seems inescapable that more complex surgical manipulations of similar, *susceptible* tissue are likely to be effective if, as seems clear from the recent findings in patients [4], CSD does indeed occur in the human brain. There is also experimental evidence that spreading depression occurs in the spinal cord [9]. What determines susceptibility, by which is meant the frequency of occurrence of CSD (rather than vulnerability to damage from depolarisations), is an important theme of this review. The factors which are currently believed to affect this are species differences, location in the brain, haemodynamic and metabolic conditions in the cortex, anaesthesia, and systemic metabolic variables (essentially – in the present state of knowledge – plasma glucose). All of these factors are best considered after we have first reviewed the basic electrophysiological, haemodynamic and metabolic properties of CSD.

The DC Potential Transient

For the purposes of a discussion focussed on brain injury, the CSD complex is best considered in its two phases, onset and recovery, since, as we shall see, it is probably deficiencies in the recovery process that underlie the differences between CSD and PIDs. When Leão measured the DC potential difference between a point on the cortex in the path of the propagating wave of depression and a remote reference point, he noted a transient negativity of some 10 to 15 mV. The observation has been repeated many times, and when sought, the DC potential transient is an invariable feature of both the CSD and PID patterns of depolarisation. The nature of the DC potential transient – presumably indicating a brief accumulation of negative charge in the cortex – is still unknown, although an increase in one or more anions, – lactate, amino-acids, or bicarbonate – has been suggested as a cause [10].

Mass Neuronal Activity: Grafstein – 1956

Studying areas of cerebral cortex isolated electrically by subpial transection but with perfusion intact, Grafstein recorded unit activity with an extracellular microelectrode, and found a short phase of intense firing at the onset of the DC potential change, followed by prolonged silence [11] (Fig. 2). There is no suggestion that this transient neuronal activity conveys any physiological information, and as we shall see it is initiated by local changes in the extracellular environment. The observation suggests that excitatory or depolarising influences on neurones – not necessarily synaptic – contribute to the initiation of the CSD event as it reaches a new locus. It is of interest for current researchers studying models of stroke that

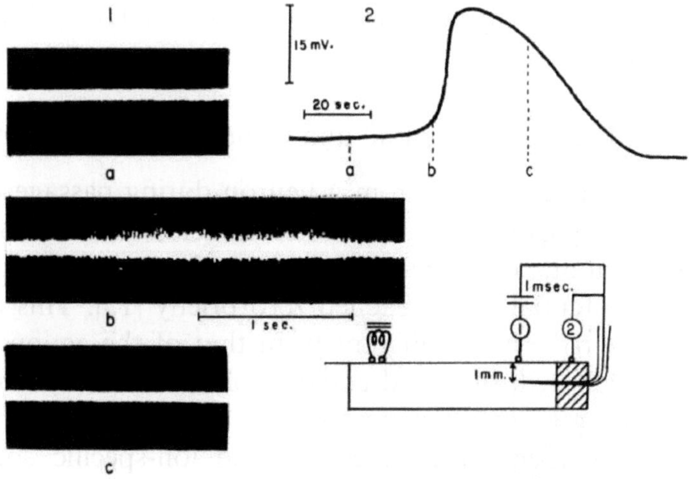

Fig. 2. Comparison of extracellular microelectrode recording (*Panel 1, a,b,c*) and simultaneous recording of DC potential (*Panel 2*) in a slab of cerebral cortex isolated electrically by subpial transections (Grafstein, 1956). The cortex was stimulated remote from the recording electrodes, initiating a wave of spreading depression. In 1a, no spontaneous neuronal firing is present prior to arrival of the wave, but as depolarisation commences there is a *brief* phase of intense neuronal firing (*b*) followed by silence (*c*) when depolarisation has begun to recover. (Reproduced with permission from Grafstein [11])

Grafstein was able to suspend and then restore resolution of the DC potential transient by first occluding and then releasing the middle cerebral artery (MCAO) in rabbits, showing that resolution of the DC potential is energy-dependent. Her proposal that potassium ion liberated by neuronal depolarisation caused subsequent depolarisation of adjacent neurons still forms the basis of current thinking on mechanisms of CSD propagation (see below).

Thus Grafstein's 1956 paper demonstrated or inferred three of the key features of CSD – mass neuronal depolarisation, mediation by potassium ion, at least in part, and the dependence of recovery on availability of perfusion and energy. The findings and inferences have remained substantially unchallenged, and form the foundation of our understanding of depolarisation events in the cerebral cortex; the paper is perhaps one of the key contributions to neuroscience in the past 50 years.

Changes in Extracellular Ion Concentrations $[K^+]_e$, $[Na^+]_e$, $[Cl^-]_e$, $[Ca^{2+}]_e$

A transient, marked increase in K_e from the normal 3 mM to 60 mM or more is a striking and regular feature of CSD, and lasts for approximately 30–40 seconds in total, often resolving with an undershoot below the baseline [12]. There are accompanying decreases in $[Na_e]$ [13], in $[Cl^-]_e$ and in $[Ca^{2+}]_e$ [14].

Changes in Membrane Potential and Conductance During CSD

The first intracellular recordings from a neuron during passage of a wave of CSD were made by Collewijn and Van Harreveld, who concluded, after allowing for the simultaneous change in extracellular potential, that neuronal membrane potential (V_m) reached zero briefly [15]. This pattern of depolarisation to zero volts is different from that of the action potential, where V_m may reach +20 mV, and could imply simultaneous opening of several or all membrane conductances in CSD; this might represent mechanical opening of membrane pores with no ion-specific conductance properties, but according to Somjen it is not necessary to postulate such special channels in order to explain the membrane potential changes in CSD [1].

Redistribution of Water: Tissue Impedance

An increase in electrical impedance of tissue is largely a measure of cell swelling, and Leão and Martins-Ferreira demonstrated an increase in impedance during CSD in 1953 [16]. Measurements of extracellular space volume using indicators such as tetramethylammonium, together with morphological evidence, support this, and the basis most probably lies in the excess of the decrease in $[Na^+]_e$ over the increase in $[K^+]_e$ [13]. This would imply a net movement of ions into cells, accompanied by osmotically obliged water, in turn raising impedance to current flow in the ECS. However, Somjen points out [1] that there is evidence that when impedance is measured some current flow is through rather than around cell membranes, perhaps more especially glia, and that there is also evidence of a marked drop in neuronal membrane resistance during SD [17, 18]. Whatever their precise nature, these changes in CSD appear closely related to the transient reduction in apparent diffusion coefficient (ADC) that can be detected in the rat [19] and cat [20] brains during CSD using magnetic resonance diffusion-weighted imaging (see also page 33: Section on Occlusive Stroke).

Mode of Propagation of CSD

Early experimental studies of propagation of spreading depression were aided by the use, principally by Martins-Ferreira and Oliveira Castro [21], of the isolated chick retina, in which the presence and propagation of spreading depression is evident to the naked eye from a transient change in optical properties. They were able to establish a "ring" of retina in which the phenomenon could be constrained to propagate in circular fashion, at a rate measured at 3.7 mm/minute – similar to that originally described by Leão, and they found that alkaline conditions, or increased K_e or Cl_e, all accelerated propagation, whereas acidification or an increase in Mg_e slowed it.

In one of her 1956 papers [11], and noting the likelihood of neuronal depolarisation (as the basis for the brief phase of spontaneous spike discharges), Grafstein suggested that the resulting liberation of *potassium* ion into the ECS could occur in sufficient concentration to cause adjacent neurones to depolarise, thus causing – or at least supporting – propagation. The simultaneous reduction in ECS volume (see above) would increase the effective $[K]_e$, thus facilitating depolarisation of neurones in the path of the wave.

The *separate* idea that potassium ion might diffuse slightly further in the ECS and cause depolarisation in *non*-contiguous neurones was explored in detail by Gardner-Medwin, who determined a rate for cortical extracellular diffusion of K^+, and showed that this was slower than that of CSD propagation [22]. A further argument against extracellular diffusion of K_e as the basis of propagation is that in CSD, no increase in K_e can be recorded in the cortex prior to the DC depolarisation (unlike PIDs, where a gradual, prior increase in K_e *does* occur [23, 24].

A second candidate agent explaining propagation is glutamate released into the extracellular space (ECS) by mass neuronal depolarisation, and in turn depolarising adjacent neurons. Van Harreveld induced CSD by application of compounds in brain extracts, one of which was glutamate [25], and he and Fifkova later demonstrated release of glutamate during CSD in the retina [26]. However, glutamate dialysed into the cortical ECS does not elicit CSD, nor does inhibition of glutamate reuptake [8, 27].

Propagation of CSD via Glial and/or Neuronal Gap Junctions

The possible roles of intercellular coupling either of neurones or of astrocytes in initiation and propagation of CSD have received much attention in the last few years. In the case of astrocytes, it is now abundantly clear that in cultures of astrocytes studied with intracellular calcium-sensitive dyes, waves of transient increase in intracellular calcium ion (Ca_i) can be initiated – by glutamate [28], nitric oxide (NO) [29] or mechanical stimulation [30] – and will then propagate across the culture at a rate very similar to that of CSD in the intact cortex [6]. Nedergaard has shown that in mixed glia-neuronal cultures, such glial waves are associated with elevations in neuronal calcium concentrations [31]. Transmission of calcium waves through glial cultures is believed to occur through glial gap junctions – specialised and specific membrane openings whose molecular structure is now well-characterised and which are usually readily permeable to ions and compounds of smaller molecular weight; examples are inositol trisphosphate (IP_3) and potassium. IP_3 is thought to mediate propagation of Ca_i waves through its role as a ligand for IP_3-receptor-Ca-conductance complexes on the endoplasmic reticulum, and glial gap junctions are also thus a probable substrate for the mechanism of "spatial buffering" of increases in K_e, as envisaged by Somjen [32]. Propagation is also mediated by an extracellular agent, ATP [33].

At least in the cell culture preparations in which glial communication has been studied, the capacity to propagate Ca_i waves seems exceptionally well supported, by a range of agents that include ATP [33], nitric oxide [29], and inositol trisphosphate (IP_3), the latter via glial gap junctions [34]. The demonstration that an intracellular calcium wave *precedes* the arrival of spreading depression [35] also lends support to the idea that CSD propagation is mediated primarily by glia. A further argument for the concept is based on the fact that halothane, which blocks glial gap junctions [36], also reduces the frequency of CSD in the gyrencephalic brain [37], and reduces MCAO infarct volume and PID frequency by an effect either on perfusion or on intrinsic PID susceptibility [38].

Other findings argue against this hypothesis. First, CSD is more readily elicited in areas of grey matter with relatively *lower* glia : neuron ratio, such as the CA1 layer of the hippocampus (in experimental studies) [39], and the occipital cortex in humans [40, 41] (if it is accepted that migraine with visual aura is a manifestation of CSD, as discussed below). Secondly, the use of specific agents toxic to glia such as fluorocitrate or fluoroacetate fails to prevent CSD [42, 43]. Third, CSD can occur in the absence of Ca_i waves [44].

The Recovery Phase of CSD, and the Responses of Cerebral Metabolism and Blood Flow to CSD

Resolution of the cation transients might in theory be due either to restitution of normal, resting distributions by active transport, or in the case of the increased $[K]_e$, to diffusion through the extracellular space (which would necessarily be slower than the observed resolution rate [22]), to spatial buffering by the astrocytes through gap junctions [32], or to passive elution through cerebral perfusion (probable only under conditions of energy failure [45]). Grafstein's experiment with MCAO described above is perhaps the earliest evidence for a role for energy-dependent active transport in the recovery phase, and evidence for the concept has steadily accumulated. Demonstration of the cation transients that are an integral feature of CSD makes it almost inevitable that restoration of resting cation distributions should necessitate a considerable increase in ATP utilisation.

Indeed CSD, and epileptic seizures, are perhaps the most extreme forms of activation challenge to reactivity of cerebral metabolism and blood flow (CBF).

Detailed studies by Rosenthal & Somjen and their colleagues of CSD in the normally perfused brain indicated transient oxidation of the mitochondrial respiratory chain [46]. In the light of subsequent work demonstrating transient *increases* in perfusion [47] and in tissue pO_2 during CSD [48], one simple interpretation of Rosenthal's work is that the redox potentials of the respiratory chain coenzymes are in equilibrium, and are determined by the balance between the rate of ATP hydrolysis and availability to mitochondria of molecular oxygen from cerebral perfusion.

Glucose Utilisation During Recovery from CSD

Studies of normal, functional activation in the human brain using positron emission transverse tomography [49] indicated for the first time that the rate of glucose utilisation increased in greater proportion to oxygen utilisation, suggesting upregulation of glycolysis rather than of oxidative glucose utilisation. The finding of transient increases in brain lactate of some 30% in experimental studies of somatosensory activation [50] supported this interpretation, and suggested a degree of dependence on glycolytic generation of ATP during activation. The very large cation shifts that occur in CSD make it highly likely that similar and greater – but still transient – changes in glycolysis would occur during repolarisation after CSD. However, an extracellular lactate transient need not necessarily mean a shift to anaerobic metabolism, and Back and colleagues showed that in the normally perfused brain CSD is accompanied by an *in*crease in partial tissue pressure of oxygen [48]; this may be attributed to the hyperaemic response to CSD which is described below.

The model of the cerebral metabolic response to activation developed by Magistretti and colleagues [51] envisages that glycolytic activity is predominantly in the astrocytic compartment (where almost all glycogen in the brain is held [52, 53]), stimulated by an increase in extracellular glutamate during functional activation. It is further proposed that astrocytes deliver lactate to neurons, which, relying on lactate dehydrogenase activity in reverse, convert lactate to pyruvate. This pyruvate is then metabolised via the tricarboxylic acid cycle. Glucose transport across the blood brain barrier is highly efficient, to the extent that total unidirectional flux into the brain under non-activated conditions is approximately twice the rate of utilisation by glycolysis [54]. This, allied with the hyperaemic response to CSD discussed below, endows the cortex with its capacity to meet the challenges of activation. It is not appropriate to pursue further this important topic in this context, and the reader is referred to work by Magistretti and colleagues [55], to a review questioning some aspects of this "compartmented glial glycolysis" model [56], and to the review by Chen and Swanson of astrocytic function and changes in brain injury [57]. Changes in glucose metabolism in focal ischaemia and during PIDs are described later in this review.

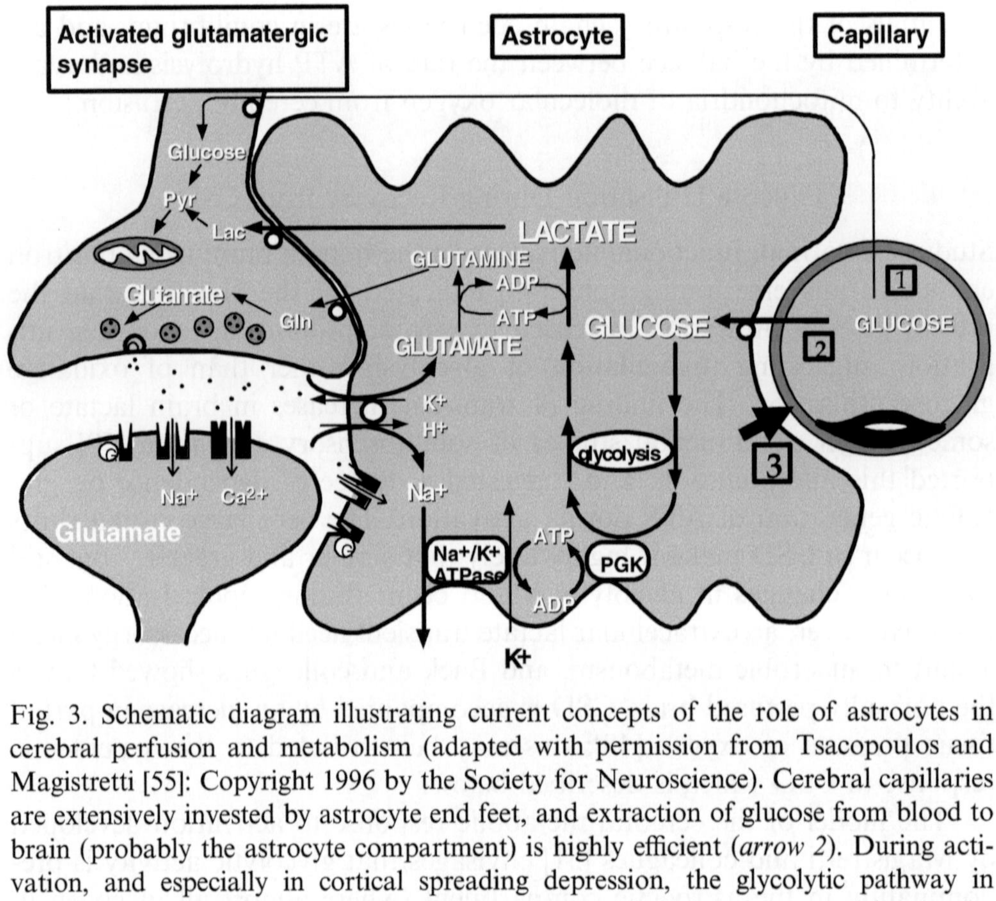

Fig. 3. Schematic diagram illustrating current concepts of the role of astrocytes in cerebral perfusion and metabolism (adapted with permission from Tsacopoulos and Magistretti [55]: Copyright 1996 by the Society for Neuroscience). Cerebral capillaries are extensively invested by astrocyte end feet, and extraction of glucose from blood to brain (probably the astrocyte compartment) is highly efficient (*arrow 2*). During activation, and especially in cortical spreading depression, the glycolytic pathway in astrocytes is upregulated, and the different kinetics of glial and neuronal lactate dehydrogenases favour net movement of lactate from astrocytes to neurons; the (limited) brain glycogen pool is located in astrocytes. Under resting conditions, glycolysis in neurons may be sufficient to meet energy demands (*arrow 1*). Neurotransmitter glutamate released into the synaptic cleft is re-accumulated into astrocytes by high-affinity cotransport with Na^+ ion, making use of the normal electrochemical gradient generated by Na^+/K^+ ATPase.

Several mechanisms regulate cerebral perfusion, with a prominent role proposed for astrocytes (*arrow 3*) [140]. First, their high membrane conductance for K^+ allows astrocytes to buffer the increased extracellular levels resulting from activation, with a direct vasodilator effect of K^+ on the microcirculation via the astrocyte cytosol and end feet. Adenosine- and nitric oxide-based mechanisms also contribute. Recent work by Zonta *et al.* now supports an additional mechanism of astrocyte-mediated vasodilation during activation [141].

Haemodynamic Response

Leão himself was the first to demonstrate hyperaemia in association with CSD; he observed a doubling in width of pial surface arterioles during CSD [58]. If CSD induced in the prefrontal region of the rat is assumed to propagate anteroposteriorly in the cerebral hemisphere at a constant rate, serial coronal sectioning of the hemisphere after it has been frozen at a single time point will provide in the section sequence a time series of the response of the brain to the propagation wave. Using autoradiography for CBF, and reasoning in this way, Lauritzen *et al.* showed that CSD is closely followed by an intense (>200%) but brief (2 minutes) transient hyperaemia [47]. An extended phase of mild hypoperfusion (80–90% control) follows, lasting for some 60 minutes. This feature of CSD was later used by the same group to allow mapping with isotope scanning of a phase of hypoperfusion associated with migraine with aura that propagated forwards in the cerebral hemisphere at a rate in accordance with that of CSD – a finding that argues quite strongly for CSD as the basis of migraine with aura [59].

Histology of the Cortex Following CSD

A careful histological study by Nedergaard & Hansen [60] found no evidence of classical ischaemic pathological changes in the cortex following CSD *in the normally perfused cortex* of rats. As will be described later, the situation is very different in focal ischaemia.

Molecular Responses to CSD

Expression after induction of CSD of some of the immediate early genes (IEG) that respond to stress has been studied extensively, principally in rats, mice and transgenic mice. The IEG responses to MCAO have also been studied. In many such MCAO studies, increases in gene expression extend to the whole hemisphere rather than remaining within the core and penumbral regions. It is generally believed that such widespread upregulation represents a response to a depolarisation event that started as a PID in the ischaemic territory but then propagated throughout the rest of the hemisphere as CSD. According to Sharp *et al.* [61] this applies to c-fos and jun-B. Cyclooxygenase-2 is also induced by CSD [62]. In some cases, the association is relatively specific: for example, the degree of induction of the mRNAs encoding brain-derived neurotrophic factor and heat-shock protein-72 in response to CSD induced in the rat is dependent on the number of CSDs [63]. It needs to be stated that in MCAO other gene expression patterns may relate more to cell damage than to CSD. Thus HSP70, a heat shock protein, behaves as a protein chaperone, increasing in the presence of denatured proteins [64], although expression in the infarct core may be limited by ATP depletion [65].

CSD as an Initiator of Inflammation

That cerebral ischaemia causes an increase in levels of interleukin-1β (IL-1β: an inflammatory cytokine) in the brain is well established [66–68]. CSD has a similar effect: Jander and colleagues recently showed that mRNA levels for IL-1β and tumour necrosis factor-α (TNF-α, also an inflammatory cytokine) are increased 24- and 60-fold respectively 4 hours after CSD induction with KCl [69]. Expression of the IL-1β protein was largely confined to microglia in the superficial cortical layers. These authors suggest that "cytokine expression following CSD forms part of a physiological stress response that contributes to the development of ischaemic tolerance in this and other preconditioning paradigms" (see below). That IL-1β can promote CNS repair has also been shown [70]. Another view of the effects of IL-1β comes from the work of Blamire *et al.*, who examined the effects of microinjection of recombinant IL-1β into the striatum of 3-week-old rats, and found significant reduction in apparent diffusion coefficient (ADC) and increases in cerebral blood volume and blood brain barrier permeability [71]. ADC reductions are usually attributed to a shift of water from extra- to intracellular compartments, but a reduction in water mobility in the intracellular compartment may occur [72]; both explanations are in keeping with an adverse effect of IL-1β.

Pre-Ischaemic Conditioning with CSD as Protection in Experimental Stroke

In experimental studies of stroke in rats, it is possible to confer a degree of protection from the effects of a period of ischaemia by prior induction of CSDs [73]. Levels of mRNAs for FOS, BDNF, and tPA, are increased by preischaemic conditioning with CSD [74]. TNF-α and IL1-β are believed to contribute to increased tolerance of ischaemia [75, 76], and an antagonist to nuclear factor κ-B (NF κ-B) blocked NF κ-B activity and reduced the pre-conditioning effect [77].

It seems very likely that one or more of the currently identified expression cascades – or other(s) still to be detected, underlie the protective effect of preischaemic conditioning with CSD, and increasing understanding of the molecular response to CSD may in time allow us to identify which of the several genes upregulated by CSD is/are responsible for the protective effects of preconditioning, and so perhaps lead to novel therapy for cerebral ischaemia, or at least better protection of the brain when some degree of prospective risk exists.

Factors Determining Ease of Induction of CSD

Species Differences and Cytoarchitecture

It has long been clear that CSD is more readily induced – and its repetition maintained – in rats than in larger experimental animals [5], with primates seen as the most "resistant" group of species. However, it is certainly possible to induce CSD in the primate brain [78]. A specific attempt to compare PID frequency in cats and squirrel monkeys after MCAO showed that PIDs do indeed occur spontaneously in a primate species, but failed to confirm a species difference in frequency of PIDs because of wide variability within both species [79]. However, the results revealed a clear dependence of PID frequency on plasma glucose level: this is discussed below in the context of PIDs.

One of the most widely canvassed explanations for species differences starts with the observation that lissencephaly is characteristic of the CSD-prone species, whereas the more resistant brains are gyrencephalic. There are however also regional differences in susceptibility within the brain of a given species, with the hippocampus particularly liable to CSD, together with – in migraineurs-with-aura – the occipital cortex. A clue to the puzzle comes from consideration of the cytoarchitecture and glial:neuronal ratios of different brain regions and in the brains of different species. Thus neurons are particularly tightly packed, glia relatively sparse, and CSD frequent, in the CA1 layer of the hippocampus. Migraine with aura typically commences with a visual aura on or near the fixation point (although auras apparently arising from the somatosensory cortex also occur), and the glia:neuron ratio in the occipital cortex is lower than elsewhere in neocortex [40, 41].

Tower and Young compared the glia:neuron ratio with brain size in a group of mammals ranging from mice to whales and elephants, and, using a log:log plot, demonstrated a convincing hierarchy in which the glia:neuron ratio increases in proportion with brain size [80]. Primates are distributed appropriately for their brain size within this hierarchy, rather than all of them possessing a high glia:neuron ratio independent of brain size, as might be predicted on "evolutionary" grounds. The issue is of interest in relation to the discussion above on mechanisms of CSD propagation, and the relationship of CSD propensity with Tower and Young's hierarchy is more in keeping with a *homœostatic* role for glia in the context of CSD than with one in which they *propagate* CSD.

Spreading depression has also been observed in experiments on the spinal cord [9, 81], and the possibility therefore arises that perilesion depolarisations might contribute to the evolution of spinal cord damage – at least in grey matter – not only in trauma but also in vascular lesions.

Drugs and Anaesthetic Agents

The reduction of CSD frequency by agents known to block glial gap junctions such as halothane and propofol has been referred to above. The role of increased K_e in initiating CSD has been referred to above, and glutamate and other excitatory amino acid agonists can also effect this [25, 47, 82]; conversely, it is widely recognised that CSD and/or PID frequencies can also be reduced by the action of some excitatory amino acid antagonists, notably antagonists of the n-methyl-D-aspartate (NMDA) class of glutamate receptors [83–86]. Such mechanisms may operate in ischaemic and traumatic brain injury, although with no proven definitive therapeutic benefit in humans to date, and are considered below in the context of PIDs.

Factors Precipitating Migraine with Aura

Evidence favouring CSD as the basis of the migraine aura has gradually accumulated since Leão and Morison first suggested this [87], and is reviewed in Migraine page 29. We can learn something of the mechanisms of CSD induction from descriptions from migraineurs (with aura) of the precipitating factors they implicate. Sometimes onset follows relaxation after a period of intense concentration or physical exercise. The onset is attributed to hunger by some migraineurs with aura, and we may speculate in the light of discussion below (page 25: Relationship of Cortical Glucose Availability with PID Frequency) that hypoglycaemia is responsible in these individuals. The various other precipitating factors do not at present appear relevant in this context.

Genotype

Familial patterns of migraine incidence and inheritance are well recognised, and there is evidence that familial hemiplegic migraine is due to a calcium channelopathy [88]. Migraine with (non-hemiplegic) aura is much more common, and appears sometimes to have a familial element. It seems likely that other gene/ion-channel abnormalities will emerge in due course. A patient with the appropriate genotype seems likely to be at increased risk of depolarisations occurring in association with stroke, subarachnoid haemorrhage or serious head injury.

Haemodynamic and Metabolic Conditions in the Cortex

The role of ischaemia, trauma, increased K_e and glucose availability to the cerebral cortex in stroke and head injury will be considered below in relation to PIDs. Disturbances of magnesium metabolism have also been invoked as an additional factor increasing migraine risk [89].

Peri-Infarct Depolarisations (PIDS)

Historical

In their 1977 paper Branston *et al.* [45] referred to spontaneous, transient increases in extracellular potassium ion concentration (K_e) which occurred in the ischaemic penumbra following experimental MCAO. Similar, spontaneous events were later reported in another MCAO preparation, also in a gyrencephalic species [90]. It was suggested then that such events, PIDs [3] or HSDs [1], were "not necessarily benign" [91], and specific studies have confirmed this page 24: Evolution of PID Patterns with Time Pathogenic Potential and Recruitment of Penumbra into Core Territory). The critical points of difference between CSD and PIDs are that CSD in completely healthy cortex requires an initiating stimulus and does not damage normally perfused and metabolising grey matter, whereas PIDs are *spontaneous* and do cause damage, and in the case of the ischaemic penumbra, appear to play a large part in recruiting this zone of tissue into the expanding core infarct until this reaches what appears to be a "predestined" size (assuming no treatment).

Detection with Electrodes, and Characteristics of PIDs in Experimental in Vivo Models

PIDs have usually been documented from recordings of the cortical DC potential, and traditionally this has been regarded as a reference detection method. Such electrodes need to be non-polarisable, and usually consist of a glass micropipette filled with physiologically neutral electrolyte and inserted into the cortex, or a chlorided silver ball placed on the cortical surface. Twin-barrelled surface contact or glass microelectrodes allow the signal from an ion-selective barrel (most often to K^+) to be compared with that from an adjacent electrode, both of them referenced to a remote ground electrode. The time course of K_e as recorded from such an electrode during a PID resembles that of CSD in respect of onset and peak amplitude, but may differ in that the recovery phase may be more prolonged. In baboons, a linear, direct relationship of K_e clearance half time with degree of ischaemia was shown, and interpreted as indicating that clearance was no longer by Na-K ATPase (energy-dependent), but relied instead on passive elution by residual perfusion [45]. Studying MCAO in rats, Gill and colleagues [85] distinguished "small" (duration ~1 minute) and "big" PIDs, both recorded with DC electrodes, the latter having much longer time courses. In the same study, this group showed that the time course of depletion of extracellular calcium mirrored that of the DC potential, indicating that "big" PIDs were associated with protracted increases in intracellular calcium, likely to be cytotoxic.

The Response of CBF to a Peri-Infarct Depolarisation

The hyperaemic response to a CSD wave is well recognised from observation of cortical vessels [58], serial section autoradiography in the rat brain (generating a time series as the event propagates along the hemisphere) [47], laser Doppler flowmetry [48], and, by inference, from monitoring of transient increases in tissue pO_2 [48]. Following MCAO, the CBF response is greatly attenuated, or even reversed; thus laser Doppler flowmetry in a deteriorating patient with an intracerebral haematoma at first revealed transient increases in perfusion coupled to probable CSD episodes, but the perfusion responses reversed to transient *hypo*perfusion as brain swelling progressed [92] (Fig. 8). Back *et al.* showed that the positive tissue hyperoxia of CSD becomes a transient *decrease* in tissue pO_2 in focal ischaemia [48].

Detection and Tracking of PIDs with Imaging

In open-skull animal models of stroke, it is usually necessary to leave electrodes at a fixed location rather than probing different cortical areas sequentially, and it is also not possible to determine the extent of propagation of a presumed PID wave with one or more electrodes in the cortex. The use of a method that acquires sequential images of the exposed core and penumbral areas offers a solution if the variable being imaged is affected by the pathophysiology. When illuminated with fluorescent light at 370 nm, the cortex will fluoresce blue, emitting light in the range 445–470 nm; the fluorochrome responsible is the reduced species of the nicotinamide adenine dinucleotide redox couple (NAD/NADH), the coenzyme for succinic dehydrogenase in the mitochondrial respiratory chain. Only NADH – the reduced species – fluoresces, so that oxidation of the couple leads to a fall in fluorescence, whereas reduction causes an increase. Interpretation of such images needs to take account of the capacity of haemoglobin, particularly when oxidised, to absorb or quench blue light (hence its colour!). This method was applied in non recovery MCAO studies in cats [93], and revealed spontaneous *increases* in 450 nm fluorescence that appeared almost always to originate near the core territory and propagate outwards into the penumbra at rates in the range 1–4 mm cortex per minute and hence very characteristic of CSD (Figs. 4–6). Propagation is invariably around the walls of a sulcus, with no evidence that the event can spread directly between gyri lying in contact at the surface. Time courses of the events could be classified into (1) fluorescence increases that did not reverse, (thus closely resembling the time course of terminal depolarisation as recorded with a K_e or with a DC-potential electrode), and which occurred on penumbral cortex close to the core, (2) more peripheral

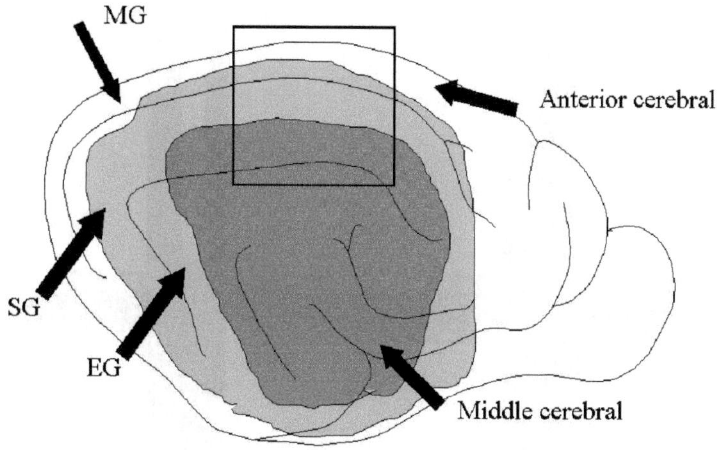

Fig. 4. Schematic diagram illustrating the concept of an ischaemic penumbra or boundary zone in experimental focal cerebral ischaemia in the cat brain, induced in this case by permanent occlusion of the right middle cerebral artery. The ectosylvian (*EG*), suprasylvian (*SG*) and marginal (*MG*) gyri lie at respectively increasing distances from the proximal Sylvian fissure. Directions of arterial inputs from the anterior and middle cerebral (*MCA*) arteries are indicated (posterior cerebral omitted for clarity). The heavily shaded area represents the *core* cortical territory associated with permanent MCA occlusion; terminal depolarisation has occurred within an hour or less of occlusion and is irreversible except by early reperfusion. The lighter shaded area (penumbra) is the site of recurrent peri-infarct depolarisations originating at the edge of the core and propagating outwards into the penumbra (see text and Fig. 5). The square area represents the field of view in each panel of Fig. 5

transient increases in fluorescence that had propagated centrifugally from cortex affected by PIDs with the first pattern, and (3) transient *decreases* in fluorescence, occurring in cortex close to the anterior cerebral artery input, and probably lying just outside penumbra (Figs. 4–6). In some cases, a single PID was seen to propagate from penumbra into anterior cerebral territory, changing its polarity from increase to decrease as an unseen interface was crossed (Fig. 5).

Increases in fluorescence may represent either reduction of the redox couple or a decrease in haemoglobin at the same locus, or a combination of the two, although also not excluding a small increase in haemoglobin outweighed by a larger NADH increase. Whichever the explanation, the observed increase in raw fluorescence indicates either vascular or metabolic compromise, and the method has been used largely to confirm propagation of the events, and to detect them. The depression in crude fluorescence grey level in normally perfused cortex *outside* the penumbra accords well with the depression of *compensated* fluorescence during CSD as shown by Rosenthal and Somjen [46].

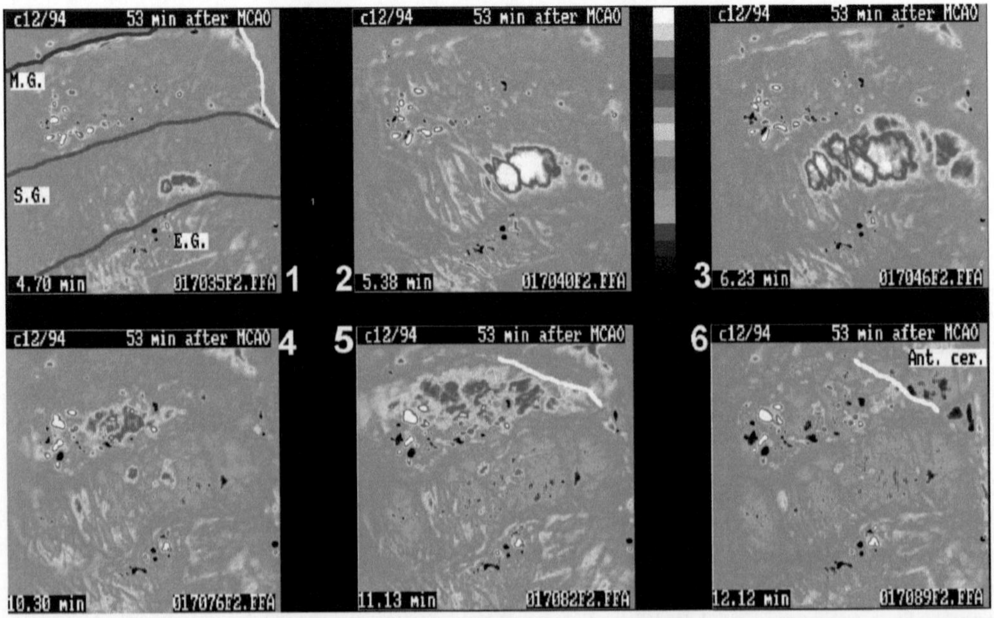

Fig. 5. *Sequence of digital images illustrating initiation and propagation of a peri-infarct depolarisation in the penumbra following experimental middle cerebral artery occlusion.* (For orientation of the image field in relation to the whole hemisphere please see Fig. 4) After exposure of the brain a sequence of grey scale fluorescence images was acquired 53 minutes after occlusion of the middle cerebral artery. The baseline image acquired at time zero was subtracted from each subsequent image and the difference image calculated and displayed in pseudocolour. Green background indicates no change in fluorescence while colours up through the rainbow spectrum to red, pink, white represent increases in fluorescence, and changes into blue, purple or black, decreases respectively. *Panel 1: EG* ectosylvian gyrus (ischaemic core). *SG* Suprasylvian gyrus (inner penumbra). *MG* Marginal gyrus (outer penumbra). Principal middle cerebral input is from lower right of the field, and anterior cerebral from upper right (*panel 6*) (see also Fig. 4). White line in panel 1 represents the anterior margin of the craniectomy exposing the cortex. Red lines represent sulci, and blue line, the line of the sagittal sinus medial to MG. Shortly before the image shown in panel 1, an area of increased fluorescence emerges from the lower sulcus and propagates outwards (from MCA input) throughout the SG (*panels 2–3*). After an interval between panels 3 and 4, the depolarisation (verified by potassium-selective electrode on posterior SG) emerges onto the MG and propagates forwards and medially (*panel 5*) but on reaching cortex perfused by anterior cerebral artery (ant. cer.), the event dissipates, represented only by a *decrease* in fluorescence in panel 6 (upper right of panel). Thus, the white line drawn on MG in panels 5 and 6 represents an apparent interface between middle and anterior cerebral territory. In this example, fluorescence has returned to baseline in the suprasylvian gyrus, but after one or more subsequent similar events, fluorescence increases on this gyrus often culminate in a permanent increase, probably indicating terminal depolarisation (Fig. 6). (Reproduced with permission from Strong *et al.* 1996 [93])

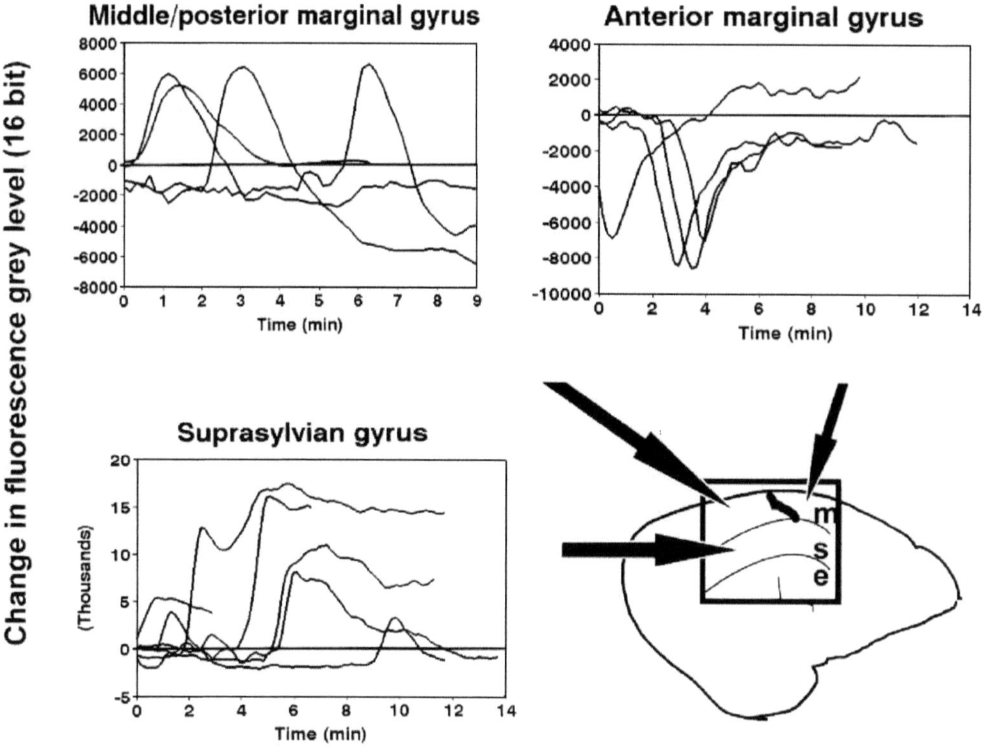

Fig. 6. Examples of time course of fluorescence events recorded from suprasylvian gyrus(s), middle and posterior marginal gyrus(m), and anterior marginal gyrus. (See also Fig. 5). On suprasylvian gyrus, the majority of fluorescence increases are sustained, probably indicating terminal depolarisation. On the middle and posterior MG, still within MCA territory but better collateralised, fluorescence increases are smaller than on SG, and not sustained. In the anterior MG, within anterior cerebral territory, fluorescence transients are all *de*creases, indicating either oxidation of the NAD/H couple, or an increase in total haemoglobin content in the parenchymal circulation, implying vasodilation. Please see also text (page 20: Detection and Tracking of PIDs with Imaging) (reproduced with permission from Strong *et al.* 1996 [93])

Initiation of PIDs

Experience with *in* vivo imaging suggests that the great majority of PIDs originate at the edge of core territory [93], and the high levels of K_e present in core areas are a probable cause [11, 27], but the same considerations apply as in CSD, and glutamate or other factors liberated from ischaemic tissue might contribute.

Terminal Depolarisation

In the core infarct territory established soon after experimental MCAO, the DC potential rapidly becomes negative, but, unlike a PID, does not then resolve, instead becoming increasingly negative and reaching a plateau that is interpreted as indicating complete depolarisation of all cellular elements. Terminal depolarisation – effectively a failure to repolarise spontaneously (as does a PID) – is commonly taken to imply complete depletion of the ATP pool required for repolarisation, and will lead inevitably to infarction unless the ATP pool can be restored promptly by reperfusion.

Evolution of PID Patterns with Time, Pathogenic Potential, and Recruitment of Penumbra into Core Territory

PIDs – however detected – recur at irregular intervals during ischaemia, and observation over 10–12 hours reveals that, at least under chloralose anaesthesia, the pattern of recurrence eventually culminates in terminal depolarisation in outer areas of penumbra, similar to the sequence that occurs earlier in more central penumbra [38]. A feature of the progression, when K_e is monitored, is that resolution of each K_e PID transient towards the pre-transient baseline becomes steadily less complete with time, leading to a gradually increasing K_e baseline. Harris *et al.* showed that in the case of K_e, there is a striking acceleration of K_e increase when it reaches 13 mmol, suggesting a specific change in a membrane conductance [24]; terminal depolarisation follows, and the area of penumbra affected is thus recruited into the core infarct. This sequence of events suggests that number or frequency of PID events in the penumbra is a principal determinant of infarct size, and three pieces of evidence support this. First, Gill *et al.* showed that when the number of PIDs was restricted with the non-competitive NMDA antagonist dizocilpine in rats subjected to MCAO, infarct size was reduced [85]. Secondly, Mies and colleagues reported findings closely similar to those of Gill's group [94]. The association of a larger infarct with increasing PID number may simply reflect the operation of a different, underlying mechanism determining both infarct size and PID frequency. However, thirdly and conclusively, Busch *et al.* were able to increase infarct size in rats by inducing CSD events *outside* the penumbra which propagated into it and caused enlargement of the definitive core infarct [95].

Arising from the original demonstration that loss of evoked potential amplitude could be reversed upon reperfusion, the initial concept of the ischaemic penumbra was of a "sleeping beauty" – a zone of cortex whose function was *reversibly* suppressed in a stable fashion, so that function could be restored at a much later time point by the magical touch of a vascular neurosurgeon carrying out an extra-intracranial vascular bypass procedure [96]. The study of PIDs and manipulations of their frequency has demonstrated instead that – without early reperfusion – the ischaemic penumbra is a maturation phenomenon in which the core infarct gradually expands into penumbra, thus "recruiting" it. The time course of this progression is probably shortest in rats – perhaps 3 hours, extending to 12 to 24 hours in cats, and is believed in humans to extend to perhaps 48 hours. The factors which might influence PID frequency and hence the rate of progression need to be considered.

Species Variations in PID Frequency

Tower and Young's observation of a relationship of cerebral cortical glia : neuron ratio with brain mass is relevant to brain injury since, as mentioned earlier, glial buffering of potassium ion concentration and uptake of neurotransmitters, especially glutamate, are important mechanisms for homeostasis of the extracellular space. It is therefore not a matter of surprise that the frequency of PIDs following MCAO in rats should be high [85], but much less so in cats [79]. CSD is also difficult to induce in monkeys [78]. Efforts to make a direct comparison of PID frequency between cats and primates were frustrated by considerable inter-experiment variability in frequency within a species, but variations in plasma glucose emerged from these experiments as a cause of this variability; this is discussed below (some page: Relationship of Cortical Glucose Availability with PID Frequency). The inference from such comparisons is that PIDs in humans might be rarer still – perhaps vanishingly so – and the relevant, new evidence is described later.

Effects of Drugs and Anaesthetic Agents on PID Frequency

The beneficial effects of NMDA-type glutamate receptor blockade on PID frequency and on infarct size have been reviewed above (page 18: Drugs and Anaesthetic Agents). The AMPA/kainate-type glutamate receptor antagonist NBQX has been shown to reduce PID frequency and volume of ATP depletion in rats subjected to MCAO [97], and this agent has also been shown to reduce ischaemic lesion volume [98]. It is of interest that, unlike MK-801, NBQX does not prevent induction of CSD in the normal brain [99]. The volatile anaesthetic agent halothane may achieve its experimental neuroprotective effect by reducing PID numbers [38], and can, like propofol, block CSD [37]. The fact that halothane also blocks gap junctions in cultures of astrocytes [36] supports the argument for a role of glial gap junctions in the propagation of CSD [100].

Relationship of Cortical Glucose Availability with PID Frequency

As CBF progressively falls in focal ischaemia, a shift to anaerobic glycolysis is inevitable once oxygen extraction is maximal. At that point, a dramatic loss in efficiency of glucose utilisation is equally inevitable, with a fall in net ATP yield per mole glucose utilised from 38 to 2 moles. Glucose utilisation increases to compensate [101]; this is possible despite presence of ischaemia, due to the remarkable effectiveness of the capillary glucose uptake/transport mechanism. This concept is based on several lines of evidence. Hansen showed that following cardiac arrest in rats, delay before terminal ischaemic depolarisation was proportional to plasma glucose, indicating an inverse relationship between depolarisation rate (the dependent variable) and glucose availability in the brain [23]. In 1986, Nedergaard and Astrup showed in rats (MCAO) that hyperglycaemia reduced the frequency of PIDs (although a plasma level in excess of 30 mmol/L was needed to achieve this) [102]. They also showed an increase in phosphorylation of [^{14}C]2-deoxyglucose (an index of metabolic rate) that was related to frequency of PIDs, and predicted that with ischaemia accompanied by PIDs the brain free glucose pool would tend towards zero as delivery and extraction from plasma would quickly become inadequate, given the high, anaerobic utilisation rate. In cats (MCAO), dependence of homœostasis on plasma glucose is demonstrable at glucose levels that are frequently encountered in clinical practice: thus Strong *et al.* showed a striking increase in PID frequency in this situation when mean post-occlusion plasma glucose fell below 4.5 mmol/L (the lower limit of normal quoted for clinical plasma glucose assays in our institution is 3.3 mmol/L) [79]. Our subsequent, unpublished work suggests that the threshold may be nearer 6.5 to 7 mmol/L. This is of potential importance for clinical management since insulin is used to control hyperglycaemia in many intensive care units, with the target range varying in different units. At least one trial of glucose and insulin (to restrict ischaemic acidosis) in acute stroke is under way [103]. There is also striking (and influential) evidence favouring the use of insulin in the intensive care of systemic critical illness [104].

In summary, the initiation of a PID appears to be a random event in which an elevated K_e level at the edge of core infarct territory causes depolarisation of neighbouring tissue because membrane homœostasis there is partially impaired. The impairment is due to a combination of factors in which reduction of glucose availability (the multiple of perfusion (absolute, ml/100 g/min) and plasma glucose levels) as ischaemia deepens becomes particularly important. It seems that reduction of glucose availability increases the probability of initiation of a PID.

The Metabolic "Signature" of PIDs

The transient hypoperfusion or reduction in tissue pO_2 that occurs in association with a PID has been described above. Given the likelihood of transient tissue glycopœnia during recovery from a PID, and the critical dependence of the ATP pool on the balance between on the one hand,

ATP utilisation for restitution of cation gradients during PID recovery, and glucose availability on the other, it becomes valuable to measure the available tissue glucose pool with sufficient time resolution to detect the effects on it of a PID. This has recently been achieved with the use of cerebral microdialysis coupled with rapid sampling of dialysate by means of an online, automated flow–injection assay [105, 106]. The technology allows enzymatic assay of microlitre dialysate samples for glucose and lactate at intervals of 30 seconds each. When dialysate was sampled from penumbral tissue closely adjacent to the core area after MCAO in cats, a PID arriving at the microdialysis probe was associated with complete disappearance of glucose from the dialysate within approximately 3 minutes. In more peripheral penumbra, PIDs were accompanied by transient, stereotyped increases in lactate and decreases in glucose, superimposed in the case of recurrent PIDs on decreasing glucose and increasing lactate baselines (Fig. 7) [107]. This reproducible combination of transient metabolite changes may be taken as a typical metabolic "signature" for a PID, of potential value for the monitoring of patients with severe TBI or acute cerebral ischaemia.

The Role of Depolarisations in Pathophysiology of CNS Disorders in Humans

Speculation and then evidence have accumulated, at first gradually [59, 108, 109], but now more steadily [4, 92, 110], that depolarisations do indeed occur in the human brain – in the functional disorder of migraine with aura as well as in acute traumatic brain injury. It seems that it will only be a matter of time before evidence emerges that they also occur in acute ischaemic or haemorrhagic lesions affecting grey matter in the CNS.

Cortical spreading depression and peri-infarct depolarisations compared:
It is appropriate at this point to summarise the similarities and differences between cortical spreading depression (CSD) and peri-infarct depolarisations (PIDs). CSD is a general, *asynchronous*, neuronal and glial depolarisation that usually commences at a focus in the cerebral cortex, and usually in response to quite vigorous efforts to induce it. It propagates radially in the cortex at 2–5 mm/minute, is accompanied by intense but transient hyperaemia, and does not result in histologically demonstrable cell damage.

A PID is a general neuronal and glial depolarisation that occurs *spontaneously* in an ischaemic boundary zone, especially when plasma glucose is mildly reduced, and propagates into adjacent boundary zone territory at the same velocity as CSD. There is little or no recruitment of perfusion, and, probably as a result of this, ischaemic damage accumulates in the

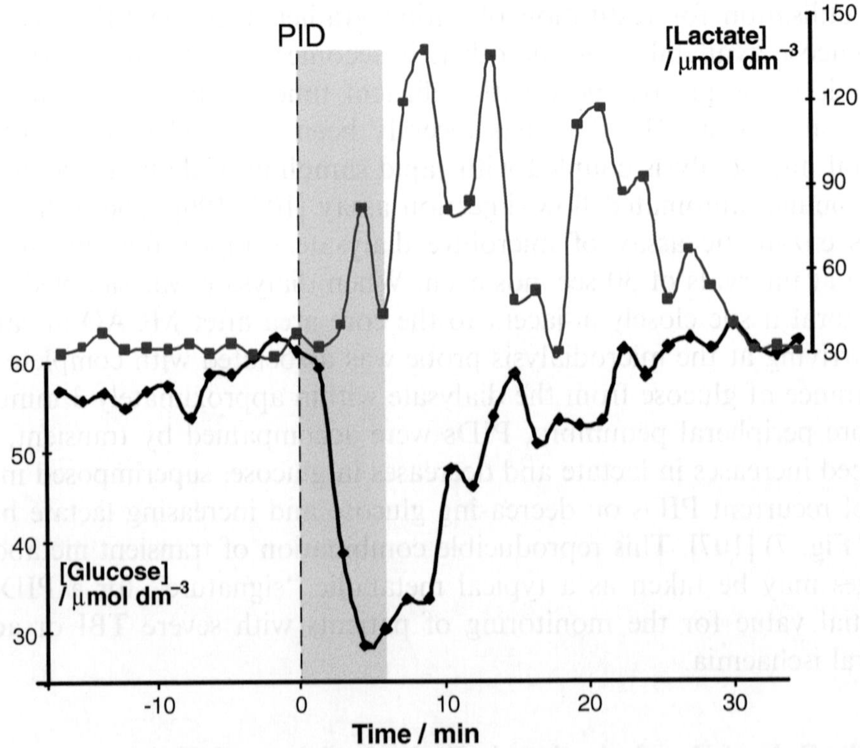

Fig. 7. Time courses of lactate (upper trace) and glucose (lower) concentrations in dialysate from a probe placed in the marginal gyrus (peripheral penumbra) of the cat brain (chloralose anaesthesia). Samples were analysed at 30-second intervals using an enzymatic flow-injection assay [106]. The data demonstrate the typical transient increase in dialysate lactate and decrease in glucose that accompany a PID; this was verified by fluorescence imaging [93]. (Reproduced with permission from Parkin *et al.* [142])

affected territory, culminating after varying periods, perhaps 24–48 hours in humans, in terminal depolarisation and complete infarction.

Spreading depolarisations and epileptic seizures compared:
The essential electrophysiology of an epileptic seizure affecting the cerebral cortex differs from that of CSD in that a degree of *synchronous* firing/depolarisation of neurons is required to generate the dipole whose presence is detected by EEG/ECoG electrodes during a seizure. The transient phase of *asynchronous* neuronal firing at onset of CSD results simply in silence at overlying electrodes. However, the apparent capacity of a Jacksonian fit (a rare event) to propagate across the cerebral cortex resembles the behaviour of CSD, and in the light of the new concepts of non-synaptic communication between different cells in grey matter, one may envisage that Jacksonian epilepsy and CSD might propagate through similar mechanisms. CSD and an epileptic fit both result in transient increases in the metabolic

load inherent in membrane repolarisation, and hyperaemia is a feature common to both. Provided the hyperaemic response is sufficient to permit prompt restoration of "resting" transmembrane cation gradients, neither CSD nor an epileptic fit should lead to any neuronal necrosis. A focal fit is a recognised complication of surgery to clip a ruptured middle cerebral artery aneurysm, and we, probably like most neurosurgeons, view such fits with concern as to their cytotoxic potential since vasospasm may attenuate the hyperaemic response that is required.

Migraine

Classical migraine, now designated migraine-with-aura, is characterised by the migraineur's experience of a visual, somatic motor or sensory symptom as the first component of a stereotyped sequence. The typical visual aura starts as a central scotoma and propagates outwards into the more peripheral visual field (usually a hemifield) as a scintillating, often multicoloured pattern. In 1941 Lashley published a description of his own visual aura, and suggested that it represented propagation of an unknown disturbance across the visual cortex at a rate which he calculated lay in the range of some 3 mm per minute [111], and Leão and Morison suggested that CSD was the basis of migraine with aura [87]. Milner [108] drew attention to the similarity of Lashley's figure for migraine aura propagation with Leão's for CSD propagation, and since then evidence has gradually accumulated that CSD is the basis of migraine with aura. For example, Lauritzen and colleagues [59] mapped CBF using the intra-arterial xenon method and reported propagation in serial images of a phase of reduced blood flow following migraine with aura – probably representing the oligaemic phase of the haemodynamic response to CSD. Woods, Iacoboni and Mazziotta achieved similar results with positron emission tomography [112], and Hadjikhani and colleagues recently described transient loss of normal magnetic resonance blood-oxygen level dependent (BOLD) responses to repetitive visual stimuli during migraine with visual aura. This inhibition propagated outwards from the occipital pole at a rate that was appropriate for CSD [110]. Gardner-Medwin and colleagues had earlier demonstrated propagation of a similar MRI change in experimental CSD [113].

Transient Global Amnesia

Transient global amnesia (TGA) is a neurological syndrome possibly arising in the hippocampus and characterised by sudden onset of complete memory loss; TGA is believed to have as its basis CSD in the hippocampus. The individual appears to be completely alert and can communicate, but enquires frequently about present events. Most episodes last around eight hours but can last for 24 hours and indeed an episode lasting seven days has been described. The onset of memory loss may occur during an emotional stimulus or physical exertion. A history of migraine is recognised in up to 25% of TGA patients. Cerebral blood flow studies using the 133-xenon inhalation method in TGA patients suggest temporary regional hypoperfusion [114]. Marked hypoperfusion in the region of the posterior cerebral arteries has been displayed with single photon emission computed tomography [115]. Diffusion-weighted magnetic resonance imaging during an episode of TGA indicated a decrease in the interstitial space and cellular oedema of the temporal lobe [116]. The induction of spreading depression by the injection of KCl in the hippocampus creates an irreversible retrograde amnesia in the rat [117, 118]. It is currently believed that the amnesic effect of CSD depends on the duration and density of the phenomenon, repetitive CSD causing a more sustained retrograde amnesia.

Trauma

Depolarisation and Concussion

The suggestion that neuronal depolarisation might account for disturbance of consciousness following head injury originates with a paper by A. Earl Walker and colleagues in 1944 [119], and there is ample, more recent experimental evidence – from use of DC potential or ion-selective electrodes in *in vivo* small-animal models of traumatic brain injury – that is compatible with this concept (some page: Recurrent Depolarisations following Experimental Traumatic Brain Injury (TBI)). However, other mechanisms may also contribute to or account for concussion. For example, there is evidence to implicate sublethal, reversible diffuse white matter shearing injury as a mechanism of concussion [120, 121], and it is beyond the scope of this review to explore this issue in detail.

Recurrent Depolarisations Following Experimental Traumatic Brain Injury (TBI)

The term "peri-*infarct* depolarisation" is not strictly applicable to a depolarisation occurring spontaneously in the periphery of a *traumatic* contusion or intracortical haematoma, but several reports (below) of depolarisations in experimental TBI raise two questions. First, do such depolarisations have the characteristics of CSD or of PID, and second, is there evidence for similar events in the injured human brain? Until it becomes clear whether or not depolarisations around a contusion have the characteristics of an (ischaemic) PID, it seems wiser not to assign the term "PID" or "CSD" to them. Although there is evidence for ischaemia surrounding such lesions in humans [122, 123], it is by no means clear that the ischaemia is distributed as widely in TBI as it is in MCAO (unless intracranial pressure is markedly elevated). Notwithstanding these uncertainties about the extent or severity of ischaemia, there is ample experimental evidence from electrode studies for the occurrence of depolarisations in the rat brain following TBI [124–127] or in association with an intracerebral haematoma [128].

Kubota and colleagues [125] and Sunami and colleagues [126] showed a relationship between severity of contusion (from fluid percussion injury) and subsequent frequency of CSD-like events; they also found marked elevations in local cerebral glucose utilisation in those hemispheres in which CSD occurred. Similar observations, suggesting hyperglycolysis and later a *hypo*metabolic state, were reported by Hovda, Lee, and Katayama and their colleagues [129]. B. Nilsson and colleagues studied the effect of mild, non-lethal acceleration head injury on cerebral blood flow and metabolism in rats, and found marked but transient increases in CBF and in brain lactate:pyruvate ratio [130, 131]. Although DC potential was not recorded, the time course of the changes closely resembles the transient hyperaemia associated with CSD described above [47]. P. Nilsson *et al.* demonstrated a relationship between CSD and neuronal damage after a weight drop injury [132].

Direct Detection and Characterisation of Depolarisations in Humans, and Their Role in Human Traumatic Brain Injury

Is there any *direct* evidence for the occurrence of depolarisations in the human brain? In the course of stereotaxic neurosurgical procedures, Sramka and colleagues were able to demonstrate CSD in deep grey matter [109]. Mayevsky and colleagues used a multimodal monitoring system located on the right frontal convexity in 14 patients [92]. In only one did they find evidence for CSD, but the findings in this individual were striking. Recurrent ECoG suppressions were seen, associated with transient increases in CBF (laser Doppler), oxidation of NADH, and elevations of K_e, a combination of features closely compatible with CSD, although – despite the authors' claims – proper verification of propagation of the events was not possible at the single monitoring point used. As brain swelling progressed, the NAD/H transients became reduction rather than oxidation events, and the CBF transients became negative – the features now of PIDs (Fig. 8). Although this group saw CSD/PID events in only one patient, it is important to note that their regular use of a right frontal monitoring site irrespective of the site of any contusion or haematoma, although standard practice at the time of the study, will have precluded detection of events confined to the margins or "traumatic penumbra" of a lesion elsewhere

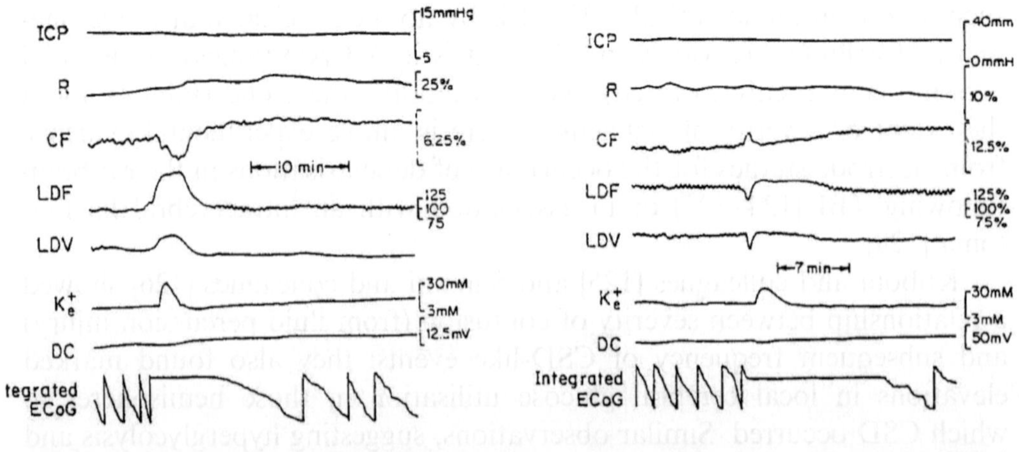

Fig. 8. Traces of intracranial pressure (*ICP*), cortical reflectance (*R*), compensated fluorescence (*CF*), laser Doppler flow (*LDF*), laser Doppler blood volume (*LDV*), extracellular potassium (*K*) DC potential (*DC*) and time-integrated cortical surface EEG activity recorded from a multiparametric probe assembly in the right frontal region of a patient with a severe left parietal contusion [92] (Reproduced with permission from: Mayevsky A et al (1995) J Cereb Blood Flow Metab 15, S1, p S34). The right panel was acquired several hours after the left, following deterioration and shortly before death. In the left-hand panel, a single event – characterised by an increase in perfusion, decrease in fluorescence, increase in extracellular potassium, and a period of electrical silence – has the characteristics of CSD. The characteristics of the event in the right panel have changed in that fluorescence now increases, but perfusion (*LDF*) shows a decrease. The characteristics now correspond more with features of PID rather than CSD

in the brain. There now appears to be wide recognition of the value of locating detection devices, at the very least for research purposes, near the edge of focal lesions.

Recently our group has undertaken a prospective, pilot study designed to detect – or exclude – CSD/PID in patients undergoing emergency craniotomy for traumatic and spontaneous intracranial haematoma [4]. Linear strips of 6 corticography electrodes were placed on the cortex *adjacent to the focal contusion*, lying over both marginal as well as healthy cortex. A working definition of CSD as "suppression of amplitude of the voltage envelope by 50% or more, occurring between one electrode pair and propagating to the next 2 adjacent sites" was adopted, and 14 patients were studied. During periods of observation that lasted up to 63 hours, 6 definite and 23 possible episodes that met this definition were seen in 14 patients (Fig. 9). A second pattern was seen 19 times in 8 patients, in which essentially synchronous suppression was seen in all channels. A proportion of these apparently synchronous events may have been due to arrival of a CSD wave propagating across the array rather than along it, but modelling the statistical distribution of a set of CSD waves reaching

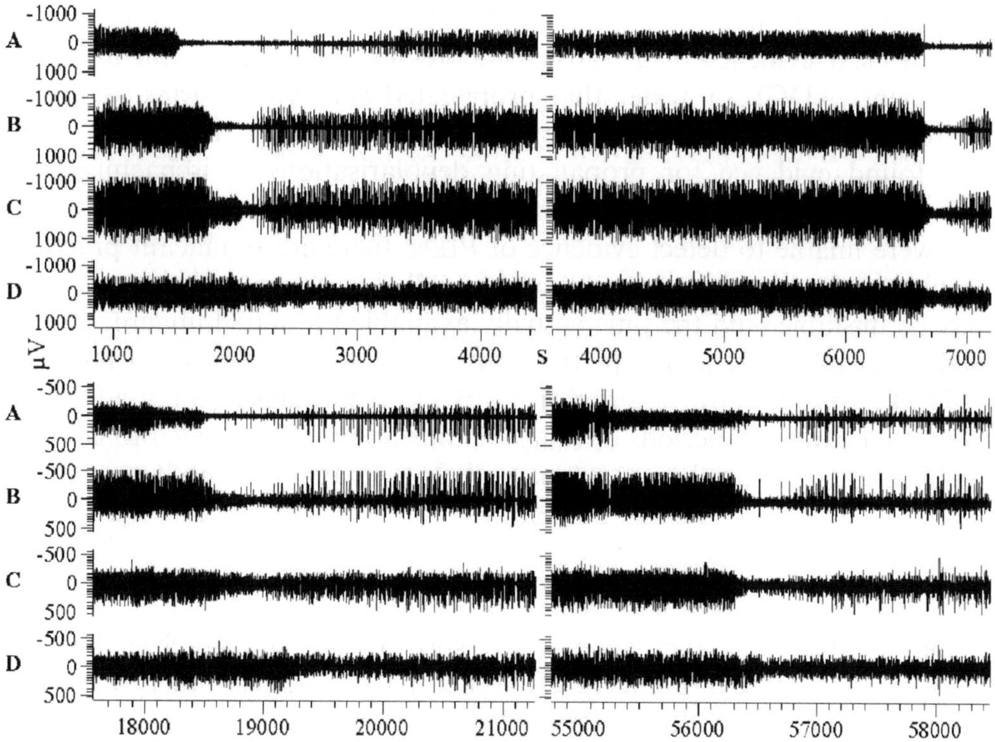

Fig. 9. Examples of two time-compressed electrocorticographic (*ECoG*) traces from two patients (Patient 1: upper, Patient 2, lower) (four bipolar traces per patient). In the first patient there is a sudden reduction of ECoG amplitude propagating sequentially to the next two adjacent channels and to a lesser extent to the third channel. Propagation rate was 2.4 and 2.3 mm per minute, corresponding closely with velocities characteristic of CSD. In the second trace from this patient, ECoG suppression occurs rapidly, recovering most slowly in the upper channel as in the left panel. Synchronous suppression suggests arrival of a wave from a site equidistant from all electrode pairs and to one side of the array, rather than, as in the first panel, propagating along the length of the electrode array. In the lower panel (Patient 2) a phase of ECoG amplitude suppression again propagates along the electrode array, with the respective time points indicating propagation rates of 1.4, 5.0 and 1.0 per minute. (Strong *et al.* 2002 [4]: reproduced with permission)

the strip from a range of angles between 0 and 90 degrees did not fully account for the frequency of synchronous events that we recorded, and the existence of a second pattern of truly synchronous event (possibly due to a partial seizure elsewhere in the cortex [133]) could not be excluded.

Cerebrovascular Disease

Occlusive Stroke

The extensive experimental evidence of PIDs in MCAO models has not as yet been mirrored by studies in patients with occlusive stroke. In 1995

Hasegawa *et al.* had shown that MR diffusion-weighted imaging of rats yielded clear evidence of a transient depression of the apparent diffusion coefficient (ADC) for water that propagated across the cortex at a rate appropriate for CSD in response to stimuli capable of inducing it; they also found evidence for propagating depolarisations in ischaemia [19]. However, Back *et al.* [134] used the same approach in patients with stroke and were unable to detect evidence of PIDs; there are significant practical problems posed by this approach when applied in patients whose condition may be unstable, and the time actually available for actual imaging in this study was relatively short. In the light of the intermittent occurrence of CSD-like episodes now reported from ECoG recordings in trauma [4], it is likely that extended periods of image acquisition will be needed to capture depolarisations in stroke or trauma with MR imaging methods.

Intracerebral Haemorrhage

Using a collagenase that generates a spontaneous intracerebral haematoma in swine and monitoring DC potential (as well as other variables) Mun-Bryce and her colleagues recorded recurrent, spontaneous CSDs originating in perilesion cortex [128].

Subarachnoid Haemorrhage (SAH)

Although there is at present no direct evidence for CSD or PIDs in patients with SAH, there is persuasive experimental evidence to suggest the likelihood of depolarisations occurring in these patients, perhaps restricted to those in intermediate or poor grade. For example, Dreier and colleagues showed that superfusion over the cortex of rats of a combination of increased K^+ and free haemoglobin (such as would arise from lysed erythrocytes in the subarachnoid space) could induce recurrent CSDs [135]. These were accompanied not by hyperaemia but by ischaemia, and thus meet the essential criteria for designation as PIDs. When haemoglobin (which scavenges nitric oxide) was replaced with the nitric oxide synthase inhibitor *N*-nitro-L-arginine, the same effect was observed. This group suggested that this mechanism might account for non-haemorrhagic deterioration in patients with SAH. The common clinical observation of fluctuations in clinical state of intermediate grade SAH patients over intervals often of less than an hour is compatible with the capricious behaviour of CSDs and PIDs in the laboratory; clearly, other explanations are possible and cannot be discounted, but, taken together, the demonstration of CSD-like events in TBI [4] and the work of Dreier and colleagues provide support for this hypothesis.

Non-Invasive Detection of Depolarisations in Ischaemic and Traumatic Brain Injury

Availability of a simple non-invasive method for detection of depolarisations would greatly aid studies of their frequency, properties and effects. Using near-infrared spectroscopy (NIRS), Wolf *et al.* characterised non-invasively the transient changes in oxidised and reduced haemoglobin (HbO, Hb) that accompany CSD in the rat brain [136]. A transient increase in HbO was accompanied by a reduction in Hb, a combination suggesting hyperaemia (Fig. 10, upper). There are clinical NIRS data from this department in one patient with TBI (hitherto unpublished: Fig. 10 lower) and from the Berlin group in two with ischaemic stroke (Fig. 11), in which the HbO and Hb transients closely resembled those seen with CSD in the laboratory. However, in neither case was it feasible to confirm depolarisation by ECoG or DC potential measurement. The potential use of serial measurements of the apparent diffusion coefficient for water using diffusion-weighted MRI has been discussed above.

Characterisation of Depolarisation Events in the Injured Human Brain

It will be clear from the distinction drawn between CSD and PIDs throughout this review that although it would seem that PIDs are invariably cytotoxic and therapy should aim at their control, this is much less certain in the case of CSD. The doubt arises from the evidence that experimental preconditioning with CSD confers protection against subsequent insults (page 16: Pre-Ischaemic Conditioning with CSD as Protection in Experimental Stroke). Since an episode of depolarisation detected by ECoG may represent either CSD or PID, it becomes important to distinguish which has occurred.

Several monitoring methods already well or partly established in clinical or research use are capable of making this distinction. The different responses of cerebral cortical tissue pO_2 to CSD and PIDs (transient increase and decrease respectively) have been well characterised by Back [48]. Mayevsky *et al.* showed that laser Doppler flow monitoring would provide similar information [92]. Dirnagl suggests that the NIRS profile of CSD-linked hyperaemia, transient increase in HbO and decrease in reduced Hb, is reversed in PID – a decrease in HbO with an increase in Hb [137]. Finally, our experimental work with rapid-sampling microdialysis in the MCAO stroke model shows that the occurrence of transient depletion of dialysate glucose and increase in lactate would indicate a PID [105] (Fig. 7). To date we have been unable to identify any comparable "signature" for CSD.

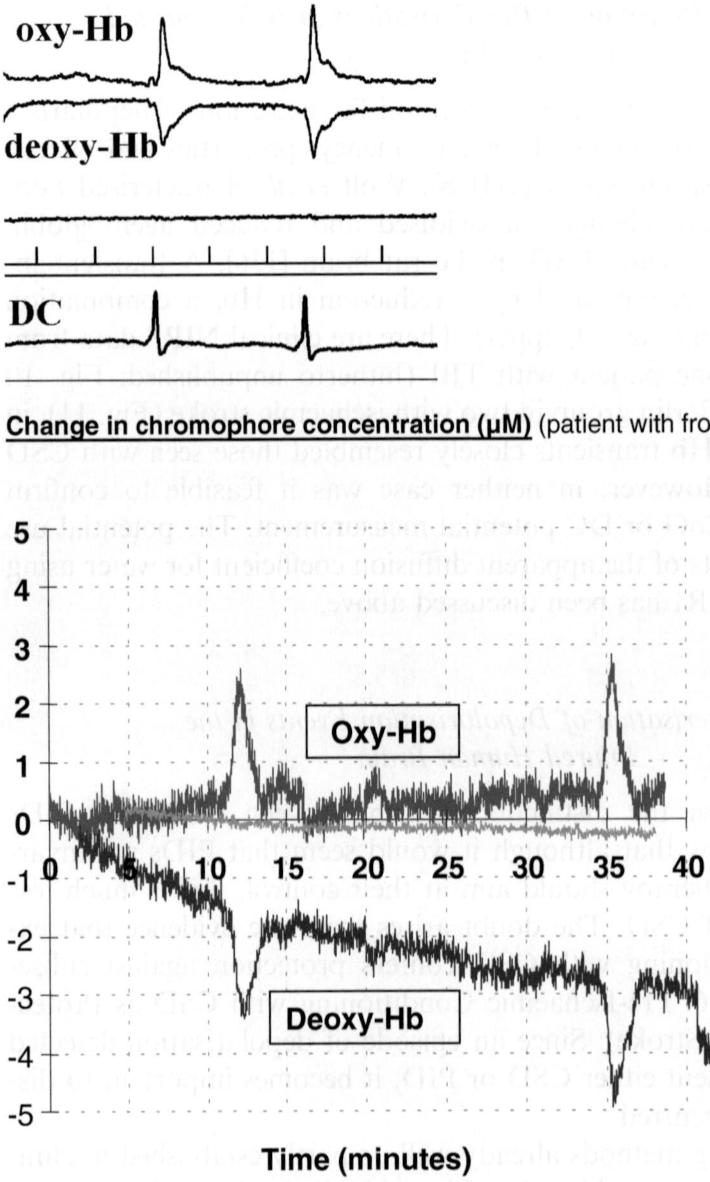

Fig. 10. (*Upper panel*) Time course of near infra-red spectroscopy (NIRS) data (oxy-haemoglobin and deoxy-haemoglobin) from the exposed rat cerebral cortex during an episode of induced CSD, verified by the changes in DC potential. Time bars are at 10 minute intervals. (Reproduced with permission from Kohl *et al.* [139]). (*Lower panel*) Time courses of changes in oxidized haemoglobin (upper trace), cytochrome oxidase (middle trace) and deoxyhaemoglobin (lower trace) in a ventilated patient following severe head injury, obtained non-invasively with NIRS (Cheng, Prowse and Strong, unpublished). The traces show stereotyped combinations of increased HbO and decreased Hb, separated by an interval of some 25 minutes, and suggesting increased oxygen availability characteristic of CSD; ECoG was not available to verify CSD (Please see (page 15: Haemodynamic Response and page 31: Direct Detection and Characterization of Depolarisations in Humans, and Their Role in Human Traumatic Brain Injury). The time course and patterns of the HbO and Hb transients

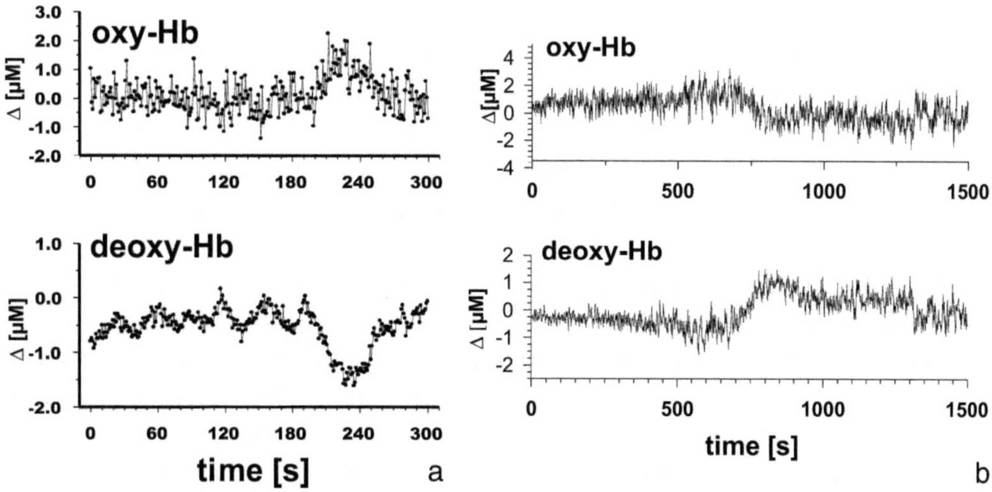

Fig. 11. NIRS traces from 2 patients during the acute phase of ischaemic stroke. (a) Transient increase in HbO with decrease in Hb, suggesting a hyperaemic response to a (unverified) CSD wave. (b) Sustained reduction in HbO signal and increase in Hb, suggesting PID, characterised by *de*crease in HbO signal, and *in*crease in the deoxy Hb signal (depolarisation or propagation not verified by electrophysiology). (Reproduced with permission from Dirnagl, 2001 [137])

The Biological Significance of CSD

Nearly 60 years after its first description by Leão, we remain uncertain of the biological role of a phenomenon that seems to be at times beneficial and a normal response of the brain, and at other times, in the case of PIDs, harmful, principally under conditions of ischaemia.

As one speculation, perhaps the apparent paradox can be explained if we see CSD as reflecting the operation of well-conserved intercellular communications in the brain – serving to protect the brain against inflammation and infection. We do well to bear in mind that our relative mastery of infection in the central nervous system – incomplete and perhaps temporary – is only very recent on the time scale on which evolution operates, and there has probably been more survival value for our own and other vertebrate species in effective responses to CNS infection before and during reproductive life than in avoidance of the cost of aberrant, deleterious operation of the same mechanism in the ageing or irretrievably injured brain.

A quite different speculative view of CSD emerges from the recent rapid growth in our knowledge of the physiological role of astrocytes in modulating synaptic function, to the extent that the synapse is now seen as a tripartite entity – pre- and post-synapse, and astrocyte [138]. Perhaps the probability of CSD (variable depending on glia : neuron ratio and other factors) is an inevitable consequence of this arrangement and of the intimate communication between astrocytes through their gap junctions.

We may also speculate that the depolarisation events that we are able to observe propagating across the cortex with current methods and stimuli may occur much more frequently than we can presently detect, but restricted to microfoci of grey matter, not propagating widely, and below the limits of the available resolution and sensitivity. Results of imaging work with glial and organotypic cultures seem to support this possibility.

As the methods of molecular biology expand, so does the range of gene responses to CSD that have been documented, and it is by no means yet certain which are the most significant. However, it does seem likely that we shall learn as much about the biological significance of CSD from greater knowledge of the expression cascades that it initiates as from the longer-established neurophysiological approaches to CSD.

For neurosurgeons studying and caring for acute brain injury, the only certainties are that *PIDs*, when identified, should be controlled, and that there is more to be learned about the effects of *CSD* on the human brain before we can reach a view on whether to attempt to control it.

Summary

1. *Cortical spreading depression* is a non-physiological global depolarisation of neurones and astrocytes that can be initiated with varying degrees of difficulty in the normally perfused cerebral cortex in the experimental laboratory. Induction is typically with electrical stimulation, needling of the cerebral cortex, or superfusion of isotonic or more concentrated potassium chloride solution. The phenomenon propagates across the cerebral cortex at a rate of 2–5 mm per minute, and is accompanied by marked but transient increases in cerebral blood flow, in local tissue oxygen tension, and most probably in metabolic rate.
2. *Peri-infarct depolarisation* is also a depolarisation event affecting neurones and glia, with an electrophysiological basis similar or identical to CSD, but occurring *spontaneously* in the ischaemic penumbra or boundary zone in focal cerebral cortical ischaemia. Most such events arise from the edge of the ischaemic core, and propagate throughout the penumbra, at a rate similar to that of cortical spreading depression.
3. Cortical spreading depression in the normally perfused cortex does *not* result in histological damage whereas peri-infarct depolarisations *augment* neuronal damage in the penumbra, and are believed by many authors to constitute an important, or the principal, mechanism by which electrophysiological penumbra progressively deteriorates, ultimately undergoing terminal depolarisation and thus recruitment into an expanded core lesion.
4. There is some experimental evidence to suggest that under some circumstances induction of episodes of cortical spreading depression can confer protection against subsequent ischaemic insults.

5. Although cortical spreading depression and peri-infarct depolarisations have been extensively studied in the experimental *in vivo* models, there is now clear evidence that depolarisations also occur and propagate in the human brain in areas surrounding a focus of traumatic contusion.
6. Whether such events in the injured human brain represent cortical spreading depression or peri-infarct depolarisation is unclear. However, invasive and probably non-invasive monitoring methods are available which may serve to distinguish which event has occurred.
7. Much further work will be needed to examine the relationship of depolarisation events in the injured brain with outcome from cerebral ischaemia or head injury, to examine the factors which influence the frequency of depolarisation events, and to determine which depolarisation events in the human brain augment the injury and should be prevented.

Acknowledgements

We thank the Medical Research Council, the Wellcome Trust, the National Lottery Charities Board (Community Fund), HeadFirst, the Golden Charitable Trust, the Patrick Berthoud Trust and GlaxoSmithkline for their support over the period covered by this review.

We are very grateful to Drs Sue Robertson and George Somjen for helpful comments on the manuscript.

Key Original Papers and Reviews

Busch E, Gyngell ML, Eis M, Hoehn Berlage M, Hossmann KA (1996) Potassium-induced cortical spreading depressions during focal cerebral ischemia in rats: contribution to lesion growth assessed by diffusion-weighted NMR and biochemical imaging. J Cereb Blood Flow Metab 16:1090–1099

Grafstein B (1956) Mechanism of spreading cortical depression. J Neurophysiol 19:154–171

Hadjikhani N, Sanchez DR, Wu O, Schwartz D, Bakker D, Fischl B *et al* (2001) Mechanisms of migraine aura revealed by functional MRI in human visual cortex. Proceedings of the National Academy of Sciences of the United States of America 98:4687–4692

Hossmann KA (1996) Periinfarct depolarizations. [Review] [81 refs]. Cerebrovasc Brain Metab Rev 8:195–208

Leao AAP (1944) Spreading depression of activity in cerebral cortex. J Neurophysiol 7:359–390

Somjen GG (2001) Mechanisms of spreading depression and hypoxic spreading depression-like depolarisation. Physiol Rev 81:1065–1096

Strong AJ, Fabricius M, Boutelle MG, Hibbins SJ, Hopwood SE, Jones R *et al* (2002) Spreading and synchronous depressions of cortical activity in acutely injured human brain. Stroke 33:2738–2743

Tsacopoulos M, Magistretti PJ (1996) Metabolic coupling between glia and neurons. J Neurosci 16:877–885

References

1. Somjen GG (2001) Mechanisms of spreading depression and hypoxic spreading depression-like depolarisation. Physiol Rev 81:1065–1096
2. Leão AAP (1944) Spreading depression of activity in cerebral cortex. J Neurophysiol 7:359–390
3. Hossmann KA (1996) Periinfarct depolarizations. [Review] [81 refs]. Cerebrovasc Brain Metab Rev 8:195–208
4. Strong AJ, Fabricius M, Boutelle MG, Hibbins SJ, Hopwood SE, Jones R *et al* (2002) Spreading and synchronous depressions of cortical activity in acutely injured human brain. Stroke 33:2738–2743
5. Marshall WH (1959) Spreading cortical depression of Leão. Physiol Rev 39:239–279
6. Martins-Ferreira H, Nedergaard M, Nicholson C (2000) Perspectives on spreading depression. [Review] [124 refs]. Brain Res – Brain Res Rev 32:215–234
7. Gorji A (2001) Spreading depression: a review of the clinical relevance. Brain Res Rev 38:33–60
8. Obrenovitch TP, Zilkha E, Urenjak J (1996) Evidence against high extracellular glutamate promoting the elicitation of spreading depression by potassium. J Cereb Blood Flow Metab 16:923–931
9. Streit DS, Ferreira Filho CR, Martins-Ferreira H (1995) Spreading depression in isolated spinal cord. J Neurophysiol 74:888–890
10. Nicholson C (1984) Comparative neurophysiology of spreading depression in the cerebellum. [Review] [40 refs]. Anais Da Academia Brasileira de Ciencias 56:481–494
11. Grafstein B (1956) Mechanism of spreading cortical depression. J Neurophysiol 19:154–171
12. Vyskocil F, Kritz N, Bures J (1972) Potassium-selective microelectrodes used for measuring the extracellular brain potassium during spreading depression and anoxic depolarization in rats. Brain Res 39:255–259
13. Muller M, Somjen GG (2000) Na(+) and K(+) concentrations, extra- and intracellular voltages, and the effect of TTX in hypoxic rat hippocampal slices. J Neurophysiol 83:735–745
14. Hansen AJ, Zeuthen T (1981) Extracellular ion concentrations during spreading depression and ischemia in the rat brain cortex. Acta Physiol Scand 113:437–445
15. Collewijn H, Van Harreveld A (1966) Membrane potential of cerebral cortical cells during spreading depression and asphyxia. Exp Neurol 15:425–436
16. Leão AAP, Martins-Ferreira H (1953) Alteraçao da impedancia electrica no decurso de depressão alastrante da atividade do cortex cerebral. Ann Acad Brasil Cienc 25:259–266
17. Czeh G, Aitken PG, Somjen GG (1993) Membrane currents in CA1 pyramidal cells during spreading depression (SD) and SD-like hypoxic depolarization. Brain Res 632:195–208

18. Snow RW, Taylor CP, Dudek FE (1983) Electrophysiological and optical changes in slices of rat hippocampus during spreading depression. J Neurophysiol 50:561–572
19. Hasegawa Y, Latour LL, Formato JE, Sotak CH, Fisher M (1995) Spreading waves of a reduced diffusion coefficient of water in normal and ischemic rat brain. J Cereb Blood Flow Metab 15:179–187
20. James MF, Smith MI, Bockhorst KH, Hall LD, Houston GC, Papadakis NG *et al* (1999) Cortical spreading depression in the gyrencephalic feline brain studied by magnetic resonance imaging. J Physiol 519 Pt 2:415–425
21. Martins-Ferreira H, de Castro GO (1966) Light-scattering changes accompanying spreading depression in isolated retina. J Neurophysiol 29:715–726
22. Gardner-Medwin AR (1983) A study of the mechanisms by which potassium moves through brain tissue in the rat. J Physiol 335:353–374
23. Hansen AJ (1978) The extracellular potassium concentration in brain cortex following ischemia in hypo- and hyperglycemic rats. Acta Physiol Scand 102:324–329
24. Harris RJ, Symon L, Branston NM, Bayhan M (1981) Changes in extracellular calcium activity in cerebral ischaemia. J Cereb Blood Flow Metab 1:203–209
25. Van Harreveld A (1959) Compounds in brain extracts causing spreading depression of cerebral cortical activity and contraction of crustacean muscle. J Neurochem 3:300–315
26. Van Harreveld A, Fifkova E (1970) Glutamate release from the retina during spreading depression. J Neurobiol 2:13–29
27. Obrenovitch TP, Zilkha E (1995) High extracellular potassium, and not extracellular glutamate, is required for the propagation of spreading depression. J Neurophysiol 73:2107–2114
28. Cornell-Bell AH, Finkbeiner SM, Cooper MS, Smith SJ (1990) Glutamate induces calcium waves in cultured astrocytes: long-range glial signaling. Science 247:470–473
29. Willmott NJ, Wong K, Strong AJ (2000) A fundamental role for the nitric oxide-G-kinase signaling pathway in mediating intercellular Ca(2+) waves in glia. J Neurosci 20:1767–1779
30. Willmott NJ, Wong K, Strong AJ (2000) Intercellular Ca(2+) waves in rat hippocampal slice and dissociated glial-neuron cultures mediated by nitric oxide. FEBS Lett 487:239–247
31. Nedergaard M (1994) Direct signalling from astrocytes to neurons in cultures of mammalian brain cells. Science 263:1768–1771
32. Somjen GG (1975) Electrophysiology of neuroglia. [Review] [174 refs]. Ann Rev Physiol 37:163–190
33. Cotrina ML, Lin JH, Nedergaard M (1998) Cytoskeletal assembly and ATP release regulate astrocytic calcium signaling. J Neurosci 18:8794–8804
34. Charles A, Giaume C (2002) Intercellular calcium waves in astrocytes: underlying mechanisms and functional significance. In: Volterra A, Magistretti P, Haydon P (eds). The Tripartite Synapse: glia in synaptic transmission, 1 edn. Oxford University Press, New York, p 110–126

35. Kunkler PE, Kraig RP (1998) Calcium waves precede electrophysiological changes of spreading depression in hippocampal organ cultures. J Neurosc 18:3416–3425
36. Mantz J, Cordier J, Giaume C (1993) Effects of general anesthetics on intercellular communications mediated by gap junctions between astrocytes in primary culture. Anesthesiology 78:892–901
37. Saito R, Graf R, Hubel K, Taguchi J, Rosner G, Fujita T *et al* (1995) Halothane, but not alpha-chloralose, blocks potassium-evoked cortical spreading depression in cats. Brain Res 699:109–115
38. Saito R, Graf R, Hubel K, Fujita T, Rosner G, Heiss WD (1997) Reduction of infarct volume by halothane: effect on cerebral blood flow or perifocal spreading depression-like depolarizations. J Cereb Blood Flow Metab 17:857–864
39. Green JD, Petsche H (1961) Hippocampal electrical activity. IV. Abnormal electrical activity. Electroenceph Clin Neurophysiol 13:868–879
40. Rockel AJ, Hiorns RW, Powell TP (1980) The basic uniformity in structure of the neocortex. Brain 103:221–244
41. Leuba G, Garey LJ (1989) Comparison of neuronal and glial numerical density in primary and secondary visual cortex of man. Exptl Brain Res 77:31–38
42. Largo C, Ibarz JM, Herreras O (1997) Effects of the gliotoxin fluorocitrate on spreading depression and glial membrane potential in rat brain in situ. J Neurophysiol 78:295–307
43. Largo C, Tombaugh GC, Aitken PG, Herreras O, Somjen GG (1997) Heptanol but not fluoroacetate prevents the propagation of spreading depression in rat hippocampal slices. J Neurophysiol 77:9–16
44. Basarsky TA, Duffy SN, Andrew RD, MacVicar BA (1998) Imaging spreading depression and associated intracellular calcium waves in brain slices. J Neurosci 18:7189–7199
45. Branston NM, Strong AJ, Symon L (1977) Extracellular potassium activity, evoked potential and tissue blood flow: relationships during progressive ischaemia in baboon cerebral cortex. J Neurol Sci 32:305–321
46. Rosenthal M, Somjen G (1973) Spreading depression, sustained potential shifts, and metabolic activity of cerebral cortex of cats. J Neurophysiol 36:739–749
47. Lauritzen M, Jorgensen MB, Diemer NH, Gjedde A, Hansen AJ (1982) Persistent oligemia of rat cerebral cortex in the wake of spreading depression. Ann Neurol 12:469–474
48. Back T, Kohno K, Hossmann KA (1994) Cortical negative DC deflections following middle cerebral artery occlusion and KCl-induced spreading depression: effect on blood flow, tissue oxygenation, and electroencephalogram. J Cereb Blood Flow Metab 14:12–19
49. Fox PT, Raichle ME, Mintun MA, Dence C (1988) Nonoxidative glucose consumption during focal physiologic neural activity. Science 241:462–464
50. Ueki M, Linn F, Hossmann KA (1988) Functional activation of cerebral

blood flow and metabolism before and after global ischemia of rat brain. J Cereb Blood Flow Metab 8:486–494
51. Magistretti PJ, Sorg O, Yu N, Martin JL, Pellerin L (1993) Neurotransmitters regulate energy metabolism in astrocytes: implications for the metabolic trafficking between neural cells. Dev Neurosci 15:306–312
52. Koizumi J (1974) Glycogen in the central nervous system. Prog Histochem Cytochem 6:1–37
53. Phelps CH (1975) An ultrastructural study of methionine sulphoximine-induced glycogen accumulation in astrocytes of the mouse cerebral cortex. J Neurocytol 4:479–490
54. Gjedde, A (1993) Relationship of unidirectional and net fluxes of glucose across the blood brain barrier. Personal Communication
55. Tsacopoulos M, Magistretti PJ (1996) Metabolic coupling between glia and neurons. J Neurosci 16:877–885
56. Chih CP, Lipton P, Roberts EL Jr (2001) Do active cerebral neurons really use lactate rather than glucose? [Review] [66 refs]. Trends Neurosci 24:573–578
57. Chen Y, Swanson RA (2003) Astrocytes and brain injury. [Review] [184 refs]. J Cereb Blood Flow Metab 23:137–149
58. Leão AAP (1944) Pial circulation and spreading depression of activity in the cerebral cortex. J Neurophysiol 7:391–396
59. Lauritzen M, Skyhoj OT, Lassen NA, Paulson OB (1983) Changes in regional cerebral blood flow during the course of classic migraine attacks. Ann Neurol 13:633–641
60. Nedergaard M, Hansen AJ (1988) Spreading depression is not associated with neuronal injury in the normal brain. Brain Res 449:395–398
61. Sharp FR, Lu A, Tang Y, Millhorn DE (2000) Multiple molecular penumbras after focal cerebral ischemia. [Review] [373 refs]. J Cereb Blood Flow Metab 20:1011–1032
62. Koistinaho J, Pasonen S, Yrjanheikki J, Chan PH (1999) Spreading depression-induced gene expression is regulated by plasma glucose. Stroke 30:114–119
63. Rangel YM, Kariko K, Harris VA, Duvall ME, Welsh FA (2001) Dose-dependent induction of mRNAs encoding brain-derived neurotrophic factor and heat-shock protein-72 after cortical spreading depression in the rat. Brain Res Molecul Brain Res 88:103–112
64. Ananthan J, Goldberg AL, Voellmy R (1986) Abnormal proteins serve as eukaryotic stress signals and trigger the activation of heat shock genes. Science 232:522–524
65. Nowak TS, Kiessling M (1999) Reprogramming of gene expression after ischemia. In: Walz W (ed) Cerebral ischemia: molecular and cellular pathophysiology. Totowa, Humana Press, NJ, p 145–216
66. Rothwell NJ, Relton JK (1993) Involvement of interleukin-1 and lipocortin-1 in ischaemic brain damage. Cerebrovasc Brain Metab Rev 5:178–198
67. Szaflarski J, Burtrum D, Silverstein FS (1995) Cerebral hypoxia-ischemia stimulates cytokine gene expression in perinatal rats. Stroke 26:1093–1100

68. Betz AL, Schielke GP, Yang GY (1996) Interleukin-1 in cerebral ischemia. Keio J Med 45:230–237
69. Jander S, Schroeter M, Peters O, Witte OW, Stoll G (2001) Cortical spreading depression induces proinflammatory cytokine gene expression in the rat brain. J Cereb Blood Flow Metab 21:218–225
70. Mason JL, Suzuki K, Chaplin DD, Matsushima GK (2001) Interleukin-1 beta promotes repair of the CNS. J Neurosci 21:7046–7052
71. Blamire AM, Anthony DC, Rajagopalan B, Sibson NR, Perry VH, Styles P (2000) Interleukin-1beta – induced changes in blood-brain barrier permeability, apparent diffusion coefficient, and cerebral blood volume in the rat brain: a magnetic resonance study. J Neurosci 20:8153–8159
72. Duong TQ, Sehy JV, Yablonskiy DA, Snider BJ, Ackerman JJ, Neil JJ (2001) Extracellular apparent diffusion in rat brain. Magn Res Med 45:801–810
73. Kobayashi S, Harris VA, Welsh FA (1995) Spreading depression induces tolerance of cortical neurons to ischemia in rat brain. J Cereb Blood Flow Metab 15:721–727
74. Kariko K, Harris VA, Rangel Y, Duvall ME, Welsh FA (1998) Effect of cortical spreading depression on the levels of mRNA coding for putative neuroprotective proteins in rat brain. J Cereb Blood Flow Metab 18:1308–1315
75. Ohtsuki T, Ruetzler CA, Tasaki K, Hallenbeck JM (1996) Interleukin-1 mediates induction of tolerance to global ischemia in gerbil hippocampal CA1 neurons. J Cereb Blood Flow Metab 16:1137–1142
76. Wang X, Li X, Currie RW, Willette RN, Barone FC, Feuerstein GZ (2000) Application of real-time polymerase chain reaction to quantitate induced expression of interleukin-1beta mRNA in ischemic brain tolerance. J Neurosci Res 59:238–246
77. Blondeau N, Widmann C, Lazdunski M, Heurteaux C (2001) Activation of the nuclear factor-kappa-B is a key event in brain tolerance. J Neurosci 21:4668–4677
78. Marshall WH, Essig CF, Dubroff SJ (1951) Relation of temperature of cerebral cortex to spreading depression of Leão. J Neurophysiol 14:153–166
79. Strong AJ, Smith SE, Whittington DJ, Meldrum BS, Parsons AA, Krupinski J *et al* (2000) Factors influencing the frequency of fluorescence transients as markers of peri-infarct depolarizations in focal cerebral ischemia. Stroke 31(1):214–222
80. Tower DB, Young OM (1973) The activities of butyrylcholinesterase and carbonic anhydrase, the rate of anaerobic glycolysis, and the question of a constant density of glial cells in cerebral cortices of various mammalian species from mouse to whale. J Neurochem 20:269–278
81. Czeh G, Somjen GG (1990) Hypoxic failure of synaptic transmission in the isolated spinal cord, and the effects of divalent cations. Brain Res 527:224–233
82. Curtis DR, Watkins JC (1961) Analogues of glutamic and gammaaminobu-

tyric acids having potent actions on mammalian neurones. Nature 191:1010–1011
83. Gorelova NA, Koroleva VI, Amemori T, Pavlik V, Bures J (1987) Ketamine blockade of cortical spreading depression in rats. Electroencephalography Clin Neurophysiol 66:440–447
84. Lauritzen M, Rice ME, Okada Y, Nicholson C (1988) Quisqualate, kainate and NMDA can initiate spreading depression in the turtle cerebellum. Brain Res 475:317–327
85. Gill R, Andine P, Hillered L, Persson L, Hagberg H (1992) The effect of MK-801 on cortical spreading depression in the penumbral zone following focal ischaemia in the rat. J Cereb Blood Flow Metab 12:371–379
86. Iijima T, Mies G, Hossmann KA (1992) Repeated negative DC deflections in rat cortex following middle cerebral artery occlusion are abolished by MK-801: effect on volume of ischemic injury. J Cereb Blood Flow Metab 12:727–733
87. Leão AAP, Morison RS (1945) Propagation of spreading cortical depression. J Neurophysiol 8:33–45
88. Ophoff RA, Terwindt GM, Vergouwe MN, Frants RR, Ferrari MD (1997) Wolff Award 1997. Involvement of a Ca2+ channel gene in familial hemiplegic migraine and migraine with and without aura. Dutch Migraine Genetics Research Group. [Review] [43 refs]. Headache 37:479–485
89. Welch KM, Ramadan NM (1995) Mitochondria, magnesium and migraine. J Neurol Sci 134:9–14
90. Strong AJ, Venables GS, Gibson G (1983) The cortical ischaemic penumbra associated with occlusion of the middle cerebral artery in the cat: 1. Topography of changes in blood flow, potassium ion activity, and EEG. J Cereb Blood Flow Metab 3:86–96
91. Strong AJ, Tomlinson BE, Venables GS, Gibson G, Hardy JA (1983) The cortical ischaemic penumbra associated with occlusion of the middle cerebral artery in the cat: 2. Studies of histopathology, water content, and in vitro neurotransmitter uptake. J Cereb Blood Flow Metab 3:97–108
92. Mayevsky A, Doron A, Manor T, Meilin S, Zarchin N, Ouaknine GE (1996) Cortical spreading depression recorded from the human brain using a multiparametric monitoring system. Brain Res 740:268–274
93. Strong AJ, Harland SP, Meldrum BS, Whittington DJ (1996) The use of in vivo fluorescence image sequences to indicate the occurrence and propagation of transient focal depolarizations in cerebral ischemia. J Cereb Blood Flow Metab 16:367–377
94. Mies G, Iijima T, Hossmann KA (1993) Correlation between peri-infarct DC shifts and ischaemic neuronal damage in rat. Neuroreport 4:709–711
95. Busch E, Gyngell ML, Eis M, Hoehn Berlage M, Hossmann KA (1996) Potassium-induced cortical spreading depressions during focal cerebral ischemia in rats: contribution to lesion growth assessed by diffusion-weighted NMR and biochemical imaging. J Cereb Blood Flow Metab 16:1090–1099
96. Lassen NA, Vorstrup S (1984) Ischaemic penumbra results in incomplete infarction: is the sleeping beauty dead? Stroke 15: 755–756, 15:755

97. Mies G, Kohno K, Hossmann KA (1994) Prevention of periinfarct direct current shifts with glutamate antagonist NBQX following occlusion of the middle cerebral artery in the rat. J Cereb Blood Flow Metab 14:802–807
98. Buchan AM, Xue D, Huang ZG, Smith KH, Lesiuk H (1991) Delayed AMPA receptor blockade reduces cerebral infarction induced by focal ischemia. Neuroreport 2:473–476
99. Lauritzen M, Hansen AJ (1992) The effect of glutamate receptor blockade on anoxic depolarization and cortical spreading depression. J Cereb Blood Flow Metab 12:223–229
100. Nedergaard M, Cooper AJ, Goldman SA (1995) Gap junctions are required for the propagation of spreading depression. J Neurobiol 28:433–444
101. Ginsberg MD, Reivich M, Giandomenico A, Greenberg JH (1977) Local glucose utilization in acute focal cerebral ischemia: local dysmetabolism and diaschisis. Neurology 27:1042–1048
102. Nedergaard M, Astrup J (1986) Infarct rim: effect of hyperglycemia on direct current potential and [^{14}C]2-deoxyglucose phosphorylation. J Cereb Blood Flow Metab 6:607–615
103. Scott JF, Robinson GM, French JM, O'Connell JE, Alberti KG, Gray CS (1999) Glucose potassium insulin infusions in the treatment of acute stroke patients with mild to moderate hyperglycemia: the Glucose Insulin in Stroke Trial (GIST). Stroke 30:793–799
104. Van den Berghe G, Wouters P, Weekers F, Verwaest C, Bruyninckx F, Schetz M *et al* (2001) Intensive insulin therapy in the critically ill patients.[comment]. New Engl J Med 345:1359–1367
105. Strong AJ, Wong C-K, Jones DA, Parkin M, Boutelle MG (2001) Detection and analysis of peri-infarct glucose and lactate transients with rapid-sampling microdialysis. J Cereb Blood Flow Metab 21(S1):86 (abstract)
106. Jones DA, Parkin MC, Langemann H, Landolt H, Hopwood SE, Strong AJ, Boutelle MG (2002) On-line neurochemical monitoring in Neurointensive care: enzyme-based assay for the simultaneous, continuous monitoring of glucose and lactate from critical care patients. J Electroanalytical Chem 238:243–252
107. Hopwood SE, Boutelle MG, Parkin MC, Bezzina EL, Strong AJ (2003) Rapid sampling of glucose and lactate using on-line microdialysis in a model of focal cerebral ischaemia. (Abstract) J Cereb Blood Flow Metab [Suppl] 1:115
108. Milner PM (1958) Notes on a possible correspondence between the scotomas of migraine and spreading depression of Leao. Electroenceph Clin Neurophysiol 10:705
109. Sramka M, Brozek G, Bures J, Nadvornik P (1977) Functional ablation by spreading depression: possible use in human stereotactic neurosurgery. Appl Neurophysiol 40:48–61
110. Hadjikhani N, Sanchez DR, Wu O, Schwartz D, Bakker D, Fischl B *et al* (2001) Mechanisms of migraine aura revealed by functional MRI in human visual cortex. Proc Nat Acad Sci USA 98:4687–4692

111. Lashley KS (1941) Patterns of cerebral integration indicated by the scotomas of migraine. Arch Neurol Psychiatry 46:331–339
112. Woods RP, Iacoboni M, Mazziotta JC (1994) Brief report: bilateral spreading cerebral hypoperfusion during spontaneous migraine headache [see comments]. N Engl J Med 331:1689–1692
113. Gardner-Medwin AR, van Bruggen N, Williams SR, Ahier RG (1994) Magnetic resonance imaging of propagating waves of spreading depression in the anaesthetised rat. J Cereb Blood Flow Metab 14:7–11
114. Crowell GF, Stump DA, Biller J, McHenry LC Jr, Toole JF (1984) The transient global amnesia-migraine connection. Arch Neurol 41:75–79
115. Tanabe H, Hashikawa K, Nakagawa Y, Ikeda M, Yamamoto H, Harada K *et al* (1991) Memory loss due to transient hypoperfusion in the medial temporal lobes including hippocampus.[erratum appears in Acta Neurol Scand 1991 Nov;84(5):463]. Acta Neurol Scand 84:22–27
116. Strupp M, Bruning R, Wu RH, Deimling M, Reiser M, Brandt T (1998) Diffusion-weighted MRI in transient global amnesia: elevated signal intensity in the left mesial temporal lobe in 7 of 10 patients. [comment]. Ann Neurol 43:164–170
117. Avis HH, Carlton PL (1968) Retrograde amnesia produced by hippocampal spreading depression. Science 161:73–75
118. Kapp BS, Schneider AM (1971) Selective recovery from retrograde amnesia produced by hippocampal spreading depression. Science 173:1149–1151
119. Walker AE, Kollros JJ, Case TJ (1944) The physiological basis of concussion. J Neurosurg 1:103–116
120. Povlishock JT (2000) Pathophysiology of neural injury: therapeutic opportunities and challenges. [Review] [37 refs]. Clin Neurosurg 46:113–126
121. Sahuquillo J, Poca MA (2002) Diffuse axonal injury after head trauma. A review. [Review] [151 refs]. Advances & Technical Standards in Neurosurgery 27:23–86
122. Bouma GJ, Muizelaar JP, Choi SC, Newlon PG, Young HF (1991) Cerebral circulation and metabolism after severe traumatic brain injury: the elusive role of ischemia. J Neurosurg 75:685–693
123. von Oettingen G, Bergholt B, Gyldensted C, Astrup J (2002) Blood flow and ischemia within traumatic cerebral contusions. Neurosurgery 50:781–788
124. Takahashi H, Manaka S, Sano K (1981) Changes in extracellular potassium concentration in cortex and brain stem during the acute phase of experimental closed head injury. J Neurosurg 55:708–717
125. Kubota M, Nakamura T, Sunami K, Ozawa Y, Namba H, Yamaura A *et al* (1989) Changes of local cerebral glucose utilization, DC potential and extracellular potassium concentration in experimental head injury of varying severity. Neurosurg Rev 12 [Suppl] 1:393–399
126. Sunami K, Nakamura T, Ozawa Y, Kubota M, Namba H, Yamaura A (1989) Hypermetabolic state following experimental head injury. Neurosurg Rev 12 [Suppl] 1:400–411
127. Katayama Y, Becker DP, Tamura T, Hovda DA (1990) Massive increases

in extracellular potassium and the indiscriminate release of glutamate following concussive brain injury. J Neurosurg 73:889–900
128. Mun-Bryce S, Wilkerson AC, Papuashvili N, Okada YC (2001) Recurring episodes of spreading depression are spontaneously elicited by an intracerebral hemorrhage in the swine. Brain Res 888:248–255
129. Yoshino A, Hovda DA, Kawamata T, Katayama Y, Becker DP (1991) Dynamic changes in local cerebral glucose utilization following cerebral contusion in rats: evidence of a hyper- and subsequent hypometabolic state. Brain Res 561:106–119
130. Nilsson B, Nordstrom C-H (1977) Experimental head injury in the rat. Part 3: cerebral blood flow and oxygen consumption after concussive impact acceleration. J Neurosurg 47:262–273
131. Nilsson B, Ponten U (1977) Experimental head injury in the rat. Part 2: regional brain energy metabolism in concussive trauma. J Neurosurg 47:252–261
132. Nilsson P, Hillered L, Olsson Y, Sheardown MJ, Hansen AJ (1993) Regional changes in interstitial K+ and Ca2+ levels following cortical compression contusion trauma in rats. J Cereb Blood Flow Metab 13:183–192
133. Alarcon G, Binnie CD, Elwes RD, Polkey CE (1995) Power spectrum and intracranial EEG patterns at seizure onset in partial epilepsy. Electroencephalography Clin Neurophysiol 94:326–337
134. Back T, Hirsch JG, Szabo K, Gass A (2000) Failure to demonstrate peri-infarct depolarizations by repetitive MR diffusion imaging in acute human stroke. Stroke (Online) 31:2901–2906
135. Dreier JP, Korner K, Ebert N, Gorner A, Rubin I, Back T *et al* (1998) Nitric oxide scavenging by hemoglobin or nitric oxide synthase inhibition by N-nitro-L-arginine induces cortical spreading ischemia when K+ is increased in the subarachnoid space. J Cereb Blood Flow Metab 18:978–990
136. Wolf T, Lindauer U, Reuter U, Back T, Villringer A, Einhaupl K *et al* (1997) Noninvasive near infrared spectroscopy monitoring of regional cerebral blood oxygenation changes during peri-infarct depolarizations in focal cerebral ischemia in the rat. J Cereb Blood Flow Metab 17:950–954
137. Dirnagl U, Obrig H, von Pannwitz W, Kohl M, Kerskens CM, Doge C, Lindauer U, Wolf T, Villringer A (2000) Cerebral blood flow, hemoglobin oxygenation, and water diffusion changes during stroke: fingerprinting with near-infrared spectroscopy and MRI. In: Fukuuchi Y, Tomita M, Koto A (eds) 6:232–240. 2001. Springer, Tokyo. Keio University, Symposia for Life Science and Medicine: Ischemic Blood Flow in the Brain
138. Volterra A, Magistretti PJ, Haydon PG (2003) The Tripartite Synapse: glia in synaptic transmission. Oxford University Press, New York
139. Kohl M, Lindauer U, Dirnagl U, Villringer A (1998) Separation of changes in light scattering and chromophore concentrations during cortical spreading depression in rats. Optics Lett 23:555–557
140. Anderson CM, Nedergaard M (2003) Astrocyte-mediated control of cerebral microcirculation. Trend Neurosci 26(7):340–344
141. Zonta M, Angulo MC, Gobbo S, Rosengarten B, Hossmann KA, Pozzan T,

Carmignoto G (2003) Neuron-to-astrocyte signalling is central to the dynamic control of brain microcirculation. Nature Neurosci 6(1):43–50
142. Parkin MC, Hopwood SE, Strong AJ, Boutelle MG (2003) Resolving dynamic changes in brain metabolism using biosensors and on-line microdialysis. Trends Anal Chem 22(9):487–497

Offprint from
Advances and Technical Standards in Neurosurgery, Vol. 30
Edited by J.D. Pickard

What is Magnetoencephalography and why it is Relevant to Neurosurgery?

F. H. Lopes da Silva

Section Neurobiology, Swammerdam Institute for Life Sciences, University of Amsterdam, Amsterdam, The Netherlands

With 3 Figures

Contents

Abstract

Magnetoencephalography (MEG) is a relatively novel technique that allows the study of the dynamic properties of cortical activity. The functional localization of brain sources of MEG signals depends on the models used and it always has a certain degree of uncertainty. Nevertheless, MEG can be very useful in assisting the neurosurgeon in planning and carrying out brain surgery in, or around, eloquent brain areas, and in epilepsy surgery in pharmaco-resistant patients. The following three areas of application of MEG in neurosurgery are reviewed: (i) Presurgical functional localization of somatomotor eloquent cortex; (ii) Presurgical evaluation of epileptic patients. (iii) Functional localization of speech relevant brain areas. The performance of MEG in comparison with EEG and fMRI is discussed.

Keywords: Magnetoencephalography; epilepsy; presurgical planning.

Introduction

Magnetoencephalography (MEG) is the technique of measuring the magnetic fields generated by brain activity. In general, an important asset of MEG/EEG is the ability to record signals generated by the brain in relation to distinct states of activity, whether determined by intrinsic processes, e.g. different states of sleep and alertness, or in relation to motor acts and sensory events. Here we will focus on the core of clinical applications with respect to neurosurgical isssues. In this context three fields of application will be reviewed: (i) Presurgical functional localization of somatomotor eloquent cortex; (ii) Presurgical evaluation of epileptic patients. (iii) Functional localization of speech relevant brain areas.

The implementation of the MEG in clinical settings has had a relatively slow start, after the pioneering studies of Cohen (1968), in comparison with the amazingly rapid development of other brain imaging techniques such as Magnetic Resonance Imaging (MRI). It should be stressed that by no means are MEG and MRI really competitors. Rather, the two techniques are complementary, since MRI provides precise static information about brain anatomy, while MEG allows the study of the dynamic properties of cortical activity. MRI yields unrivalled images of brain structures at all levels, whereas MEG gives information about dynamics of the activities of large populations of neurons of the cerebral cortex. Nevertheless MEG has a more direct competitor in the more recently developed functional MRI (fMRI) technique that is based on measurements of changes of the ratio between oxy- and reduced-haemoglobin, which are related to neuronal activity. Be as it may the time resolution of MEG is unrivalled. Below we analyse more specifically the comparison between MEG and fMRI regarding the problem of localizing cortical areas bordering the central sulcus.

A basic limitation of MEG, as much as of Electroencephalography (EEG), is that the neuronal signals are recorded from the scalp and that there is no unique solution to the question of where, within the brain, are localized the sources of those signals. This means that there is no solution to what is called the inverse problem. The common approach to overcome the non-uniqueness of the inverse problem in MEG/EEG is to introduce constraints in the solutions, in order to exclude all solutions except that one that is most suitable to describe the data. Thus the functional localization of brain sources of MEG/EEG signals depends on the models used and on the corresponding assumptions, and it always has a certain degree of uncertainty. This contrasts with the accuracy of MRI brain scans. It explains partly why the development of clinical applications of MEG has been slow, since the research on how to optimise the solutions of the inverse problem has been arduous and only recently it is reaching the stage at which consensual strategies may be advanced. These inherent difficulties, along with the fact that MEG needs rather costly facilities, account for the restraint of medical specialists in promoting this new methodology and the hesitation of hospital administrators in supporting the necessary investments in material and human resources.

Some Basic Notions: From Applied Physics to Biophysics

After the pioneer measurement of MEG signals by Cohen (1968) using a simple induction coil magnetometer, this method became practical only after new technologies based on the principle of superconduction became available. In the course of the last decades this led to the construction of whole-head devices that allow measuring simultaneously from more than 150 sensors. In Amsterdam we have a CTF/VSM apparatus, manufactured in Vancouver (Canada). To give a rough idea of the financial aspects, an MEG apparatus costs about 2.5×10^6€, while the yearly running costs, including salaries, are in the order of 6×10^5€. It is most important that a small staff consisting of physicists, software specialists and technicians (at least $1\frac{1}{2}$ positions), will give the necessary support and collaboration in the development of acquisition protocols and signal analysis. I should add that besides the exploration of magnetic signals produced by the brain, the same technique has proved valuable in the study of the heart, both of adults and foetuses. This can be illustrated by two recent studies: it was shown that recordings of the adult heart magnetic field (Magnetocardiogram or MCG) in the depolarization process has the potential to detect subtle myocardial ischemia induced by exercise (Kanzaki *et al.* 2003). The analysis of foetal MCG (FMCG) recordings may also contribute significantly to a better understanding of the heart function of the foetuses, and thus may help improve perinatal morbidity and mortality (Anastasiadis *et al.* 2003).

A few basic notions of applied physics and biophysics may be useful as a short technical introduction into the realm of MEG.

First, we may ponder about some basic notions of applied physics. In general, magnetic fluxes can be measured using induction coils. However, very weak magnetic fields are not measurable using normal wires since the induced currents dissipate as heat by the electrical resistance of the wires. The discovery of the principle of superconduction, i.e. of materials that have essentially no electrical resistance at extremely low temperatures, opened up the possibility of measuring tiny magnetic fields as those produced by the brain. These superconductors are usually known by the abbreviation SQUID (superconducting quantum interference devices). To assure that SQUIDs work properly they have to be maintained at a very low temperature. This is achieved by immersing the SQUIDs in liquid helium contained in an insulated vessel known as a Dewar (−269 °C). The SQUID may be considered a device for transforming a time-varying magnetic field to a time-varying voltage, since a magnetic flux passing perpendicular to a superconducting coil induces an electrical current in the coil. The latter can be further amplified using appropriate electronics. Thus the field of 'low temperature physics' created new devices that allowed measuring very weak magnetic fields such as those produced by neurons.

Second, we may review some basic principles of biophysics. Neurons generate time-varying electrical currents when activated. Longitudinal intra-cellular currents flowing along dendrites or axons generate magnetic fields around them, just as it happens in a wire, according to the well known right-hand rule of electromagnetism. Pyramidal neurons of the cortex, with their long apical dendrites oriented perpendicular to the cortical surface, if activated with a certain degree of synchrony, generate coherent magnetic fields. In this way we may say that these neurons behave as 'current dipoles', the activity of which can be detected by SQUIDs placed at a small distance from the skull. We should note that the resulting MEG signals depend on the orientation of the neurons with respect to the skull.The MEG 'sees' only those magnetic fields that are perpendicular to the skull. These magnetic fields are generated by neuronal currents that are oriented *tangentially* to the skull. In contrast, those that are oriented *radially* to the skull do not generate a magnetic field outside the head.

Third, we have to note that in order to record MEG signals it is necessary to use specialized recording conditions, since these signals are very weak. Even when thousands of cortical pyramidal neurons are synchronously active the resulting magnetic field at the head surface has a very small magnitude, in the order of 10^{-12} Tesla (1 fT $= 10^{-15}$ Tesla), which is much smaller than the earth magnetic field and urban magnetic noise fields. This poses a hard detection problem, equivalent to that of detecting, at a distance, the voice of one single individual in a large noisy crowd. To accomplish this strenuous task the MEG is generally recorded in a magnetic and radiofrequency shielded room with walls made of mu-metal and aluminium, what adds appreciably to the cost of the system.

Clinical Applications of MEG in a Neurosurgical Setting

In this overview we will examine the evidence available regarding the use of MEG in the clinical settings where the emphasis is on assisting the neurosurgeon in planning and carrying out brain surgery in, or around, eloquent brain areas, and in pharmaco-resistant epileptic patients. The following three areas will be considered, as indicated in the Introduction: (i)

Presurgical functional localization of eloquent cortex; (ii) Presurgical evaluation of epileptic patients. (iii) Functional localization of speech relevant brain areas.

Magnetic Functional Source Imaging of the Sensorimotor Strip

The central sulcus is an important anatomical reference in order to localize the sensorimotor cortical strip. It is important to identify precisely the functional anatomy of these cortical areas with the aim of performing surgical resections in and around these areas without harming the eloquent cortex. In particular space-occupying lesions, as tumours and vascular malformations may deform the cortical anatomy such that pre-operative MRI scans may not be appropriate for the identification of the central sulcus. In these cases a functional method is necessary. The central sulcus (CS) marks the border between the agranular motor cortex in the pre-central gyrus and the granular somatosensory cortex in the post-central gyrus (Zilles, Schlaug *et al.* 1995). The question is whether MEG may provide reliable information regarding the localization of the somato-sensory and motor cortical areas.

Functional Localization of Somatosensory Cortex

A number of studies showed that MEG provides a powerful method for the functional mapping of the somatosensory cortex (Hari, Reinikerinen *et al.* 1984), (Wood, Cohen *et al.* 1985), (Sutherling, Crabdall *et al.* 1988), (Buchner, Fuchs *et al.* 1994), (Rezai, Hund *et al.* 1996), (Ossenblok, Luiten *et al.* 2003). This is very relevant to guide the neurosurgeon in interventions where it is necessary to map the somatosensory cortex around the central sulcus. It is well established that the electrical stimulation of the median nerve generates an evoked field in the SI hand area located in the post-central gyrus. Most commonly the component of the cortical electric evoked field that is taken as reference is the negative peak at about 20 ms latency (Okada, Tanenbaum *et al.* 1984; Allison, McCarthy *et al.* 1989).This fits well with the results obtained by way of direct cortical recordings showing (reviewed by Allison (1991)) that the sources of early cortical somatosensory evoked potentials are located in the parietal cortex, more specifically in the primary somatosensory area of the posterior bank of the central sulcus. The source modelling studies of electrical and magnetic fields, along with the cortical recordings, agree that the N20 field is generated by a dipolar source tangential to the cortical surface. This dipolar field is produced by the activity of neurons of area 3b evoked by cutaneous inputs. The localization of this dipolar field is nowadays the standard method to estimate the localization of the central sul-

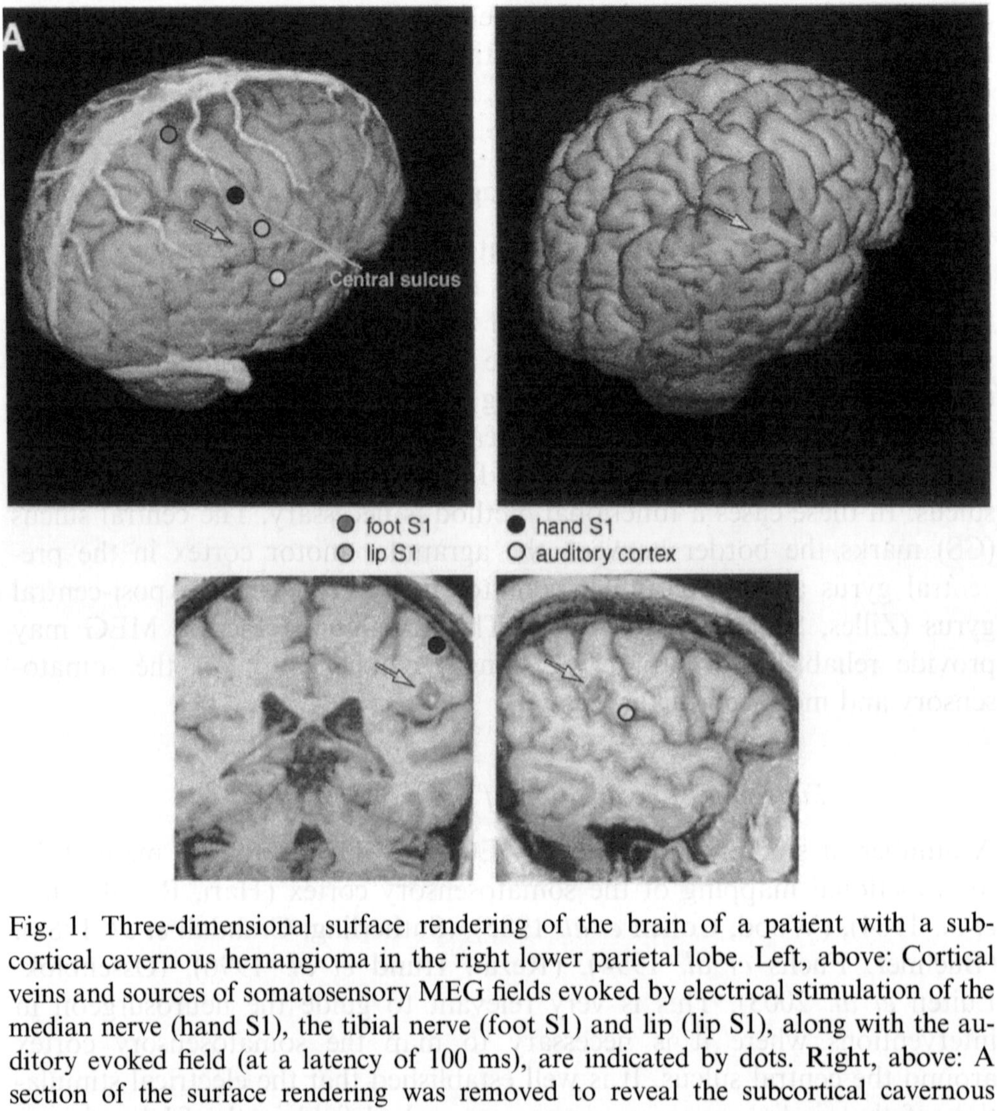

Fig. 1. Three-dimensional surface rendering of the brain of a patient with a subcortical cavernous hemangioma in the right lower parietal lobe. Left, above: Cortical veins and sources of somatosensory MEG fields evoked by electrical stimulation of the median nerve (hand S1), the tibial nerve (foot S1) and lip (lip S1), along with the auditory evoked field (at a latency of 100 ms), are indicated by dots. Right, above: A section of the surface rendering was removed to reveal the subcortical cavernous hemangioma, and the sulcal route (arrow) used for its removal. Below: on the left a coronal MRI section showing the source of the median nerve evoked field (dot) and on the right a sagittal section showing the source of the auditory evoked field (dot); The arrows show the approximate location of the tumour, that is below the median nerve source and posterior to the auditory source. (Adapted with permission from (Mäkelä, Kirveskari *et al.* 2001))

cus (Fig. 1). A thorough analysis of this and other components of the somatosensory cortex evoked potentials is given by Mauguiere (1998) and of the corresponding magnetic fields by Hari and Forss (1999). In this context it is important to note that the MEG is to be preferred to the EEG because the former is less sensitive to the different conductivities of the tissues which, in pathological cases may be even more complex than in normal cases, due to local changes in conductivity caused by tumour masses, oedema and/or skull defects. In a clinical application (Ganslandt, Fahlbusch *et al.* 1999) during surgery, the pre- and postcentral gyri were identified by neuronavigation and, in addition, the central sulcus was localized using intraoperative recording of somatosensory evoked potentials. In all cases MEG localizations of the sensory or motor cortex were correct. These authors conclude that the method of incorporating functional data into neuronavigation systems is a useful tool that can be used to lessen morbidity around eloquent brain areas.

Functional Localization of Motor Cortex

The studies described above had the aim of localizing the posterior bank of the central sulcus using the evoked responses of the somatosensory cortex. In addition a number of studies have focused on the functional localization of the primary motor cortex on the anterior bank of the central sulcus. This may provide important complementary information. Rezai *et al.* (1996) recorded movement-related magnetic fields to estimate the location of the cortical motor strip. The observation that MEG rhythmic activity of the motor cortex at about 15–30 Hz shows a significant coherence with the EMG of arm muscles during isometric contractions by Salenius and collaborators (Salenius, Portin *et al.* 1997; Hari and Salenius 1999) has yield a new method to estimate the localization of the motor strip. Mäkelä *et al.* (2001) combined information obtained using magnetic fields evoked by median and tibial nerves and by lip stimulation, with that obtained by estimating the maximal coherence of MEG oscillatory cortical activity with the electromyogram of hand and foot muscles in order to estimate the precentral motor and the post-central somatosensory cortices. The neuromagnetic information was displayed on 3D-surface reconstructions of the individual brains. These data were verified using intraoperative corticography. The sources of the somatosensory evoked fields located the posterior bank of the central sulcus correctly in all patients, whereas information about the localization of the motor strip was obtained in 8 of 12 patients. This study shows that this complementary information, obtained pre-operatively, can be useful in guiding the neurosurgeon so that possible damage to eloquent cortical areas of the sensorimotor strip may be substantially reduced.

Comparison Between MEG and fMRI Regarding the Functional Localization of Somatomotor Cortex

Above we noted that fMRI can compete with MEG/EEG to some extent with respect to the functional localization of the cortical areas around the central sulcus. In this context which of these techniques has the best performance? The fMRI measures changes in magnetic signals related to spatio-temporal haemodynamic events that are associated with neuronal activity evoked by a stimulus. These basic haemodynamic events (Malonek, Dirnagl *et al.* 1997) consist of increased neuronal O2 consumption (latency: 100 ms), and associated increased blood volume (latency: 300–500 ms) and increased blood flow (latency: 500–1500 ms) in surrounding vascular compartments. These haemodynamic changes are reflected in variations of the ratio oxy-/reduced-haemoglobin that can be picked up, to some extent, by fMRI. We should note that there is, however, a time delay between the occurrence of changes of neuronal activity, that in the case considered here is evoked by stimulation of the median nerve, and the associated haemodynamic changes. This implies that there is a physiological dead-time that does not allow to detect the earliest changes of neuronal activity by way of haemodynamic measurements. In addition there are also technical limitations, since the sensitivity of the current MRI apparatuses (1.5 T) is not sufficient to pick up the weak haemodynamic changes occurring at the shortest latencies. This performance may improve with more powerful MRI apparatuses.

What is the correspondence between fMRI signals and neuronal activity evoked by peripheral stimulation in the somatosensory cortex? This question was addressed, in a comprehensive way, by Disbrow *et al.* (2000). Cortical activation maps obtained after somato-sensory stimulation of hand, face and forelimb in monkey, were measured using standard fMRI (1.5 T) and electrocorticography (microelectrode electrophysiology) and compared. In 55% of the maps the centroids of fMRI maps co-localized with those obtained electrophysiologically. In 45%, however, they did not, and the mean distance between the two kinds of centroids was 1 cm. These differences were related to how the cortical maps are oriented in relation to the cortical blood vessels. The cortical maps that are oriented parallel to the blood vessels display a good overlap between fMRI and electrophysiological signals, whereas those that are perpendicular do not overlap, or overlap only partially. This means that in addition to the differences in timing, there are also spatial differences between fMRI and electrical/magnetic signals, determined mainly by the orientation of the surrounding blood vessels that condition the fMRI signals, while they do not affect the MEG.

A direct comparison of fMRI and MEG in their capacity to localize the central sulcus was carried out by Inoue *et al.* (1999) in a neurosurgical setting. They found that fMRI and MEG coincided in defining the central sulcus in all 24 hemispheres of volunteers and all 10 examined non-affected hemispheres of patients. The percentage of concordance, however, decreased to 82% for the affected hemispheres of 11 patients with brain tumors. All MEG localizations were confirmed by intra-operative recordings of somatosensory evoked potentials from the cortical surface. In those few cases where fMRI was not reliable, this was probably due to venous flow changes by tumor compression. These authors conclude that for precise functional assessment of the brain affected by intracranial tumors, the combination of fMRI and MEG is recommended. The same conclusion was reached by Kober *et al.* (2001) who estimated the distances between MEG and fMRI activation sites in a number of neurosurgical patients. The central sulcus could be identified by MEG and fMRI in 33 of 34 cases. However, MEG and fMRI localization results showed significantly different activation cortical sites for motor and sensory tasks with a distance of 10 and 15 mm, respectively. These authors conclude that both modalities are useful for the estimation of the somatomotor cortex, but a single modality may err in the exact topographical labelling of the somatomotor cortex. In some unclear cases a combination of both methods should be used in order to avoid post-surgical neurological deficits.

In short, the MEG, whether alone or in combination with fMRI, is an important tool for functional localization of the cortical areas surrounding the central sulcus.

MEG in Epilepsy: Identification of Epileptiform Inter-Ictal Foci

With the aim of localizing epileptogenic areas particularly in order to plan surgical resections of these areas it is important to identify where in the brain epileptic seizures start. This cannot be easily done using MEG since it is not practical to monitor epileptic patients during prolonged MEG recordings while waiting for a spontaneous seizure to occur. However, very useful information can also be obtained through the identification of the areas where epileptiform inter-ictal activity, e.g. spikes are present (Sutherling, Levesque *et al.* 1991; Ebersole 1997; Ossenblok, Fuchs *et al.* 1999; Baumgartner, Pataraia *et al.* 2000; Stefan, Hummel *et al.* 2000). Both EEG and MEG can accomplish this task efficiently, but the latter has some advantages since it yields more localized fields and it allows a better differentiation of multiple sources. In our experience this applies particularly to neocortical sources. However, temporal lobe epilepsies tend to yield too few interictal spikes in the MEG, likely due to the distance between the sensors and the hippocampal region. In any case the signal-to-noise ratio of individual spike discharges is usually relatively low. Therefore especial techniques have to be applied to enhance it. Most often the MEG/EEG recordings of inter-ictal epileptiform activity are complex signals, with artefacts and other non-epileptiform transients that have to be avoided using appropriate methods (Gotman and Wang 1991), along with real epileptiform spikes that may arise from multiple brain sources (Engel

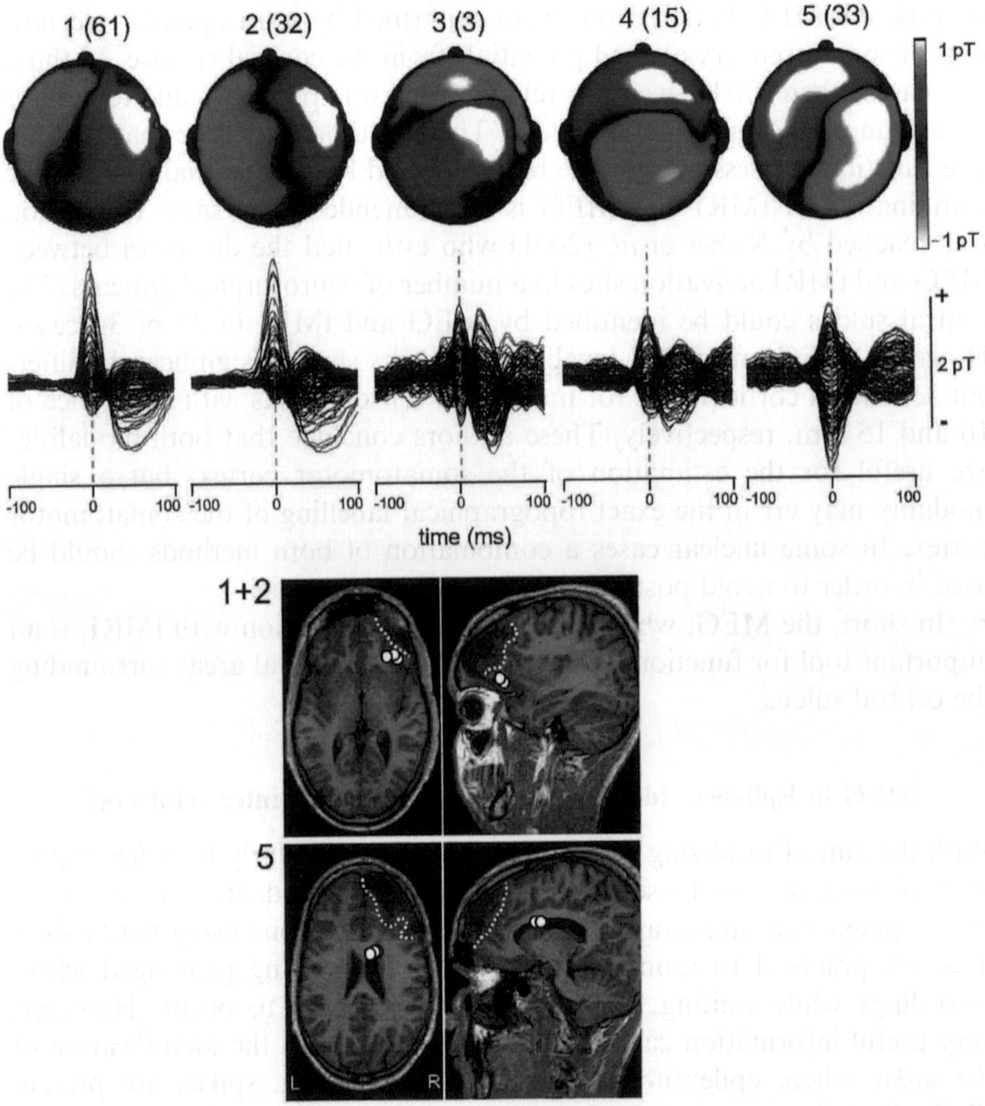

Fig. 2. Above: MEG Topographic maps and overlay of MEG signals recorded from different sensors for 5 epileptiform spike clusters in a patient with a 10-year history of a seizure disorder that became manifest after drainage of an intracerebral abscess in the right frontal lobe. The seizures proved unresponsive to medical treatment so that the patient was evaluated for epilepsy surgery. Above each map the cluster number and the number of spikes it contains (within brackets) is indicated. The field maps represent the magnetic field distribution at the time of the marker indicated by the vertical dashed line, set at time zero. Below: position of equivalent dipoles in MRI slices for clusters 1 and 2 (1st row) and for cluster 5 (2nd row). Clusters 3 and 4 are not indicated because they did not yield residual errors less than 10%. The boundary of the structural lesion in the right frontal lobe is marked by the dotted line. Note that the dipoles of clusters 1 and 2 were located at the posterior border of the lesion in the right frontal lobe. Those of cluster 5 were also located in the same hemisphere but more posterior, likely indicating propagated epileptiform activity from the main more anterior sources. The patient underwent a right frontal lobectomy. Pathology of the

Jr 1993), (Lantz, Wahlberg *et al.* 1998), (Iwasaki, Nakasato *et al.* 2002). This implies that the MEG/EEG epileptiform spikes should be grouped into distinct clusters before estimating the corresponding brain sources. Several methods have been proposed to achieve this objective (Lantz, Wahlberg *et al.* 1998), (Wilson, Turner *et al.* 1999; Wahlberg and Lantz). Recently an automated algorithm for clustering of epilepiform spikes was developed in our group (Van't Ent, Manshanden *et al.* 2003) that was applied to MEG recordings from epileptic patients with localization-related epilepsy. Inverse computations applied to selected averages of spike clusters, in order to identify the corresponding brain sources, yielded source solutions that agreed with the presumed site of seizure initiation and were associated with the structural lesions seen in MRI scans (Fig. 2). Even in patients with no evident brain lesion in the MRI this approach helped to obtain insight in the characteristics of a possible neocortical epileptogenic area.

In conclusion these recent advances indicate that objective cluster analysis of large numbers of MEG epileptiform spikes yields useful clinical information regarding the delineation of an epileptogenic zone. This information can converge with that obtained using classic scalp video/EEG seizure monitoring, even in the absence of a structural lesion in MRI.

Functional Localization of Speech Relevant Brain Areas

In addition to the applications described above, which are directly relevant for neurosurgical applications, the MEG is becoming an important tool for the analysis of dynamical brain activities in relation to neuro-cognitive functions (Momjian, Seghier *et al.* 2003). In this context the functional mapping of speech relevant areas is also clinical useful since in neurosurgery it is of the uttermost importance to avoid damage to these eloquent areas of the brain. This is usually done by way of the invasive Wada test (Wada and Rasmussen 1960). The question is whether the MEG may be an alternative non-invasive method to the Wada test. A number of studies, particularly of the Helsinki school, have shown that MEG signals evoked by speech related tasks can be recorded and localized to speech relevant brain areas (Fig 3). A picture naming task in normal subjects revealed that MEG fields evoked by the visual presentation of pictures to be named appear first in the occipital areas and evolve in time from the posterior to frontal areas (Salmelin, Hari *et al.* 1994). These MEG signals

excised brain tissue revealed the presence of extensive gliosis secondary to the intracerebral drainage performed earlier. The patient has remained seizure free, on anti-epileptic medication, 14 months after surgery. (Adapted with permission from (van't Ent, Manshanden *et al.* 2003))

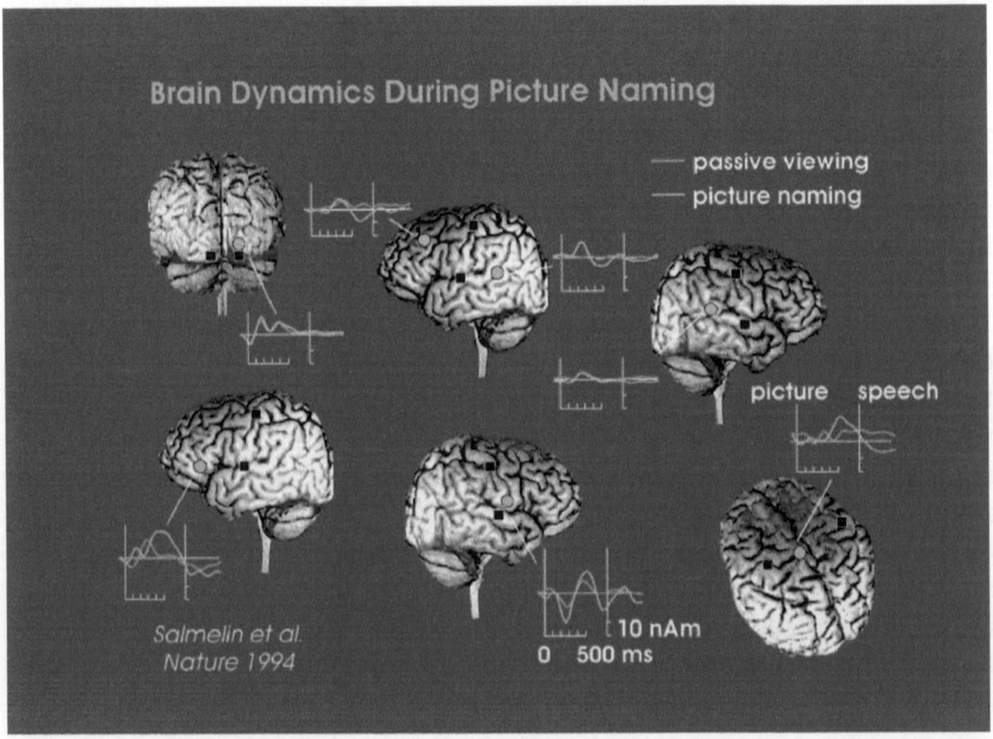

Fig. 3. Brain activation in the course of a naming-aloud experiment (orange traces) of pictures presented on a computer screen once every 5 s. The early visual MEG evoked response (source 1, picture upper left) at the back of the head was followed by more lateral posterior signals in both hemispheres (sources 2 and 3, upper middle and right), suggesting a contribution from Wernicke's area and its right-sided counterpart. Within the same time window (grey rectangles), a left frontal site (4, below left) close to the cortical face representation area showed activation, as in preparation for mouth movements. About 500 ms after the picture had been presented, signals emerged from a site anterior to the face representation area in the motor cortex, reflecting activation of Broca's area (5, below middle) and its counterpart in the right hemisphere. These signals were immediately followed by activity at the top of the brain (6, below right), probably generated in the vicinity of the supplementary motor area. MEG responses during passive viewing (white traces) and quiet naming (green traces) are also shown. The locations of visual (*V*), auditory (*A*), and somatosensory (*S*) projection cortices are marked by yellow squares. (Adapted with permission from (Salmelin, Hari *et al.* 1994))

are bilateral but over the left hemisphere they are usually stronger and leading in time. In a more specific task Gootjes, Raij *et al.* (1999) performed a MEG study in 11 healthy right-handed subjects to estimate individual hemispheric dominance for speech sounds. The auditory stimuli comprised binaurally presented vowels, tones, and piano notes in groups of two or four stimuli. In the left hemisphere, vowels evoked significantly stronger (37–79%) responses than notes and tones, whereas in the right hemisphere the responses to different stimuli did not differ significantly. Specifically, in the two-stimulus task, all 11 subjects showed left-hemisphere dominance in the vowel vs tone comparison. This simple paradigm may be helpful in non-invasive evaluation of language lateralization. Other MEG studies using different forms of signal analysis are congruent with these findings. Papanicolaou *et al.* (1999) reported that they could identify the dominant hemisphere by inter-hemispheric comparison of the number of dipolar sources, and the same applies to the study of Kober (2001) who used a laterality index calculated from the strength of current density distributions after spatial filtering.

More well controlled studies in larger groups of subjects are, of course, necessary. Nevertheless the results obtained until now allow us to conclude that MEG is a valuable method to investigate the dynamics of speech processing. Furthermore MEG merits being explored as a potential alternative to the Wada test for non-invasive presurgical localization of speech eloquent brain areas and the identification of the language dominant hemisphere.

Discussion and Future Developments

The MEG has been slow in being accepted as a major method of diagnosis due to its high cost relative to the well established EEG. Nevertheless the advantages of MEG over EEG are becoming accepted in a number of applications as indicated above. We discuss here in more general terms the main advantages and disadvantages of MEG versus EEG. Since the MEG is the newcomer we may start by discussing whether it has advantages with respect to the older EEG method. We may distinguish three main advantages of the MEG.

First, the EEG is a relative measurement; it needs a reference electrode, while the MEG does not. This makes the analysis of correlations between signals recorded from different sites over the head, using the MEG, easier to interpret. This is particularly relevant in dynamical investigations of MEG signals during the performance of cognitive tasks, such as in the analysis of speech processing.

Second, the MEG field generated by a given cortical population is more tight than the corresponding EEG field. This is due to the fact that the EEG depends mainly on volume currents that are smeared out by the isolating shell of the skull, whereas the latter do not contribute to the MEG. This implies also that in the case that several generators are simultaneously present in the cortex, the MEG is capable of resolving the corresponding sources more accurately than the EEG.

Third, the MEG is much less affected by the electric properties of the shells that surround the brain, namely the skull and the scalp. This makes the determination of brain sources of MEG activities more accurate than of EEG sources. There are, of course, also some disadvantages of the MEG. Long recordings such as it may be necessary to be made in sleep research are more difficult to obtain, although not impossible (Simon *et al.* 2000). It is unpractical to record epileptic seizures since the patient may be hurt due to the head confinement in the MEG container. In addition a non-trivial disadvantage is the price of the MEG installation. This is the main reason why there are relatively few MEG facilities around the world, while most of these are still primarily dedicated to fundamental research.

One problem that has hindered the more generalized acceptance of MEG as a clinical tool is the fact that a consensus about how to process MEG signals has not yet emerged. Different methods are being used, which very often are variations of the basic equivalent dipolar source model. Indeed a general approach is to estimate from a given MEG signal, after adequate processing, the equivalent current dipole model for a series of time samples within a time-window of interest. Dipolar solutions should be accepted only when the residual error between measured and computed magnetic fields is below a certain threshold (typically values: 10 or 5%). The dipolar solutions should be presented on the corresponding MRI slices along the three planes (horizontal, sagittal and coronal). Thereafter the dipolar distributions on these MRI slices have to be interpreted as representing reliable sources, either in the form of one single or multiple clusters. Alternative methods have been proposed such as those based on spatial filtering to estimate current distributions (Robinson and Rose 1992). Among other applications this methodology has been used to estimate the dynamics of speech processing both in normal subjects and patients. Be as it may, the field of MEG analysis has entered the phase of maturity such that more comprehensive studies can reasonably be expected.

Besides advances in the creation of comprehensive software for MEG analysis that may be more widely accepted, we may expect, in the near future, also technical developments in hardware, with the implementation of novel high-temperature SQUIDs operating at the temperature of liquid nitrogen, such that the MEG sensors may be brought closer to the scalp improving the signal-to-noise ratio. Accordingly information from brain sources of MEG signals lying deeper under the scalp may become more accessible. New technological improvements may also lead to a decrease of the cost of the MEG installations, what would be most desirable.

References

1. Allison T, McCarthy G *et al* (1989) Human cortical potentials evoked by stimulation of the median nerve. J Neurophysiol 62: 694–710

2. Allison T, McCarthy G *et al* (1991) Potentials evoked in human and monkey cerebral cortex by stimulation of the median nerve: a review of scalp and intracranial recordings. Brain 114: 2465–2503
3. Anastasiadis PG, Kotini A, Anninos P, Adamopoulos A, Sigalas J, Koutlaki N (2003) Chaotic and periodic analysis of fetal magnetocardiogram recordings in growth restriction. Prenat Diagn 23(5): 405–409
4. Baumgartner C, Pataraia E *et al* (2000) Magnetoencephalography in focal epilepsy. Epilepsia 41 [Suppl] 3: S39–S47
5. Buchner H, Fuchs M *et al* (1994) Source analysis of median nerve and finger stimulated somatosensory evoked potentials: multichannel simultaneous recording of electric and magnetic fields combined with 3D MRI tomography. Brain Topography 6: 299–310
6. Cohen D (1968) Magnetoencephalography: evidence of magnetic field produced by alpha-rhythm currents. Science 161: 784–786
7. Disbrow EA, Slutsky DA *et al* (2000) Functional MRI at 1.5 tesla: a comparison of the blood oxygenation level-dependent signal and electrophysiology. Proc Natl Acad Sci USA 97(17): 9718–9723
8. Ebersole JS (1997) Defining epileptogenic foci: past, present and future. J Clin Neurophysiol 14: 470–483
9. Engel J Jr (1993) Intracerebral recordings: organization of the human epileptogenic region. J Clin Neurophysiol 10: 90–98
10. Ganslandt O, Fahlbusch R *et al* (1999) Functional neuronavigation with magnetoencephalography: outcome in 50 patients with lesions around the motor cortex. J Neurosurg 91(1): 73–79
11. Gootjes L, Raij T *et al* (1999) Left-hemisphere dominance for processing of vowels: a whole-scalp neuromagnetic study. Neuroreport 10(14): 2987–2991
12. Gotman J, Wang IY (1991) State-dependent spike detection: concepts and preliminary results. Electroenceph Clin Neurophysiol 79: 11–19
13. Hari R, Forss N (1999) Magnetoencephalography in the study of human somatosensory cortical processing. Phil Trans R Soc Lond B 354: 1145–1154
14. Hari R, Reinikerinen K *et al* (1984) Somatosensory evoked cerebral magnetic fields from SI and SII in man. Electroenceph Clin Neurophysiol 57: 254–263
15. Hari R, Salenius S (1999) Rhythmical corticomotor communication. NeuroReport 10: R1–R10
16. Inoue T, Shimizu H *et al* (1999) Accuracy and limitation of functional magnetic resonance for identification of central sulcus: comparison with magnetoencephalography in patients with brain tumours. Neuroimage 10: 738–748
17. Iwasaki M, Nakasato N *et al* (2002) Surgical implications of neuromagnetic spike localization in temporal lobe epilepsy. Epilepsia 43: 415–424
18. Kanzaki H, Nakatani S, Kandori A, Tsukada K, Miyatake K (2003) A new screening method to diagnose coronary artery disease using multichannel magnetocardiogram and simple exercise. Basic Res Cardiol 98(2): 124–132
19. Kober H, Moller M *et al* (2001) New approach to localize speech relevant brain areas and hemispheric dominance using spatially filtered magnetoencephalography. Hum Brain Mapp 14(4): 236–250
20. Kober H, Nimsky C *et al* (2001) Correlation of sensorimotor activation with

functional magnetic resonance imaging and magnetoencephalography in presurgical functional imaging: a spatial analysis. Neuroimage 14(5): 1214–1228
21. Lantz G, Wahlberg P *et al* (1998) Categorization of inter-ictal epileptiform potentials using a graphic-theoretic method. Electroenceph Clin Neurophysiol 107: 323–331
22. Mäkelä JP, Kirveskari E *et al* (2001) Three-dimensional integration of brain anatomy and function to facilitate intraoperative navigation around the sensorimotor strip. Human Brain Mapping 12: 180–192
23. Malonek D, Dirnagl U *et al* (1997) Vascular imprints of neuronal activity: relationships between the dynamics of cortical blood flow, oxygenation, and volume changes following sensory stimulation. Proc Natl Acad Sci USA 94(26): 14826–14831
24. Mauguiere F (1998) Somatosensory evoked potentials: normal reposnses, abnormal waveforms and clinical applications in neurological diseases. 1014–1058
25. Momjian S, Seghier M *et al* (2003) Mapping of the neuronal networks of human cortical brain functions. Adv Tech Stand Neurosurg 28: 91–142
26. Okada YC, Tanenbaum R *et al* (1984) Somatotopic organization of the human somatosensory cortex revealed by neuromagnetic measurements. Exp Brain Res 56(197–205)
27. Ossenblok P, Fuchs M *et al* (1999) Source analysis of lesional frontal lobe epilepsy. IEEE Eng Med Biol Mag 18: 67–77
28. Ossenblok P, Luiten FSS *et al* (2003) Magnetic source imaging contributes to the presurgical identification of sensorimotor cortex in patients with frontal lobe epilepsy. Clin Neurophysiol 114: 212–232
29. Papanicolaou AC, Simos PG *et al* (1999) Magnetoencephalographic mapping of the language specific cortex. J Neurosurgery 90: 85–93
30. Rezai AR, Hund M *et al* (1996) The interactive use of magnetoencephalography in stereotaxic image-guided neurosurgery. Neurosurgery 39(1): 92–102
31. Robinson SE, Rose DG (1992) Current source image estimation by spatial filtered MEG
32. Salenius S, Portin K *et al* (1997) Cortical control of human motorneuron firing during isometyric contractions. J Neurophysiol 77: 3401–3405
33. Salmelin R, Hari R *et al* (1994) Dynamics of brain activation during picture naming. Nature 368: 463–465
34. Stefan H, Hummel C *et al* (2000) Magnetoencephalography in extratemporal epilepsy. J Clin Neurophysiol 17: 190–200
35. Sutherling WW, Crabdall P *et al* (1988) The magnetic and electric fields agree with intracranial localization of somtaosensory cortex. Neurology 38: 373–381
36. Sutherling WW, Levesque MF *et al* (1991) Localization of partial epilepsy using magnetic and electric measurements. Epilepsia 32 [Suppl] 5: S29–S40
37. Van't Ent D, Manshanden I *et al* (2003) Spike Cluster analysis in neocortical localization related epilepsy achieves clinically significant source localization results in MEG. Clin Neurophysiol (in press)

38. Van't Ent D, Manshanden I *et al* (2003) Spike cluster analysis in neocortical localization related epilepsy yields clinically significant equivalent source localization results in magnetoencephalogram (MEG). Clin Neurophysiol
39. Wada J, Rasmussen T (1960) Intracarotid injection of sodium amytal for the lateralization of cerebral speech dominance. J Neurosurgery 17: 266–282
40. Wahlberg P, Lantz G (2000) Methods for robust clustering of epileptic EEG spikes. IEEE Trans Biomed Eng 47: 857–868
41. Wilson SB, Turner CA *et al* (1999) Spike detection. II. Automatic, perception-based detection and clustering. Clin Neurophysiol 110: 404–411
42. Wood CC, Cohen D *et al* (1985) Electrical sources in human somatosensory cortex: identification by combined magnetic and potential recordings. Science 227: 1051–1053
43. Zilles K, Schlaug G *et al* (1995) Mapping of human and macaque sensorimotor areas by integrating architectonic, transmitter receptor, MRI and PET data. J Anat 187(Pt 3): 515–537

Offprint from
Advances and Technical Standards in Neurosurgery, Vol. 30
Edited by J.D. Pickard

Basic and Clinical Aspects of Olfaction

B. N. Landis[1,2], T. Hummel[2], and J.-S. Lacroix[1]

[1] Unité de Rhinologie-Olfactologie, Service d' Oto-Rhinologie-Laryngologie, Hopitaux Universitaires de Genève, Genève, Switzerland
[2] Smell and Taste Clinic, Department of Otolaryngology, University of Dresden Medical School, Dresden, Germany

With 1 Figure

Contents

Abstract

Disturbances of olfaction are a common occurrence in many neurological and neurosurgical patients and their correct diagnosis might be helpful in management and enhancement of quality of life. However, olfaction is seldom checked in most neurosurgical units and the "smell bottles" are often either absent or out of date. This chapter reviews systematically recent advances in our understanding of the anatomy, physiology (olfactory coding) and measurement of olfactory function in the human. The causes and symptoms of smell disorders, risk of damage to the olfactory system by various surgical procedures and, finally, the natural history of recovery and treatment of smell disorders, for example after trauma, are discussed.

Keywords: Olfaction; smell disorder; anatomy; physiology; olfactory coding; measurement of olfactory function; craniotomy.

Anatomy

Although this review focuses on the olfactory system, it is necessary to mention, at least briefly, other sensory channels involved in chemosensory perception. All the senses can be stimulated by chemicals, which in fact, typically activate not only one but several of the "chemical senses". For example, nicotine not only activates the olfactory nerves, but also produces activation of the intranasal chemosensory trigeminal system.

Main Olfactory System

Olfactory perception starts at the level of the olfactory epithelium in the roof of the nasal cavity. Olfactory receptor neurons (ORN) are embedded within the respiratory epithelium and send their axons through the cribriform plate towards the olfactory bulbs. ORN carry olfactory receptors (OR) which are the key to olfactory information processing (see below). In the olfactory bulb ORN axons synapse with second order neurons, the mitral cells. The wiring between the olfactory epithelium and the olfactory bulb is characterized by a convergence of ORN axons. Specifically, all ORN carrying the same OR converge in the same site within the bulb, called "glomerulus". Axons from the mitral cells follow the olfactory tract and divide into two bundles. Most fibers directly project to the pyriform and entorhinal cortices as well as to the amygdalae (all structures formerly subsumed under the term "limbic system") whereas a minority of fibers project through the thalamus towards the orbito-frontal cortex [1].

Compared to other sensory modalities the olfactory system has some particularities. First, the majority of the olfactory fibers do not cross but project ipsilaterally in the brain. Second, most olfactory fibers bypass the thalamus and project very rapidly and directly in the pyriform cortex, amygdalae, and entorhinal cortex which are implicated in emotional and memory processing [2]. This difference in central anatomy has been claimed to be partly responsible for the emotional load olfactory memories can carry [3]. In contrast to other sensory modalities, no main olfactory cortex has yet been found. Numerous works indicate the orbitofrontal cortices to be an important relay in olfactory information processing [4].

Trigeminal System

The trigeminal system provides the somato-sensory innervation to the nasal mucosa. Somato-sensory input from the skin, the nasal and oral cavities, respectively, is mediated by the trigeminal system. Since most odorous compounds stimulate trigeminal nerve endings, at least at higher concentrations, this system is almost always co-activated in the perception of odors. With few exceptions almost all odorants have been shown to exhibit trigeminal activation to some extent [5] (e.g., mint has a somewhat fruity odor, but also evokes a typical cooling effect which is mainly trigeminally mediated). The main modalities supplied by the trigeminal system are temperature, pain, touch, and irritation. Testing the chemosensory intranasal trigeminal system psychophysically is more complex than olfactory testing. Since olfactory thresholds are always lower than the trigeminal thresholds for a given substance, olfactory biases are obvious.

The olfactory system is unable to localize the site of stimulation when one nostril receives clean air and the other nostril simultaneously receives an odor at the same time. In contrast, trigeminal stimulation can be localized. This difference is used to solve the bias inherent to trigeminal testing. Using lateralisation paradigms, trigeminal function can be easily and rapidly measured in a clinical context [6, 7]. Since the olfactory and trigeminal systems are so closely related anatomically and physiologically, there is a strong interaction between the two systems [8, 9]. In patients with olfactory loss, the trigeminal function is also weakened [10, 11]. Older literature on trigeminal trans-sections also discusses its impact on olfactory

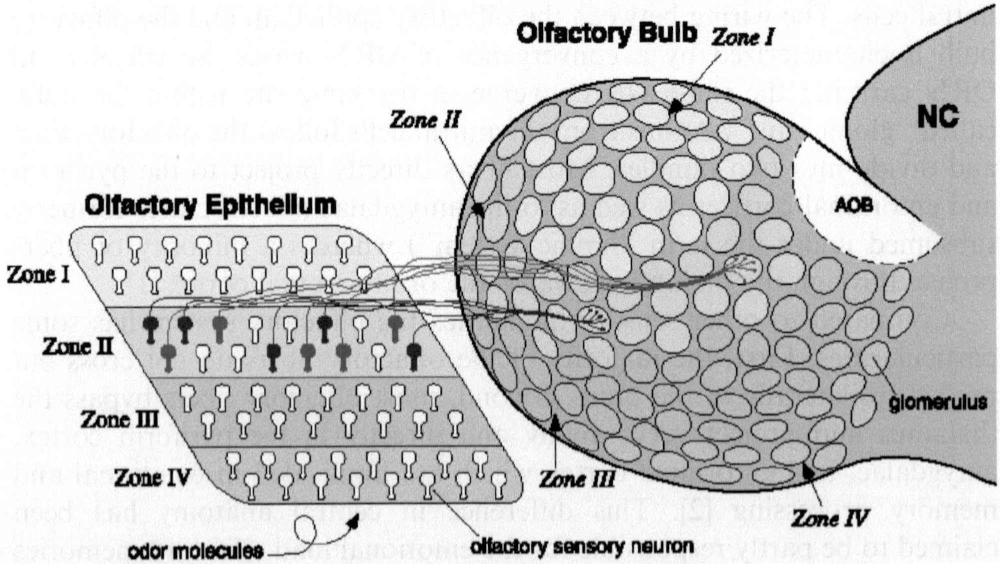

Fig. 1. Schematic diagram illustrating the axonal connectivity pattern between the nose and the MOB. The OE in mice is divided into four zones (zones I through IV) that are defined by the expression of odorant receptors. Olfactory sensory neurons in a given zone of the epithelium project to glomeruli located in a corresponding zone (*zones I* through *IV*) of the MOB. Axons of sensory neurons expressing the same odorant receptor (red or dark blue) converge to only a few defined glomeruli. *NC* Neocortex; *AOB* accessory olfactory bulb. Reprinted (abstracted/excerpted) with permission from Mori K, Nagao H, Yoshihara Y (1999). The olfactory bulb: coding and processing of odor molecule information. Science 286 (5440): 711–715. Copyright (1999) AAAS

function [12]. However, patients with complete trigeminal loss are extremely rare and no large study has so far been conducted to investigate the effects of trigeminal loss on olfactory function.

Gustatory System

The gustatory system provides the five basic tastes; sweet, sour, salty, bitter, and umami (glutamate). The latter, which resembles mainly the taste of chicken soup, has long been claimed in the Asian literature to be a basic taste quality [13, 14], whereas the western scientific community considered umami mainly as a "taste enhancer". This controversy was resolved when monosodium glutamate receptors were found on the tongue surface acting as specific taste receptors [15]. Molecular biological knowledge about taste receptors started to emerge a few years ago. Most basic taste qualities are not mediated by just one receptor type; several receptors act, for example, as sweet receptors. Many other taste modalities have been postulated (metallic taste, fat taste) and are currently under investigation. Future research is expected to clarify the coding mechanisms in taste perception.

Taste receptors are located within the taste buds, which are situated on all papillae except the filiform type. The highest densities of taste buds are found on the tongue and palate but they are also found throughout the entire oral cavity, hypopharynx and subglottic larynx [16]. Like ORNs, gustatory sensory receptor cells have the ability to regenerate [17]. Neural supply for these cells is provided by the facial, glossopharyngeal, and vagal nerves. The facial nerve innervates the anterior two thirds of the tongue, while the glossopharyngeal and vagal nerves innervate the posterior third of the tongue, as well as the hypopharynx and larynx. Like olfaction, taste fibers project ipsilaterally into the basal ganglia and brain stem. All gustatory fibers (facial, glossopharyngeal and vagus) innervating the oral-pharyngeal cavity converge into the nucleus solitarius within the brain stem.

Vomeronasal System

In 1813, Jacobson described a mucosal organ located on each side of the nasal septum and which was subsequently named "Jacobson's organ" [18, 19]. Later, the vomeronasal organ (VNO, consisting of vomeronasal epithelium, nerve, and accessory olfactory bulb) was shown to mediate effects attributed to pheromones [20]. A pheromone is a chemical molecule or compound which is secreted by one member of a species and, as soon as it is perceived by another member of the same species, elicits physiological, behavioral, or endocrinological effects [21, 22]. According to the original definition of Karlson and Lüscher [20] such effects must be species-specific. While the functionality and biological relevance in most animals is well documented, there is ongoing debate about the functional significance of the vomeronasal pouch in humans. Some authors claim to find neuronal activity within the vomeronasal epithelium [23] while many other studies suggest that the vomeronasal duct is nonfunctional in humans, with some vomeronasal nerves missing and lack of accessory olfactory bulbs in adults [24–27]. Furthermore, a vomeronasal duct is not always present in humans; recent investigations revealed that approximately 60% of humans have one [25]. Nevertheless, a few papers indicate that pheromonal-like effects occur in humans [28, 29] and several vomeronasal-like receptor genes have been found in the human genome, one of which is expressed in the olfactory epithelium (V1r) [30]. It is not yet known whether these receptors are functional or not. Their expression, however, indicates that putative "human pheromones" may act via the main olfactory system.

Olfactory Coding

The question of olfactory information encoding has been a concern for a long time. One main problem was to find a theory or model that would predict the odorous properties of a given molecule. Although the fragrance industry spends a lot of money on the creation of new – and hopefully smelly – molecules, no model exists which could predict the smell of any given molecule. The search for new odorants is still a very expensive procedure based on trial and error. Therefore, a universal model of stereochemical – odor interaction would greatly assist the search for new odorants.

Several models have been proposed to explain how the olfactory system discriminates between odorants. In the early sixties Mozell hypothesized that the chromatography of a molecule would determine its processing [31, 32]. According to Mozell, the olfactory receptors, which are located on the cilia of the olfactory neurons, are covered by a mucous layer and odorants have to cross this mucus before reaching the receptor cell. His theory was based on experiments using frog olfactory epithelium. Although no clear evidence has been presented that absorption of odorants is irrelevant to its interaction with the receptors, this theory has received less attention during recent years. Nevertheless, recent work on humans suggests that absorption could have implications for olfactory perception [33]. Another model indicates that olfactory recognition is mainly based on a few basic odors and that combination of these odors encodes the olfactory information [34, 35]. This model claims that olfaction works according to physiological principles similar to those governing vision. This assumption was mainly based on experiments on specific anosmias to isovaleric acid. Further experiments with other odorants were not able to confirm this model. Another theory receiving interest from the media is an old idea [36] reactivated by Turin [37]. According to this theory, olfactory coding could be based on vibration properties of the odorants. Recent work, however, indicates that this model can not predict the olfactory characteristics of a given molecule [38].

Since odorants are chemical structures, the existence of a ligand–receptor interaction has been claimed for many years, and was finally substantiated in 1991 by the discovery of a large family of seven transmembrane receptor proteins, expressed exclusively in the olfactory neuroepithelium. These olfactory receptors (ORs) are encoded by approximately 1000 genes in the mouse, or approximately 1% of its genome [39]. While the mouse expresses approximately 850 of these genes, the rest being pseudogenes, humans have far fewer functional ORs (approximately 350) [40]. Although this seems to indicate a loss in function, the simple equation "less receptors = less function" is currently under debate and some studies argue that humans have a very high preservation rate for specially important ORs [41–43].

The discovery of the OR superfamily led to a renaissance of olfactory research. During the last decade, potential OR binding sites [44] and the topographical organization and distribution of the ORs within the olfactory epithelium have been partly identified [45–47]. A recent finding has been the astonishingly high degree of organization found within the peripheral olfactory system. The first striking observation was that, among all the potentially expressible OR, every ORN expresses only one single OR gene [48, 49]. Furthermore, axons from all ORN expressing the same OR, whatever their location within the olfactory epithelium, project into two glomeruli in each olfactory bulb. This organization is called glomerular convergence [50]. Thus, a large glomerular map in the bulb, containing hundreds of glomeruli, will correspond to all OR expressed in the olfactory neuroepithelium. Molecular and electrophysiological studies revealed that OR are not selective for only one odorant, but numerous molecules bind with varying affinities to a certain OR. A given receptor may bind to a molecule with a given carbon chain length, but may lose binding affinity as the agonist's chain length increases. Similarly, the OR binding affinity for a molecule may dramatically change upon modification of the functional groups (aldehydes, ketones, acids, esters, alcohols, etc) of this molecule [51, 52]. In addition, every odorant is recognized not by one but by several ORs simultaneously, depending on its particular chemical properties. At the level of the glomerular map this leads to a specific activation pattern for each odorant [53]. This odor-specific activation pattern is believed to be responsible for the recognition of and distinction between different odorants [54].

However, as previously mentioned and although the olfactory receptor theory adequately explains how olfactory coding could work, olfactory research is still a long way away from predicting the odor of a molecule based solely on the stereo-chemical properties of the latter.

Measurement of Olfactory Function

Similar to other sensory modalities, olfactory testing procedures will yield information which is either based on subjects' insights ("subjective" or "psychophysical" tests) or on more "objective" techniques less biased by the subjects' observations. Since the subjects' self assessment of olfactory function is unreliable, testing of olfactory function is necessary [55].

Psychophysical Methods of Olfactory Testing

The basic principle of psychophysical testing of olfaction is to expose a subject to an olfactory stimulus and to interpret the responses or reactions of the tested subject.

This procedure has numerous advantages in clinical application, but also important limitations. The most valuable advantage compared to objective testing methods in daily clinical life is the rapidity which allows psychophysical tests to serve as quick screening tools for olfactory dysfunction

[56]. More extensive testing sets, which can also be used for clinical research, allow graduation of the olfactory disorder. Fundamentally, every collection of odors is a potential olfactory test. Nevertheless, research during the last three decades [57, 58] has ruled out the importance of a well validated and reliable testing device. Whatever a clinical test consists of, it should reliably distinguish between anosmic, hyposmic, and normosmic subjects. Thus, the test should be based on normative data acquired and validated on large samples of healthy and diseased subjects. This includes comparison of the results with other validated tests and a good test-retest-reliability. These requirements apply to only a few olfactory tests available worldwide [57–63], since many tests of olfactory function do not comply with these criteria [64].

The best-validated olfactory tests include the UPSIT (University of Pennsylvania Smell Identification Test) [57, 59], the CCCRC-test (Connecticut Chemosensory Clinical Research Center) [58], and the "Sniffin' Sticks" [61, 62]. The latter one is a European test, while the first two were created in North America.

Most tests are based on a forced choice paradigm. An odorant is presented at supra-threshold concentration and the subject has to identify the odor from a list of descriptions of odors (e.g. the subject gets rose odor to smell, and is asked whether the perceived odor was "banana," "anis," "rose," or "lilac"). This forced-choice procedure controls the subjects' response bias. It also (potentially) allows the detection of malingerers since even anosmic subjects will produce a few "correct" answers provided in a random selection of items. However, this method is unreliable for medico-legal investigations since well-read or hyposmic malingerers may overcome these pitfalls. The result of the test corresponds to the sum of the correctly identified items. This test design is called a smell identification test, and is the most widely used way of testing [57–63, 65] probably because it is the most easy to understand. Most tests are based on the identification of 16 to 40 odors – the more items tested the more reliable the results. Identification tests are known to have a cultural connotation. Tests used in North America, for example, are composed of odors many of which are unfamiliar to continental Europeans or Asians (e.g., root beer, or wintergreen). The odors tested should therefore be adapted to the patients cultural background [66] in order to avoid unfamiliarity.

The two other widely used test designs are threshold tests and tests of odor discrimination. The idea of threshold tests is to expose a subject repeatedly to ascending and descending concentrations of the same odorant and to identify the least detectable concentration for this individual odor. Other designs are based on logistic regression [67, 68]. Discrimination tests mainly consist of a 3-alternative forced choice technique. Two of the administered odors are identical, one is different. The subjects' task is to detect the different one. In principal, tests for odor threshold/odor discrimination are non-verbal. In addition, they can be used repetitively – which is more difficult with odor identification tests.

Generally, identification and discrimination tests are believed to reflect central olfactory processing while thresholds are thought rather to reflect peripheral olfactory function. Accordingly, it has been claimed by several authors [69–74] that patients with diseases of the central nervous processing of odorous information exhibit selective disturbances of discrimination and identification while threshold results are normal. Although this idea of a certain pattern pathognomonic for "central" olfactory disturbances seems attractive, the vast majority of studies have yet failed to confirm such typical pathology-associated patterns [75, 76]. The only, so far reliable and recurrent test pattern in olfactory disturbance is a low threshold and normal identification and discrimination in patients with chronic sinunasal problems [77].

Besides the solid body of literature and its clinical convenience, the psychophysical tests have one main limitation. As soon as the patient's collaboration is not guaranteed, interpretation of test results becomes difficult or even impossible. Such cases include mainly willful non-collaboration in cases of malingering, or for demented, unconscientious, or inexperienced patients. In order to acquire olfactory information in such cases, more objective testing methods have been developed which rely less on the subjects' cooperation.

Electrophysiological/Imaging Techniques Used to Test Olfactory Testing

Electro-Olfactogram (EOG)

Electro-olfactograms (EOG) are electrical potentials of the olfactory epithelium that occur in response to olfactory stimulation. The EOG represents the sum of generator potentials of ORN. While this response has been used extensively in olfactory research in animals (e.g., [78]), there are only a handful of reports describing the properties of the human EOG. Among other results, EOGs have been used to provide evidence for the dominant role of the central nervous system in olfactory desensitisation [79], for the functional characterisation of the olfactory epithelium [80], the specific topographical distribution of ORN, the expression of ORN in response to exposure to odorants [81], and the characterisation of certain odorants as OR antagonists [82]. However, the EOG so far has not been systematically used in patients with olfactory dysfunction. This is partly due to the topographical specificity of EOG responses, meaning that EOGs to certain odorants may be recorded only at certain epithelial sites. Thus, the subjects' odorous impressions may not always be reflected by the presence of an EOG response. In addition, the presence of an EOG may not always represent an odorous sensation. Specifically, EOGs can be recorded in subjects with congenital anosmia [83], or EOGs are present at certain threshold levels when the subjects do not yet perceive an odor [84]. Having said this, EOGs may be extremely helpful in terms of elucidating pathological processes at the mucosal level [85].

Chemosensory Event-Related Potentials (CSERP)

Event-related potentials are EEG-derived poly-phasic signals. They are caused by the activation of cortical neurons which generate electro-magnetic fields [86]. As the EEG is a noisy signal which contains activity from many cortical neurons, ERP need to be extracted from this background activity. In other words, the signal-to-noise ratio needs to be improved. The classical approach to this problem involves averaging of individual responses to olfactory stimuli such that random activity would cancel itself out while all non-random activation would remain. In addition, stimuli are typically presented with a steep onset (<20 ms) in an extremely well-controlled, monotonous environment such that stimulus onset synchronizes the activity of as many cortical neurons as possible.

Olfactory ERP (1) are direct correlates of neuronal activation, unlike the signals that are seen, for example, in functional MR imaging, (2) have an extremely high temporal resolution in the range of micro-seconds, (3) allow the investigation of the sequential processing of olfactory information, and (4) can be obtained independently of the subject's response bias, i.e., they allow the investigation of subjects who have difficulties to respond properly (e.g., children, aphasic patients).

Based on a system developed by Kobal [87, 88], odors are applied intranasally. Presentation of odorous stimuli does not simultaneously activate mechano- or thermo-receptors in the nasal mucosa since odor pulses are embedded in a constantly flowing air stream. In contrast to audition or vision, to date no early ERP have been recorded in response to olfactory stimuli (for review see [89]) but only late near-field ERP, i.e. responses from cortical neurons. Peaks of the late near-field ERP fall into two groups. Earlier peaks like N1 encode a greater number of exogenous stimulus characteristics than of later, so-called endogenous components. That is, earlier components encode stimulus intensity or stimulus quality (e.g., "What is the nature of this stimulus?"), whereas later components are more related to the frequency, or the salience of the stimulus ("What is the meaning of this stimulus?") [86, 90–92].

Olfactory ERP are recorded all over the scalp. In terms of the topographic distribution of olfactory ERP, amplitudes exhibit characteristic patterns with a centro-parietal maximum for both amplitudes N1 and P2 [93] (compare [94–96]). Using magneto-encephalographic techniques [97]

Kobal and co-workers conducted a series of experiments which addressed the question of the generation of olfactory ERP. Cortical generators of the responses to trigeminal stimulation with CO_2 were localized in the secondary somato-sensory cortex [98]. Other work indicated [99, 100] that olfactory stimuli activate anterior-central parts of the insula, the para-insular cortex, and the superior temporal sulcus [101].

Clinical testing with chemosensory ERP [89] typically includes the recording of responses to olfactory (e.g., hydrogen sulfide, and phenyl ethyl alcohol) and trigeminal (e.g., CO_2) stimuli. This procedure has been adopted by the working group "Olfaction and Gustation" of the German ENT Society [102]. So far, in all investigated anosmic patients, intranasal trigeminal ERPs could be obtained after stimulation with CO_2 – although with significantly smaller amplitudes than in healthy controls [10]. In contrast, no olfactory ERPs could be detected in anosmic patients after stimulation with the odorants hydrogen sulfide and vanillin [103, 104]. Results from ERP investigations provide significant information in the testing of malingering patients. In a recent study, olfactory short-term recognition memory was assessed in patients with unilateral temporal lobe epilepsy and stereotactic electroencephalography (SEEG) recordings prior to surgery. Such recordings from the amygdala indicated the presence of chemosensory evoked potentials [267].

FMRI, PET, and MSI

Recent progress in the field of imaging presented the opportunity to study the functional topography of the human olfactory system in detail [106–108]. There are three major techniques being used: positron emission tomography (PET) [109–111], functional magnetic resonance imaging (FMRI) [112–114], and magnetic source imaging (MSI) based on magneto-encephalography [99, 101]. While bio-magnetic fields directly reflect electrophysiological events, PET and FMRI reflect either changes in blood flow or changes in metabolism which are epiphenomena of neuronal activity. Other major differences between these techniques relate to the temporal and spatial resolution. All three techniques have been used extensively to perform basic research, e.g., on olfactory induced emotions, odor memory, mechanisms of sniffing [109], or age- and sex-related differences in terms of olfactory function [115]. However, in order to become relevant for routine clinical investigations [116], these intriguing techniques await further standardization.

Causes and Symptoms of Smell Disorders

Since olfactory disorders or even total olfactory loss are far less of a handicap than blindness or deafness for the person concerned, there have not been many attempts to estimate the percentage of people with olfactory problems. Initial surveys were done with questionnaires and rapid smell tests. They revealed that approximately 1–3% of the population has an olfactory problem [117, 118]. Since most of the decrease in olfactory function, like any other sensory function, is due to aging [119], this high incidence was not so astonishing in an aging society with an increasing mean age. In these first attempts to evaluate the epidemiology of olfactory problems, olfactory function was tested rather rudimentarily. Consecutive studies yielded much higher percentages of the population concerned by olfactory dysfunctions [120]. Importantly, olfactory disorders seem to affect more younger people than previously thought and most anosmic or hyposmic subjects either do not realize that they have an olfactory disorder or are simply not sufficiently handicapped to consult a physician [121, 122]. Current consensus is that approximately 5 percent of the general population suffer from anosmia, unrelated to chronic nasal problems. Although the highest incidence is found in the age group above 65 years, anosmia is astonishingly frequent in subjects between 45 and 65 years of age. Results are similar for the distribution of hyposmia, with a mean percentage of approximately 20% of the general population exhibiting mild or moderate smell dysfunction [120–122]. Recent studies underlined the potential alteration of quality of life consecutive to olfactory impairment [123–125]. Although not all patients with olfactory impairment seek medical help due to decreased quality of life, some may experience hazardous events in daily life like eating spoiled food or undetected smoke or gas leaks [126].

Most Common Causes

Several reports have been published on the frequency of the diverse origins of olfactory dysfunction. A recent survey conducted in Austria, Switzerland and Germany [127] revealed that approximately 50% of patients with olfactory dysfunction seen in ENT clinics are due to sinunasal problems. Further frequent causes of olfactory dysfunction are related to traumatic and post-URTI events.

Olfactory Loss Following Infections of the Upper Respiratory Tract (URTI)

As mentioned above, epidemiological questions surrounding olfactory disorders within the general population have only recently been addressed. Previous contributions to the epidemiology of olfactory loss included retrospective analyses of specialized "Smell and Taste Centers" on their respective data bases [128–130]. With some minor differences, these reports show similar findings about the main causes of olfactory disorders. Apart from posttraumatic and sinunasal origin, post-URTI olfactory loss is among the major causes of olfactory dysfunction.

The patient's history typically starts with a cold, during which he loses his sense of smell. Not particularly bothered during the cold, the patient becomes suspicious about the smell loss when, one or two months after all sinunasal symptoms have abated, normal olfactory function does not return. This is usually the moment when the patient seeks medical advice, either from their general practitioner or from an ENT specialist. Unfortunately, very few studies focused on the epidemiology and prognostic outcome of post-URTI olfactory disorders [131–134]. Currently, no good data indicate which agent in such upper tract respiratory infections (URTI) leads to olfactory lesions. It is not even clear whether toxicity originates from a virus or bacteria, or from the immune response directed against olfactory neuroepithelium. Some authors claim that viral rather than bacterial infections are responsible for olfactory disturbances, and observe a higher incidence of dysosmias after spring and summer URTI [132]. Furthermore, women above 45 years of age seem to be affected at a higher percentage than men [132, 135, 136] – which brings up the potential olfactory protective effect of estrogens [137]. Nevertheless, the effect of estrogen on olfactory function remains an open debate [122, 138].

Clinically, it is important to know, and to inform patients with post-URTI olfactory anosmia or hyposmia, about the possibility of parosmia. Parosmia (also termed troposmia), the unpleasant distortion of odorous sensations, tends to occur two to three month after the URTI, although it appears sometimes to occur directly after the URTI. The real frequency of parosmia is probably higher than previously believed, in particular because patients do not always mention it to their physician [139, 140]. According to our clinical experience, up to 25% of subjects with URTI olfactory dysfunction experience parosmia or phantosmia.

Posttraumatic Olfactory Loss

Posttraumatic olfactory disorders represent approximately 20% of the patients seen in "Smell and Taste Clinics" [124]. Most posttraumatic olfactory dysfunctions are said to occur after occipital trauma, although no clear data on olfactory dysfunctions after lateral impacts exist. The current explanation is that "coup-contre-coup" lesions or tearing of the filae olfactoriae leads to anosmia or hyposmia. Although the entity of posttraumatic olfactory loss had already been described by the end of the last century it has, like most olfactory disorders received little systematic attention [141]. This might also be due to the modest olfactory complaints of severely poly-traumatized patients during their hospitalization. Olfactory loss seems to correlate with the severity of the trauma [142–145], although several authors pointed out the fact that there is considerable individual variability in terms of the vulnerability of olfactory structures [145, 146]. Thus, even minor trauma can lead to anosmia whereas severe brain injuries may not alter olfaction. Probably, the injured parts of the olfactory system are most often the filae olfactoriae which cross the cribriform plate. However, central structures such as the orbitofrontal cortex and gyrus rectus have also been found to be affected after head trauma [142]. Similar to post-URTI olfactory impairment, these patients are prone to develop parosmia and phantosmia several months after the trauma. Clinical experience shows that most patients with posttraumatic olfactory disturbance typically become aware of the alteration after some delay. It is usually several weeks after the injury, when the major health problems have resolved and patients are discharged from the hospital, that they begin to complain of taste or smell loss. This is probably due to increased attention to olfaction once the general health status improves.

Sinunasal Causes

The third large group of patients who seek counseling for olfactory problems are patients suffering from concomitant sinunasal problems. Approximately 20% of all patients in smell and taste consultations have lost or impaired olfactory function due to a nasal problem [124]. Nasal polyposis has been known for a long time to decrease olfactory abilities due to the mechanical obstruction of nasal cavity restricting the airflow to the olfactory cleft [77, 129, 147–151]. During the last two decades, as a result of better olfactory tests, mild olfactory impairments could also be identified in other groups of patients with sinunasal diseases such as allergic and uncomplicated chronic rhinosinusitis [77, 152, 153]. In contrast to posttraumatic and post-URTI olfactory dysfunctions, these patients rarely exhibit parosmia or phantosmia.

Neurodegenerative Causes

Olfactory loss is common in patients with idiopathic Parkinson's disease (IPD) [154–156]. While a decreased sniff volume seems to contribute to the diminution of olfactory function [157], electrophysiological recordings in response to passive olfactory stimulation clearly established the presence of olfactory impairment in IPD [158, 159]. This olfactory deficit is so reliable that it can be used as a marker of IPD [75]. In other words; if a patient with normal olfactory function presents with IPD symptoms the diagnosis should be re-investigated [160, 161]. It can also be assumed that olfactory loss precedes the onset of motor symptoms by 4–6 years [162, 163] so that IPD may be the reason for "idiopathic olfactory loss" in some patients. Olfactory loss is also observed regularly in Alzheimer's disease, but at a much lower frequency and is less pronounced in multiple system atrophy, Huntington's disease, and motor neuron disease [161]. Little or no olfactory deficit is seen in cortico-basal degeneration, progressive supranuclear palsy, or essential tremor [161, 164].

Idiopathic

According to the retrospective epidemiological studies of Taste and Smell Clinics, the diagnosis of idiopathic – unknown – origin of smell impairment accounts for almost 20% of the cases, with the sampling bias discussed above. This seems simply to reflect the poor understanding of factors interfering with olfaction. With further insight and research this percentage should logically decrease. A considerable number of these idiopathic causes might be due to sinunasal disease, post URTI dysosmias following an almost undetected URTI, or neurodegenerative diseases [165].

Less Frequent Causes

Endocrine Diseases

Diabetes is probably one of the best investigated endocrine diseases concerning olfactory disorders [121, 166–168]. Most studies reveal slight olfactory deficiencies in diabetic patients especially at threshold levels indicating a peripheral patho-mechanism compatible with a possible diabetic microangiopathy or peripheral polyneuropathy. However, olfactory impairment in diabetes is relatively mild. Two recent studies conducted with identification tests in large study samples did not find that diabetic patients exhibit a decreased ability to identify odors compared to healthy controls [121, 122]. Several other endocrine diseases such as hypothyroidism [169, 170], adrenocortical insufficiency (Addison's disease) [171] or pseudohypoparathyroidism [172], have been reported to cause olfactory disorders. Many endocrine diseases have been reported to cause hyposmia but rarely lead to anosmia.

Epilepsy

Epileptic patients have been repeatedly tested with all possible olfactory testing modalities, and the general findings were that epileptic patients perform similar to controls with regard to odor thresholds [173, 174]. In contrast, more centrally believed tasks such as odor identification, discrimination or memory tests revealed that epileptic patients have olfactory impairments predominating on the side of the epileptic focus [74, 174–176].

Furthermore, olfactory evoked potentials have been shown to be altered in epileptic patients [177]. This latter study showed increased latencies in olfactory ERP ipsilateral to the epileptic lesion. These latencies were even longer when the lesion was right-sided. Taken together the data indicate that decreased olfactory function in epileptic patients is primarily due to centrally altered olfactory structures whereby the temporal lobe is the main lesion site. Studies on olfactory function in patients with frontal epileptic lesions, however, are lacking.

General Pathologies

Long lists of general pathologies causing olfactory disorders can be found in most reviews and textbooks of smell and taste disorders [172, 178]. Nevertheless, only few studies on specific pathologies have been conducted, sometimes on small sample sizes using unreliable olfactory tests, and sometimes with contradictory outcomes. Besides the above mentioned endocrine diseases, metabolic disorders such as kidney [72] and liver [122, 179, 180] affections have been associated with decreased olfactory function. Olfactory disturbances in those patients are especially interesting, since they are discussed as a potential cause of malnutrition with a more general impact on the patients' health [181].

Post-Surgery/Anesthesia

Anosmia may occur after general anesthesia during the course of surgical interventions not necessarily associated with nasal surgery [122, 182]. Further research should clarify whether surgery under general anesthesia presents a risk of anosmia. For surgical interventions in the sinu-nasal region, anosmia as complication has been estimated to occur in 1% of the cases [183] although this risk has probably been overestimated as indicated by the results of two large studies [184, 185].

Drug-Induced/Toxic

Numerous toxins have been implicated as causes of olfactory disorders [186]. Nevertheless, this information has been mainly accumulated on the basis of case reports. Knowledge about drugs inducing smell and taste disorders is also mainly based on case reports, but several major groups of drugs have been identified as likely to cause problems. Among these, cardiovascular drugs [187], anti-hypertensive drugs [188, 189], and antibiotics [172, 190] are the most frequently mentioned. Usually, the chemosensory side effects disappear when the medication is discontinued.

Congenital

Currently we distinguish between congenital anosmia occurring as an isolated defect or occurring within the context of a syndrome [191]. Isolated congenital anosmia seems to occur more often than previously believed. Apart from the typical patient history of no odor memories, only MR imaging leads to a more definitive diagnosis [192, 193]. In the frontal imaging planes just tangential to the eye bulbs, hypoplastic or aplastic olfactory bulbs can be visualized. This plane also allows an evaluation of the olfactory sulcus which is flattened if the olfactory bulb is absent or aplastic. This is a useful indicator of congenital anosmia, especially since the bulb is not always easy to identify. Among cases of congenital anosmia as part of a syndrome, the Kallmann-Syndrom [194] is the disorder in which it is most frequently encountered. This is an anosmia associated with hypogonadotropic hypogonadism clinically characterized by infertility and anosmia, where infertility can be reversed by substitution of gonadotropins [195].

Congenital anosmia is typically discovered during early puberty. It is a matter of speculation whether olfaction starts to be more important in this period compared to younger years.

Symptoms

Although this distinction is a matter of debate, the discrimination between qualitative and quantitative olfactory disorder have proven helpful in clinical practice. This distinction is mainly based on the patient's history and psychophysical test results.

Quantitative Olfactory Disorders

Normosmia/Hyposmia/Anosmia: Normosmia is the subjectively perceived normal olfactory function, usually defined as the ability to detect the great majority of tested odors in a given olfactory test. Hyposmia means the decrease of this olfactory function and anosmia the total loss of any olfactory function. Beside total anosmia, specific anosmias have been described, where only certain odors are not perceived and most odors are smelt normally [196]. The term functional anosmia was chosen since many subjects with severe olfactory loss appear to be able to still perceive a few single odors. Nevertheless, those rare and weak olfactory impressions are too poor to be of any help to these patients in daily life.

Qualitative Olfactory Disorders

The term "qualitative olfactory disorder" reflects the qualitatively changed perception of odorous sensation. They are frequently, but not necessarily, associated with quantitative olfactory disorders.

Parosmia describes the distorted perception of smells in presence of an odor source. In other words, parosmias are triggered by odors. This is a symptom occurring particularly often in post-URTI or posttraumatic olfactory disorders. Mostly odors are distorted into unpleasant odors (although some exceptions seem to exist: TH, personal communication). For example, to parosmic patients, coffee smells like burnt plastic. The exact explanation of the molecular modifications leading to parosmia is as yet unknown. Even the site of parosmia generation (olfactory epithelium, olfactory bulb, or other central-nervous olfactory structures) is not clear. Important clinically, is the observation that most parosmic impressions tend to diminish over months and finally disappear after years.

Phantosmia describes the distorted perception of smells in the absence of an odor source. Most often, phantosmias occur after trauma or URTI and consist of unpleasant odors occurring without being elicited through environmental odor sources. Phantosmias are rarely triggered but menstruation- and stress-related phantosmias have been reported [197]. Similar to parosmia, there is no exact explanation as yet of the molecular modifications leading to phantosmia; also, the site of its generation remains unclear. Phantosmias also have a tendency to disappear over the course of years.

Surgical Risks to the Olfactory System

Endoscopic Sinus/Transnasal Surgery

Chronic rhinosinusitis is the most common chronic inflammatory disease and is frequently associated with impaired sense of smell [198, 199]. When symptomatic patients do not improve on medical treatment, endoscopic sinus surgery (ESS) may be proposed. Nasal polyposis is considered as the ultimate stage of chronic rhinosinusitis for which the mainstay of treatment is medical, but in which ESS plays a part in the majority of cases resistant to medication. Assessment of preoperative olfactory function is important since patients suffering from chronic rhinosinusitis are not always aware of their olfactory dysfunction, and occurrence of olfactory loss or disorders after endonasal surgery has been reported to be as high as 1% [183, 200, 201]. Nevertheless, this may be an overestimation, as recent studies suggested [184, 185]. Regarding bilateral choanal atresia, surgical repair at relatively advanced ages (8–10 years) was not associated with olfactory improvement [202]. This observation suggests that early sensory exposure could be important for the normal development of olfactory function.

In most cases, ESS is associated with significant improvement of rhinosinusitis symptoms and olfactory function [184, 185]. However, absence or deterioration of olfactory detection thresholds in patients with chronic rhinosinusitis after ESS have been reported [203, 204]. Post-ESS olfactory dysfunction could be due to several mechanisms with persistent mucosal inflammation/edema in the region of the olfactory epithelium being one possible explanation [205]. In addition to post-operative edema, local polyp recurrence, scar tissue, or granulation could also contribute to the absence of improvement in the sense of smell [206].

The olfactory mucosa of patients suffering from long lasting chronic rhinosinusitis could be altered by a variety of toxic inflammatory mediators. In parallel, repetitive URTIs probably alter the neuroepithelium even further [131]. Furthermore, the olfactory epithelium can degenerate in chronic rhinosinusitis and may be replaced by the respiratory epithelium [207]. Furthermore, all surgeons performing ESS should be aware of the risk of iatrogenic injuries of the olfactory epithelium associated with extensive ethmoidectomy [208].

Craniotomy

This paragraph focuses only on the interventions with access to the anterior fossa, since these are most likely to affect olfaction. As stated by Passagia [209], the olfactory structures constitute a natural obstacle to the exploration of the anterior fossa. Therefore, anosmia is a frequent complication of surgical approaches to this region [209]. Nevertheless, techniques have been described which potentially preserve olfaction [210, 211]. One crucial point in preservation of olfactory structures is to respect the blood supply to the olfactory bulb [209]. Whereas leaks of cerebrospinal fluid can be treated without destruction of olfactory structures, oncologic surgery for ethmoidal adenocarcinoma or esthesioneuroblastoma usually leads to anosmia [212, 213]. Meningiomas, which preferably grow in midline structures and especially within the olfactory groove region, are potentially dissectible with preservation of olfaction [209]. However, most reports on olfactory impairment after surgery of the anterior fossa have been conducted on small samples [214] and olfactory function has rarely been measured properly [209]. Welge-Luessen *et al.* [215] have recently published a study focusing on the olfactory outcome after meningeoma surgery. They pointed out that preservation of olfaction ipsilateral to the tumor is extremely difficult. They also showed a correlation between preserved postoperative olfactory function and tumor size. Overall, it seems that preservation of olfactorily eloquent structures might be possible when the tumor size is small. Nevertheless, olfactory function seems to be very vulnerable and seems sometimes altered even though the surgeon did not touch the olfactory structures. This corroborates findings by Delank [146] on posttraumatic cadavers, that olfactory tracts and bulbs in certain people are severed even after minor tearing.

Recovery of Smell Disorders

Age-related and congenital anosmia do not usually exhibit recovery. Sinunasal smell disorders are treatable and will be extensively discussed in the next paragraph. Toxic- and drug-induced smell disorders may recover once the drug intake is interrupted [190]. Two of the most important causes of olfactory dysfunction, post-URTI and posttraumatic causes, have received relatively little attention concerning their recovery rate. This is partly due to difficulties obtaining reliable epidemiological data on the real frequency of post-URTI olfactory disorders. Most patients with transitory or recovered post-URTI smell disorders probably do not seek medical help. The following recovery data apply to patients seen in Smell and Taste Clinics and are usually the ones with the most tenacious smell disorders. Several authors described recovery rates for post-URTI and posttraumatic disorders to be highest within the first year [133, 216–218]. According to this literature post-URTI disorders have a slightly better prognosis compared to posttraumatic disorders, mainly because they often cause hyposmia rather than anosmia. Total recovery is observed in approximately 5% of the cases, while up to 60% of all patients experience partial recovery of some olfactory function over the following years. The remaining patients do not have any improvement of chemosensory function. Although olfactory neurons have the ability to regenerate [17, 219], the exact mechanisms favoring such spontaneous recovery are not understood.

It is currently impossible to predict an individual outcome with regard to recovery. Clinically, one has the impression that younger patients might have better recovery rates but no solid data support this hypothesis [136]. For quite a while the presence of parosmia and phantosmia has been interpreted as a sign of plasticity and regeneration within the olfactory system. Recent retrospective data, however, do not support this clinical impression [220]. In contrast to the quantitative olfactory disorders, the qualitative disorders have a far better prognosis of spontaneous disappearance. Parosmias tend to decrease to a bearable level after approximately one year [139]. However, recent work revealed that more than 50% of the parosmias are still present after 2 years [220]. Over time, parosmia seems to lose its devastating effect on quality of life. To summarize, the best current therapeutic attitude towards post-URTI and posttraumatic olfactory disorders is to correctly inform the patient, without removing all hope of recovery, but not promising a quick and complete recovery. The patients should receive satisfactory olfactory testing. Follow-up investigations give both the physician and the patient the possibility to observe improvements.

Treatment of Olfactory Disorders

Surgical

The effect of surgery on quantitative olfactory disorders has already been mentioned above. Beside the routine surgery indicated in advanced and medication resistant nasal polyposis, ESS has also been proposed in very particular cases of qualitative olfactory disorders [221, 222]. Leopold was the first to describe the selective excision of the olfactory epithelium in patients suffering from very handicapping phantosmias. These cases, less than two dozen so far, have been carefully selected, and ESS in phantosmia is far from being routinely indicated. Interestingly, the histological analysis of these epithelia revealed numerous neuromas within the olfactory epithelium. Whether these neuromas are the substrate of the phantosmia is not clear. One report also treated parosmia with selective resection of the olfactory bulb [223] and a recent paper rediscovered the technique used by Leopold to treat parosmia [224]. These latter authors were unable to analyze the excised tissue and apparently ignored the existence of Leopold's work. This underlines the fact that this procedure should be reserved to experienced surgeons and is far from being a routine operation.

Conservative/Medication

Conservative Therapy of SND Related Olfactory Loss

Antibiotics: Putrid acute sinusitis is most frequently the result of infection by streptococcus pneumoniae, haemophilus influenzae, and moraxella catarrhalis which are relatively sensitive to antibiotic therapy. However, in the chronic form of putrid sinusitis, staphylococcus aureus and pseudomonas aeruginosa are much more important. Whenever possible, antibiotic therapy should only be started after the bacteria have been identified and tested for resistance to antibiotics. It is important to note that in chronic putrid sinusitis antibiotic treatment is not always successful.

Steroids: Among many other effects corticosteroids act as anti-inflammatory drugs, the anti-inflammatory effects being produced via a number of different pathways including inhibition of phospholipase A2 through induction of lipocortin [225]. They reduce submucosal edema and mucosal hypersecretion and thereby increase nasal patency. Systemically administered steroids are of help in many sinu-nasal disease (SND) patients [129, 226–228]. For example, Stevens reported that systemic administration of steroids was effective in 12 of 24 patients with SND-related olfactory loss [229]. In addition to the anti-inflammatory activity it has been postulated that corticosteroids directly improve olfactory function [230, 231] by modulating the function of ORN through effects on olfactory Na, K-ATPase [225]. In fact, also based on our own experience, systemic steroids are often helpful even in patients without nasal obstruction due to polyps or obvious inflammatory changes (compare [229, 232]).

Steroids may be administered systemically or topically. With regard to idiopathic olfactory dysfunction, systemic administration is often applied for diagnostic purposes [233]. If systemic steroids improve olfactory function, treatment is typically continued with locally administered steroids. Although systemic steroids are usually more effective than locally administered steroids [230, 234], prescription of systemic steroids over an extended period of time is rarely warranted due to their side effects [150, 232]. While there are no exact recommendations, it is possible, however, to repeatedly administer short courses of systemic steroids with an interval of 6–12 months between courses.

A number of studies indicate the usefulness of topical steroids [153, 226, 228, 235]; however, the role of topical steroids in the treatment of SND related olfactory loss has been questioned [230, 233, 234, 236–239]. So far, no factors predicting a favorable response to topical steroids have been identified. It is not entirely clear why systemic steroids have a higher therapeutic efficacy compared to topical steroids [129, 234]. One reason may relate to the deposition of the spray in the nasal cavity. In fact, it has been shown that only a small amount of nasally applied drugs reaches the olfactory epithelium which is situated in an effectively protected area of the nasal cavity [240–242]. This situation can be slightly improved by the application of sprays in "head-down-forward position" [230, 239].

Other treatments: In addition to the use of steroids there are other therapeutic approaches to restoration of olfactory loss. They include the use of anti-leukotrienes [243], saline lavages [244], or approaches which have received less vigorous scientific investigation, e.g., dietary changes [245], acupuncture [246], anti-allergy immunotherapy [247] and herbal treatments.

Conservative Therapy of Post-URTI/Posttraumatic Olfactory Loss

Post-URTI smell dysfunction seems to be due to an impairment of ORN, both in function and in numbers [248, 249]. While numerous treatments have been tried in post-URTI anosmia (e.g., zinc, vitamin A; see below), no pharmacological therapy has been established so far (see [250–252]). The situation is similar for posttraumatic olfactory loss where therapeutic options are lacking. The absence of conservative treatment for certain forms of olfactory dysfunction is underlined by the fact that, when "parosmia" is present [253, 254], in some patients surgical removal of the olfactory epithelium may be considered as a cure [255].

Having said this, there are still numerous candidates for the pharmacological treatment of olfactory dysfunction, one being *alpha-lipoic acid* (aLA) which is used in the treatment of diabetic neuropathy [256]. The effect of aLA is well described both in experimental animals and in humans (for review see [257]). It is known to stimulate the expression of nerve growth factor, substance P, and neuropeptide Y [258–260]. It enhances motor nerve conduction velocity as well as microcirculation [261, 262]. Further, due to its potent anti-oxidative effects, aLA also has neuroprotective capabilities indicating that aLA is suited to treat neural damage involving free radicals [263]. Preliminary work has already indicated that it may be useful in post-URTI olfactory loss when administered at a dose of 600 mg/d over a period of 4–7 months [136]. Other encouraging pilot studies have been performed with the NMDA-antagonist *caroverine* [135] administered at a dose of 120 mg/d for 4 weeks. Potential mechanisms for the hypothesized effect included reduced feedback inhibition in the olfactory bulb as a consequence of NMDA-antagonistic actions, or antagonism of an excitotoxic action of glutamate.

Although frequently mentioned as a therapeutic option, studies on *zinc* treatment for olfactory dysfunction have produced negative results [135, 250] (see also [264]). It may, however, be of therapeutic value in patients with severe zinc deficiency, e.g., in hemodialysis. In studies in postmenopausal women *estrogens* have been reported to provide a certain protection against olfactory disturbances [130]. However, as mentioned above, recent studies [138] indicate that estrogens are probably ineffective in the treatment of olfactory loss. Finally, although discussed frequently, the potential therapeutic use of orally administered *vitamin A* [251, 265] is questionable unless appropriate double-blinded studies become available.

A different approach to the treatment of olfactory disorders is the detection and treatment of underlying causes. This approach may also involve the replacement of drugs suspected of affecting the sense of smell [172, 266, 267]. Other possible treatments may include, for example, acupuncture [246].

Acknowledgements

The authors would like to thank Professor Ivan Rodriguez for his helpful comments on the basic olfactory sections (one and two). This work was partly supported by a grant from the Deutsche Forschungsgemeinschaft (DFG HU441/2-1) to TH and by a grant from the Swiss National Fund for Scientific Research FNSRS (n° 3100A0-100621-1) to JSL.

References

1. Dodd J, Castellucci VF (1991) Smell and taste: The chemical senses. Principles of neural sciences. In: Kandel ER, Schwartz JH, Jessel TM (eds) Elsevier Science Publishing Co, New York, p 512–529
2. Turner BH, Mishkin M, Knapp M (1980) Organization of the amygdalope-

tal projections from modality-specific cortical association areas in the monkey. J Comp Neurol 191: 515–543
3. Herz RS (2000) Scents of time. The Sciences 34–39
4. Zatorre RJ *et al* (1992) Functional localization and lateralization of human olfactory cortex. Nature 360: 339–340
5. Doty RL *et al* (1978) Intranasal trigeminal stimulation from odorous volatiles: psychometric responses from anosmic and normal humans. Physiol Behav 20: 175–185
6. Hummel T *et al* (2003) Effects of olfactory function, age, and gender on trigeminally mediated sensations: a study based on the lateralization of chemosensory stimuli. Toxicol Lett 140–141: 273–280
7. Hummel T (2000) Assessment of intranasal trigeminal function. Int J Psychophysiol 36: 147–155
8. Cain WS (1974) Contribution of the trigeminal nerve to perceived odor magnitude. Ann NY Acad Sci 237: 28–34
9. Cain WS, Murphy CL (1980) Interaction between chemoreceptive modalities of odour and irritation. Nature 284: 255–257
10. Hummel T *et al* (1996) Loss of olfactory function leads to a decrease of trigeminal sensitivity. Chem Senses 21: 75–79
11. Gudziol H, Schubert M, Hummel T (2001) Decreased trigeminal sensitivity in anosmia. ORL J Otorhinolaryngol Relat Spec 63: 72–75
12. Krause F (1895) Die Physiologie des Trigeminus nach Untersuchungen an Menschen, bei denen das Ganglion Gasseri entfernt worden ist (Schluss). Munch Med Wochenschr 27: 628–631
13. Ikeda I (1909) On a new seasoning. J Tokyo Chem Soc 30: 820–836
14. Ikeda K (2002) New seasonings. Chem Senses 27: 847–849
15. Chaudhari N, Landin AM, Roper SD (2000) A metabotropic glutamate receptor variant functions as a taste receptor. Nat Neurosci 3: 113–119
16. Witt M, Reutter K, Miller IJ Jr (2003) Morphology of the peripheral taste system. Handbook of olfaction and gustation. In: Doty RL (ed) Marcel Dekker, Inc, New York, p 651–678
17. Beidler LM, Smallman RL (1965) Renewal of cells within taste buds. J Cell Bio 27: 263–272
18. Jacobson L, Trotier D, Døving KB (1998) Anatomical description of a new organ in the nose of domesticated animals by Ludvig Jacobson (1813). Chem Senses 23: 743–754
19. Cuvier G (1811) Rapport – Fait à l'Institut, sur un Mémoire de M. Jacobson, intitulé: Descripion anatomique d'un organe observé dans les mammifères. Annales du Muséum d'Histoire Naturelle Paris. Tome 18: 412–424
20. Karlson P, Lüscher M (1959) "Pheromones": a new term for a class of biologically active substances. Nature 183: 55–56
21. Schaal B *et al* (2003) Chemical and behavioural characterization of the rabbit mammary pheromone. Nature 424: 68–72
22. Dorries KM, Adkins-Regan E, Halpern BP (1997) Sensitivity and behavioral responses to the pheromone androstenone are not mediated by the vomeronasal organ in domestic pigs. Brain Behav Evol 49: 53–62

23. Monti-Bloch L, Grosser BI (1991) Effect of putative pheromones on the electrical activity of the human vomeronasal organ and olfactory epithelium. J Steroid Biochem Molec Biol 39: 573–582
24. Witt M *et al* (2002) On the chemosensory nature of the vomeronasal epithelium in adult humans. Histochem Cell Biol 117: 493–509
25. Knecht M *et al* (2001) Frequency and localization of the putative vomeronasal organ in humans in relation to age and gender. Laryngoscope 111: 448–452
26. Witt M *et al* (2000) Characterization of the adult human vomeronasal organ using immunohistochemical and electrophysiological measures. Chem Senses 25: 668
27. Knecht M *et al* (2003) Assessment of olfactory function and androstenone odor thresholds in humans with or without functional occlusion of the vomeronasal duct. Behav Neurosci 117: 1135–1141
28. Stern K, McClintock MK (1998) Regulation of ovulation by human pheromones. Nature 392: 177–179
29. Savic I *et al* (2001) Smelling of odorous sex hormone-like compounds causes sex-differentiated hypothalamic activations in humans. Neuron 31: 661–668
30. Rodriguez I *et al* (2000) A putative pheromone receptor gene expressed in human olfactory mucosa. Nat Genet 26: 18–19
31. Mozell MM (1964) Evidence for sorption as a mechanism of the olfactory analysis of vapours. Nature 203: 1181–1182
32. Mozell MM, Jagodowicz M (1973) Chromatographic separation of odorants by the nose: retention times measured across in vivo olfactory mucosa. Science 181: 1247–1249
33. Sobel N *et al* (1999) The world smells different to each nostril. Nature 402: 35
34. Amoore JE (1967) Specific anosmia: a clue to the olfactory code. Nature 214: 1095–1098
35. Henning H (1916) Der Geruch. Johann Ambrosius Barth, Leipzig
36. Dyson GM (1938) The scientific basis of odour. Chem Ind 57: 647–651
37. Turin L (1996) A spectroscopic mechanism for primary olfactory reception. Chem Senses 21: 773–791
38. Keller A, Vosshall LB (2004) A psychophysical test of the vibration theory of olfaction. Nat Neurosci 7: 337–338
39. Buck L, Axel R (1991) A novel multigene family may encode odorant receptors: a molecular basis for odor recognition. Cell 65: 175–187
40. Gilad Y *et al* (2003) Human specific loss of olfactory receptor genes. Proc Natl Acad Sci USA 28: 28
41. Menashe I *et al* (2003) Different noses for different people. Nat Genet 34: 143–144
42. Gilad Y *et al* (2003) Natural selection on the olfactory receptor gene family in humans and chimpanzees. Am J Hum Genet 73: 489–501
43. Gilad Y *et al* (2004) Loss of olfactory receptor genes coincides with the acquisition of full trichromatic vision in primates. PLoS Biol 2: E5

44. Man O, Gilad Y, Lancet D (2004) Prediction of the odorant binding site of olfactory receptor proteins by human-mouse comparisons. Protein Sci 13: 240–254
45. Strotmann J *et al* (1994) Olfactory neurones expressing distinct odorant receptor subtypes are spatially segregated in the nasal neuroepithelium. Cell Tissue Res 276: 429–438
46. Vassar R, Ngai J, Axel R (1993) Spatial segregation of odorant receptor expression in the mammalian olfactory epithelium. Cell 74: 309–318
47. Ressler KJ, Sullivan SL, Buck LB (1993) A zonal organization of odorant receptor gene expression in the olfactory epithelium. Cell 73: 597–609
48. Vassar R *et al* (1994) Topographic organization of sensory projections to the olfactory bulb. Cell 79: 981–991
49. Nef P *et al* (1992) Spatial pattern of receptor expression in the olfactory epithelium. Proc Natl Acad Sci USA 89: 8948–8952
50. Mombaerts P *et al* (1996) Visualizing an olfactory sensory map. Cell 87: 675–686
51. Zhao H *et al* (1998) Functional expression of a mammalian odorant receptor. Science 279: 237–242
52. Araneda RC, Kini AD, Firestein S (2000) The molecular receptive range of an odorant receptor. Nat Neurosci 3: 1248–1255
53. Uchida N *et al* (2000) Odor maps in the mammalian olfactory bulb: domain organization and odorant structural features. Nat Neurosci 3: 1035–1043
54. Malnic B *et al* (1999) Combinatorial receptor codes for odors. Cell 96: 713–723
55. Landis BN *et al* (2003) Ratings of overall olfactory function. Chem Senses 28: 691–694
56. Hummel T *et al* (2001) Screening of olfactory function with a four-minute odor identification test: reliability, normative data, and investigations in patients with olfactory loss. Ann Otol Rhinol Laryngol 110: 976–981
57. Doty RL, Shaman P, Dann M (1984) Development of the University of Pennsylvania Smell Identification Test: a standardized microencapsulated test of olfactory function. Physiol Behav 32: 489–502
58. Cain WS *et al* (1988) Evaluation of olfactory dysfunction in the Connecticut Chemosensory Clinical Research Center. Laryngoscope 98: 83–88
59. Doty RL *et al* (1984) University of Pennsylvania Smell Identification Test: a rapid quantitative olfactory function test for the clinic. Laryngoscope 94: 176–178
60. Kobal G *et al* (1996) "Sniffin' sticks": screening of olfactory performance. Rhinology 34: 222–226
61. Hummel T *et al* (1997) 'Sniffin' sticks': olfactory performance assessed by the combined testing of odor identification, odor discrimination and olfactory threshold. Chem Senses 22: 39–52
62. Kobal G *et al* (2000) Multicenter investigation of 1,036 subjects using a standardized method for the assessment of olfactory function combining tests of odor identification, odor discrimination, and olfactory thresholds. Eur Arch Otorhinolaryngol 257: 205–211

63. Kondo H *et al* (1998) A study of the relationship between the T&T olfactometer and the University of Pennsylvania Smell Identification Test in a Japanese population. Am J Rhinol 12: 353–358
64. Lecanu JB *et al* (2002) Valeurs normatives du test olfactométrique Biolfa. Ann Otolaryngol Chir Cervicofac 119: 164–169
65. Briner HR, Simmen D (1999) Smell diskettes as screening test of olfaction. Rhinology 37: 145–148
66. Ho WK *et al* (2002) Change in olfaction after radiotherapy for nasopharyngeal cancer – a prospective study. Am J Otolaryngol 23: 209–214
67. Lotsch J, Lange C, Hummel T (2004) A simple and reliable method for clinical assessment of odor thresholds. Chem Senses 29: 311–317
68. Linschoten MR *et al* (2001) Fast and accurate measurement of taste and smell thresholds using a maximum-likelihood adaptive staircase procedure. Percept Psychophys 63: 1330–1347
69. Hawkes CH, Shephard BC (1993) Selective anosmia in Parkinson's disease? Lancet 341: 435–436
70. Koss E *et al* (1987) Olfactory detection and recognition in Alzheimer's disease. Lancet 1: 622
71. Koss E *et al* (1988) Olfactory detection and identification performance are dissociated in early Alzheimer's disease. Neurology 38: 1228–1232
72. Frasnelli JA *et al* (2002) Olfactory function in chronic renal failure. Am J Rhinol 16: 275–279
73. Hornung DE *et al* (1998) The olfactory loss that accompanies an HIV infection. Physiol Behav 15: 549–556
74. Jones-Gotman M, Zatorre RJ (1988) Olfactory identification deficits in patients with focal cerebral excision. Neuropsychologia 26: 387–400
75. Mesholam RI *et al* (1998) Olfaction in neurodegenerative disease: a meta-analysis of olfactory functioning in Alzheimer's and Parkinson's diseases. Arch Neurol 55: 84–90
76. Daum RF *et al* (2000) Riechprüfung mit "Sniffin' Sticks" zur klinischen Diagnostik des Morbus Parkinson. Nervenarzt 71: 643–650
77. Klimek L *et al* (1998) Lateralized and bilateral olfactory function in patients with chronic sinusitis compared with healthy control subjects. Laryngoscope 108: 111–114
78. Ottoson D (1956) Analysis of the electrical activity of the olfactory epithelium. Acta Physiol Scand 35: 1–83
79. Hummel T, Knecht M, Kobal G (1996) Peripherally obtained electrophysiological responses to olfactory stimulation in man: electro-olfactograms exhibit a smaller degree of desensitization compared with subjective intensity estimates. Brain Res 717: 160–164
80. Leopold DA *et al* (2000) Anterior distribution of human olfactory epithelium. Laryngoscope 110: 417–421
81. Wang L, Chen L, Jacob T (2004) Evidence for peripheral plasticity in human odour response. J Physiol 554: 236–244
82. Spehr M *et al* (2004) The HOR17-4 signalling system – one receptor, dual capacity. Chem Sens (in press)

83. Rawson NE *et al* (1995) Functionally mature olfactory receptor neurons from two anosmic patients with Kallmann syndrome. Brain Res 681: 58–64
84. Hummel T, Mojet J, Kobal G (1997) Electro-olfactograms are present when odorous stimuli have not been perceived. Chem Senses 22: 196
85. Knecht M, Hummel T (2004) Recording of the human electro-olfactogram. Physiol Behav 83: 13–19
86. Picton TW, Hillyard SA (1988) Endogenous event-related potentials. EEG-handbook, revised series, vol. 3. In: Picton TW (ed) Elsevier, Amsterdam, p 361–426
87. Kobal G, Plattig KH (1978) Methodische Anmerkungen zur Gewinnung olfaktorischer EEG-Antworten des wachen Menschen (objektive Olfaktometrie). Z EEG-EMG 9: 135–145
88. Kobal G (1981) Elektrophysiologische Untersuchungen des menschlichen Geruchssinns. Thieme Verlag, Stuttgart
89. Kobal G, Hummel T (1991) Olfactory evoked potentials in humans. Smell and taste in health and disease. In: Getchell TV *et al* (eds) Raven Press, New York, p 255–275
90. Pause BM *et al* (1996) The nature of the late positive complex within the olfactory event-related potential. Psychophysiology 33: 168–172
91. Krauel K *et al* (1998) Attentional modulation of central odor processing. Chem Senses 23: 423–432
92. Donchin E *et al* (1986) Cognitive psychophysiology and human information processing. Psychophysiology: systems, processes and applications. In: Coles MGH, Donchin E, Porges SW (eds) Guilford Press, New York
93. Kobal G, Hummel T, Van Toller S (1992) Differences in chemosensory evoked potentials to olfactory and somatosensory chemical stimuli presented to left and right nostrils. Chem Senses 17: 233–244
94. Pause B, Sojka B, Ferstl R (1996) The latency but not the amplitude of the olfactory event-related potential (OERP) varies with the odor concentration. Chem Senses 21: 485
95. Murphy C *et al* (1998) Age effects on central nervous system activity reflected in the olfactory event-related potential. Evidence for decline in middle age. Ann NY Acad Sci 855: 598–607
96. Lorig TS *et al* (1996) The effects of active and passive stimulation on chemosensory event-related potentials. Int J Psychophysiol 23: 199–205
97. Williamson SJ, Kaufman L (1987) Analysis of neuromagnetic signals. Handbook of electroencephalography and clinical neurophysiologgy, volume 1, Methods of brain electrical and magnetical signals. In: Gevins AS, Rèmond AA (eds) Elsevier, Amsterdam, p 405–448
98. Huttunen J *et al* (1986) Cortical responses to painful CO2-stimulation of nasal mucosa: a magnetencephalographic study in man. Electroenceph Clin Neurophysiol 64: 347–349
99. Kettenmann B *et al* (1996) Magnetoencephalographical recordings: separation of cortical responses to different chemical stimulation in man. Funct Neurosci [EEG Suppl] 46: 287–290

100. Kettenmann B *et al* (1997) Multiple olfactory activity in the human neocortex identified by magnetic source imaging. Chem Senses 22: 493–502
101. Ayabe-Kanamura S *et al* (1997) Measurement of olfactory evoked magnetic fields by a 64-channel whole-head SQUID system. Chem Senses 22: 214–215
102. Welge-Lussen A (1999) Chemosensorisch evozierte Potentiale – Anwendung und Bedeutung im klinischen Alltag. HNO 47: 453–455
103. Kobal G, Hummel T (1998) Olfactory and intranasal trigeminal event-related potentials in anosmic patients. Laryngoscope 108: 1033–1035
104. Peters JM *et al* (2003) Olfactory function in mild cognitive impairment and Alzheimer's disease: an investigation using psychophysical and electrophysiological techniques. Am J Psychiatry 160: 1995–2002
105. Hudry J *et al* (2003) Olfactory short-term memory and related amygdala recordings in patients with temporal lobe epilepsy. Brain 126: 1851–1863
106. Savic I (2002) Imaging of brain activation by odorants in humans. Curr Opin Neurobiol 12: 455–461
107. Zald DH, Pardo JV (2000) Functional neuroimaging of the olfactory system in humans. Int J Psychophysiol 36: 165–181
108. Kettenmann B, Hummel T, Kobal G (2001) Functional imaging of olfactory activation in the human brain. Methods and frontiers in chemosensory research. In: Simon SA, Nicolelis MAL (eds) CRC Press, Baco Raton, Florida, USA, p 477–506
109. Small DM *et al* (1997) Flavor processing: more than the sum of its parts. Neuroreport 8: 3913–3917
110. Kareken DA *et al* (2004) Olfactory system activation from sniffing: effects in piriform and orbitofrontal cortex. Neuroimage 22: 456–465
111. Savic I, Berglund H (2004) Passive perception of odors and semantic circuits. Hum Brain Mapp 21: 271–278
112. Sobel N *et al* (2000) Time course of odorant-induced activation in the human primary olfactory cortex. J Neurophysiol 83: 537–551
113. Poellinger A *et al* (2001) Activation and habituation in olfaction – an fMRI study. Neuroimage 13: 547–560
114. Anderson AK *et al* (2003) Dissociated neural representations of intensity and valence in human olfaction. Nat Neurosci 6: 196–202
115. Yousem DM *et al* (1999) The effect of age on odor-stimulated functional MR imaging. Am J Neuroradiol 20: 600–608
116. Henkin RI, Levy LM, Lin CS (2000) Taste and smell phantoms revealed by brain functional MRI (fMRI). J Comput Assist Tomogr 24: 106–123
117. Wysocki CJ, Gilbert AN (1989) National Geographic Smell Survey: effects of age are heterogenous. Ann NY Acad Sci 561: 12–28
118. Hoffman HJ, Ishii EK, MacTurk RH (1998) Age-related changes in the prevalence of smell/taste problems among the United States adult population. Results of the 1994 disability supplement to the National Health Interview Survey (NHIS). Ann NY Acad Sci 855: 716–722
119. Doty RL *et al* (1984) Smell identification ability: changes with age. Science 226: 1441–1443

120. Murphy C *et al* (2002) Prevalence of olfactory impairment in older adults. JAMA 288: 2307–2312
121. Brämerson A *et al* (2004) Prevalence of olfactory dysfunction: the skovde population-based study. Laryngoscope 114: 733–737
122. Landis BN, Konnerth CG, Hummel T (2004) A study on the frequency of olfactory dysfunction. Laryngoscope 114: 1764–1769
123. Miwa T *et al* (2001) Impact of olfactory impairment on quality of life and disability. Arch Otolaryngol Head Neck Surg 127: 497–503
124. Temmel AF *et al* (2002) Characteristics of olfactory disorders in relation to major causes of olfactory loss. Arch Otolaryngol Head Neck Surg 128: 635–641
125. Hummel T, Nordin S (2004) Olfactory disorders and their consequences for quality of life – A review. Acta Oto-Laryngologica, in press
126. Santos DV *et al* (2004) Hazardous events associated with impaired olfactory function. Arch Otolaryngol Head Neck Surg 130: 317–319
127. Damm M *et al* (2004) Riechstörungen – Epidemiologie und Therapie in Deutschland, Österreich und der Schweiz. HNO 52: 112–120
128. Quint C *et al* (2001) Patterns of non-conductive olfactory disorders in eastern Austria: a study of 120 patients from the Department of Otorhinolaryngology at the University of Vienna. Wien Klin Wochenschr 113: 52–57
129. Seiden AM, Duncan HJ (2001) The diagnosis of a conductive olfactory loss. Laryngoscope 111: 9–14
130. Deems DA *et al* (1991) Smell and taste disorders: a study of 750 patients from the University of Pennsylvania Smell and Taste Center. Arch Otorhinolaryngol Head Neck Surg 117: 519–528
131. Jafek BW *et al* (1990) Postviral olfactory dysfunction. Am J Rhinol 4: 91–100
132. Sugiura M *et al* (1998) An epidemiological study of postviral olfactory disorder. Acta Otolaryngol [Suppl] 538: 191–196
133. Faulcon P *et al* (1996) Anosmie secondaire à une rhinite aiguë: sémiologie et évolution à propos d'une série de 118 patients. Ann Otolaryngol Chir Cervicofac 116: 351–357
134. Duncan HJ, Seiden AM (1995) Long-term follow-up of olfactory loss secondary to head trauma and upper respiratory tract infection. Arch Otolaryngol Head Neck Surg 121: 1183–1187
135. Quint C *et al* (2002) The quinoxaline derivative caroverine in the treatment of sensorineural smell disorders: a proof-of-concept study. Acta Otolaryngol 122: 877–881
136. Hummel T, Heilmann S, Huttenbrink KB (2002) Lipoic acid in the treatment of smell dysfunction following viral infection of the upper respiratory tract. Laryngoscope 112: 2076–2080
137. Dhong HJ, Chung SK, Doty RL (1999) Estrogen protects against 3-methylindole-induced olfactory loss. Brain Res 824: 312–315
138. Hughes LF *et al* (2002) Effects of hormone replacement therapy on olfac4-tory sensitivity: cross-sectional and longitudinal studies. Climacteric 5: 140–150

139. Portier F *et al* (2000) Sémiologie, étiologie et évolution des parosmies: à propos de 84 cas. Ann Otolaryngol Chir Cervicofac 117: 12–18
140. Frasnelli J, Hummel T (2004) Olfactory dysfunction and daily life. Eur Arch Otorhinolaryngol 5: 5
141. Legg JW (1873) A case of anosmia following a blow. Lancet 2: 659–660
142. Yousem DM *et al* (1999) Posttraumatic smell loss: relationship of psychophysical tests and volumes of the olfactory bulbs and tracts and the temporal lobes. Acad Radiol 6: 264–272
143. Zusho H (1982) Posttraumatic anosmia. Arch Otolaryngol 108: 90–92
144. Frank Greiffenstein M, John Baker W, Gola T (2002) Brief report: anosmia and remote outcome in closed head injury. J Clin Exp Neuropsychol 24: 705–709
145. Sumner D (1964) Post-traumatic anosmia. Brain 87: 107–120
146. Delank KW, Fechner G (1996) Zur Pathophysiologie der posttraumatischen Riechstörungen. Laryngorhinootologie 75: 154–159
147. Fein BT, Kamin PB, Fein NN (1966) The loss of sense of smell in nasal allergy. Ann Allergy 24: 278–283
148. Doty RL (1997) Olfaction. Nasal Polyposis – An inflammatory disease and its treatement. In: Mygind N, Lildholdt T (eds) Munksgaard, Copenhagen, p 153–159
149. Landis BN *et al* (2003) Retronasal olfactory function in nasal polyposis. Laryngoscope 113: 1993–1997
150. Hotchkiss WT (1956) Influence of Prednisone on nasal polyposis with anosmia. Arch Otolaryngol 64: 478–479
151. Klimek L *et al* (1997) Olfactory function after microscopic endonasal surgery in patients with nasal polyps. Am J Rhinol 11: 251–255
152. Apter AJ *et al* (1995) Allergic rhinitis and olfactory loss. Ann Allergy Asthma Immunol 75: 311–316
153. Stuck BA *et al* (2003) Mometasone furoate nasal spray improves olfactory performance in seasonal allergic rhinitis. Allergy 58: 1195
154. Ansari KA, Johnson A (1975) Olfactory function in patients with Parkinson's disease. J Chron Dis 28: 493–497
155. Ward CD, Hess WA, Calne DB (1983) Olfactory impairment in Parkinson's disease. Neurology 33: 943–946
156. Doty RL, Deems D, Steller S (1988) Olfactory dysfunction in Parkinson's disease: A general deficit unrelated to neurologic signs, disease stage, or disease duration. Neurology 38: 1237–1244
157. Sobel N *et al* (2001) An impairment in sniffing contributes to the olfactory impairment in Parkinson's disease. Proc Natl Acad Sci USA 98: 4154–4159
158. Barz S *et al* (1997) Chemosensory event-related potentials in response to trigeminal and olfactory stimulation in idiopathic Parkinson's disease. Neurology 49: 1424–1431
159. Hawkes CH, Shephard BC (1998) Olfactory evoked responses and identification tests in neurological disease. Ann NY Acad Sci 855: 608–615
160. Hawkes CH, Shephard BC, Daniel SE (1999) Is Parkinson's disease a primary olfactory disorder? QJM 92: 473–480

161. Hawkes C (2003) Olfaction in neurodegenerative disorder. Mov Disord 18: 364–372
162. Berendse HW *et al* (2001) Subclinical dopaminergic dysfunction in asymptomatic Parkinson's disease patients' relatives with a decreased sense of smell. Ann Neurol 50: 34–41
163. Sommer U *et al* (2004) Detection of presymptomatic Parkinson's disease: combination of olfactory tests, transcranial sonography, and 123 I-FP-CIT-SPECT. Mov Disord, in press
164. Müller A *et al* (2002) Olfactory function in idiopathic Parkinson's disease (IPD): results from cross-sectional studies in IPD patients and long-term follow-up of de-novo IPD patients. J Neural Transm 109: 805–811
165. Heilmann S, Huettenbrink KB, Hummel T (2004) Local and systemic administration of corticosteroids in the treatment of olfactory loss. Am J Rhinol 18: 29–33
166. Jorgensen MB, Buch NH (1961) Studies on the sense of smell and taste in diabetics. Arch Otolaryngol 53: 539–545
167. Weinstock RS, Wright HN, Smith DU (1993) Olfactory dysfunction in diabetes mellitus. Physiol Behav 53: 17–21
168. Le Floch JP *et al* (1993) Smell dysfunction and related factors in diabetic patients. Diabetes Care 16: 934–937
169. Doty RL (1986) Gender and endocrine-related influences on human olfactory perception. Clinical Measurement of Taste and Smell. In: Meiselman R (ed) MacMillan, New York, p 377–413
170. McConnell RJ *et al* (1975) Defects of taste and smell in patients with hypothyroidism. Am J Med 59: 354–364
171. Henkin RI, Bartter FC (1966) Studies on olfactory thresholds in normal man and in patients with adrenal cortical insufficiency: the role of adrenal cortical steroids and of serum sodium concentration. J Clin Invest 45: 1631–1639
172. Schiffman SS (1983) Taste and smell in disease (first of two parts). N Engl J Med 308: 1275–1279
173. Campanella G, Filla A, De Michele G (1978) Smell and taste acuity in epileptic syndromes. Eur Neurol 17: 136–141
174. Eskenazi B *et al* (1986) Odor perception in temporal lobe epilepsy patients with and without temporal lobectomy. Neuropsychologia p 553–562
175. Jones-Gotman M *et al* (1997) Contribution of medial versus lateral temporal-lobe structures to human odour identification. Brain 120: 1845–1856
176. Kohler CG *et al* (2001) Olfactory dysfunction in schizophrenia and temporal lobe epilepsy. Neuropsychiatry Neuropsychol Behav Neurol 14: 83–88
177. Hummel T *et al* (1995) Chemosensory event-related potentials in patients with temporal lobe epilepsy. Epilepsia 36: 79–85
178. Schiffman SS (1983) Taste and smell in disease (second of two parts). N Engl J Med 308: 1337–1343
179. Henkin RI, Smith FR (1971) Hyposmia in acute viral hepatitis. Lancet 1: 823–826

180. Kleinschmidt EG, Kramp B, Schwager A (1976) Functional study on the sense of smell in patients with chronic liver disease. Z Gesamte Inn Med 31: 853–856
181. Reaich D (1997) Odour perception in chronic renal disease. Lancet 350: 1191
182. Adelman BT (1995) Altered taste and smell after anesthesia: cause and effect? Anesthesiology 83: 647–649
183. Kimmelman CP (1994) The risk to olfaction from nasal surgery. Laryngoscope 104: 981–988
184. Damm M *et al* (2003) Olfactory changes at threshold and suprathreshold levels following septoplasty with partial inferior turbinectomy. Ann Otol Rhinol Laryngol 112: 91–97
185. Briner HR, Simmen D, Jones N (2003) Impaired sense of smell in patients with nasal surgery. Clin Otolaryngol 28: 417–419
186. Hastings L, Miller ML (1997) Olfactory loss to toxic exposure. Taste and smell disorders. In: Seiden AM (ed) Thieme, New York, p 88–106
187. Doty RL *et al* (2003) Influences of antihypertensive and antihyperlipidemic drugs on the senses of taste and smell: a review. J Hypertens 21: 1805–1813
188. Kharoubi S (2003) Anosmie toxi-médicamenteuse à la nifédipine. Presse Med 32: 1269–1272
189. Levenson JL, Kennedy K (1985) Dysosmia, dysgeusia, and nifedipine. Ann Intern Med 102: 135–136
190. Welge-Luessen A, Wolfensberger M (2003) Reversible anosmia after amikacin therapy. Arch Otolaryngol Head Neck Surg 129: 1331–1333
191. Jafek BW *et al* (1990) Congenital anosmia. Ear Nose Throat J 69: 331–337
192. Abolmaali ND *et al* (2002) MR evaluation in patients with isolated anosmia since birth or early childhood. AJNR Am J Neuroradiol 23: 157–164
193. Yousem DM *et al* (1996) MR evaluation of patients with congenital hyposmia or anosmia. Am J Radiol 166: 439–443
194. Kallmann FJ, Schoenfeld WA, Barrera SE (1944) The genetic aspects of primary eunuchoidism. Am J Ment Defic 48: 203–236
195. Wustenberg EG *et al* (2001) Normosmie bei Kallmann Syndrom – Ein Fallbericht. Laryngorhinootologie 80: 85–89
196. Amoore JE (1991) Specific anosmias. Smell and taste in health and disease. In: Getchell TV *et al* (eds) Raven Press, New York, p 655–664
197. Kaufman MD, Lassiter KR, Shenoy BV (1988) Paroxysmal unilateral dysosmia: a cured patient. Ann Neurol 24: 450–451
198. Lanza DC, Kennedy DW (1997) Adult rhinosinusitis defined. Otolaryngol Head Neck Surg 117: S1–7
199. Togias A (1999) Mechanisms of nose-lung interaction. Allergy 54[Suppl] 57: 94–105
200. Stevens CN, Stevens MH (1985) Quantitative effects of nasal surgery on olfaction. Am J Otolaryngol 6: 264–267
201. Haddad FS *et al* (1985) Intracranial complications of submucous resection of the nasal septum. Am J Otolaryngol 6: 443–447
202. Gross-Isseroff R *et al* (1989) Olfactory function following late repair of choanal atresia. Laryngoscope 99: 1165–1166

203. Hosemann W *et al* (1993) Olfaction after endoscopic endonasal ethmoidectomy. Am J Rhinology 7: 11–15
204. Rowe-Jones JM, Mackay IS (1997) A prospective study of olfaction following endoscopic sinus surgery with adjuvant medical treatment. Clin Otolaryngol 22: 377–381
205. Downey LL, Jacobs JB, Lebowitz RA (1996) Anosmia and chronic sinus disease. Otolaryngol Head Neck Surg 115: 24–28
206. Min YG *et al* (1995) Recovery of nasal physiology after functional endoscopic sinus surgery: olfaction and mucociliary transport. ORL J Otorhinolaryngol Relat Spec 57: 264–268
207. Lee SH *et al* (2000) Olfactory mucosal findings in patients with persistent anosmia after endoscopic sinus surgery. Ann Otol Rhinol Laryngol 109: 720–725
208. Jafek BW, Murrow B, Johnson EW (1994) Olfaction and endoscopic sinus surgery. Ear Nose Throat J 73: 548–552
209. Passagia JG *et al* (1999) Surgical approaches to the anterior fossa, and preservation of olfaction. Adv Tech Stand Neurosurg 25: 195–241
210. Spetzler RF *et al* (1993) Preservation of olfaction in anterior craniofacial approaches. J Neurosurg 79: 48–52
211. Suzuki J, Yoshimoto T, Mizoi K (1981) Preservation of the olfactory tract in bifrontal craniotomy for anterior communicating artery aneurysms, and the functional prognosis. J Neurosurg 54: 342–345
212. Dulguerov P, Allal AS, Calcaterra TC (2001) Esthesioneuroblastoma: a meta-analysis and review. Lancet Oncol 2: 683–690
213. Dias FL *et al* (2003) Patterns of failure and outcome in esthesioneuroblastoma. Arch Otolaryngol Head Neck Surg 129: 1186–1192
214. Bakay L, Cares HL (1972) Olfactory meningiomas. Report on a series of twenty-five cases. Acta Neurochir (Wien) 26: 1–12
215. Welge-Luessen A *et al* (2001) Olfactory function in patients with olfactory groove meningioma. J Neurol Neurosurg Psychiatry 70: 218–221
216. Bonfils P, Corre FL, Biacabe B (1999) Sémiologie et étiologie des anosmies: à propos de 306 patients. Ann Otolaryngol Chir Cervicofac 116: 198–206
217. Costanzo RM, DiNardo LJ, Zasler ND (1995) Head injury and olfaction. Handbook of Olfaction and Gustation. In: Doty RL (ed) Marcel Dekker, New York, p 493–502
218. Murphy C, Doty RL, Duncan HJ (2003) Clinical disorders of olfaction. Handbook of olfaction and gustation. In: Doty RL (ed) Marcel Dekker, New York, p 461–478
219. Gradziadei PPC, Monti-Graziadei GA (1978) Continuous nerve cell renewal in the olfactory system. Handbook of sensory physiology, vol. IX. In: Jacobson M (ed) Springer, Wien, New York, p 55
220. Hummel T *et al* (2004) Qualitative olfactory dysfunction: frequency and prognostic significance. Abstractbook AchemS p 89 (Poster 339)
221. Leopold DA *et al* (1991) Successful treatment of phantosmia with preservation of olfaction. Arch Otolaryngol Head Neck Surg 117: 1402–1406

222. Leopold DA, Loehrl TA, Schwob JE (2002) Long-term follow-up of surgically treated phantosmia. Arch Otolaryngol Head Neck Surg 128: 642–647
223. Markert JM, Hartshorn DO, Farhat SM (1993) Paroxysmal bilateral dysosmia treated by resection of the olfactory bulbs. Surg Neurol 40: 160–163
224. Bonfils P *et al* (2004) Traitement chirurgical d'une parosmie post-rhinitique. Ann Otolaryngol Chir Cervicofac 121: 47–50
225. Fong KJ *et al* (1999) Olfactory secretion and sodium, potassium-adenosine triphosphatase: regulation by corticosteroids. Laryngoscope 109: 383–388
226. Golding-Wood DG *et al* (1996) The treatment of hyposmia with intranasal steroids. J Laryngol Otol 110: 132–135
227. Tos M *et al* (1998) Efficacy of an aqueous and a powder formulation of nasal budesonide compared in patients with nasal polyps. Am J Rhinol 12: 183–189
228. Mott AE *et al* (1997) Topical corticosteroid treatment of anosmia associated with nasal and sinus disease. Arch Otolaryngol Head Neck Surg 123: 367–372
229. Stevens MH (2001) Steroid-dependent anosmia. Laryngoscope 111: 200–203
230. Mott AE, Leopold DA (1991) Disorders in taste and smell. Med Clin North Am 75: 1321–1353
231. Klimek L, Eggers G (1997) Olfactory dysfunction in allergic rhinitis is related to nasal eosinophilc inflammation. J Allergy Clin Immunol 100: 159–164
232. Jafek BW *et al* (1987) Steroid-dependent anosmia. Arch Otolaryngol Head Neck Surg 113: 547–549
233. Heilmann S, Hüttenbrink KB, Hummel T (2004) Local and systemic administration of corticosteroids in the treatment of olfactory loss. Am J Rhinol 18: 29–33
234. Ikeda K *et al* (1995) Efficacy of systemic corticosteroid treatment for anosmia with nasal and paranasal sinus disease. Rhinology 33: 162–165
235. Meltzer EO *et al* (1998) Subjecive and objective assessments in patients with seasonal allergic rhinitis: effects of therapy with mometasone furoate nasal spray. J Allergy Clin Immunol 102: 39–49
236. El Naggar M *et al* (1995) Effect of Beconase nasal spray on olfactory function in post-nasal polypectomy patients: a prospective controlled trial. J Otolaryngol 109: 941–944
237. Blomqvist EH *et al* (2003) Placebo-controlled, randomized, double-blind study evaluating the efficacy of fluticasone propionate nasal spray for the treatment of patients with hyposmia/anosmia. Acta Otolaryngol 123: 862–868
238. Heilmann S *et al* (2004) Untersuchung der Wirksamkeit von systemischen bzw. topischen Corticoiden und Vitamin B bei Riechstörungen. Laryngo-Rhino-Otol 86: 1–6
239. Benninger MS *et al* (2004) Techniques of intranasal steroid use. Otolaryngol Head Neck Surg 130: 5–24
240. Hardy JG, Lee SW, Wilson CG (1985) Intranasal drug delivery by spray and drops. J Pharmacy Pharmacol 37: 294–297

241. Newman SP, Moren F, Clarke SW (1987) Deposition pattern from a nasal pump spray. Rhinology 25: 77–82
242. McGarry GW, Swan IR (1992) Endoscopic photographic comparison of drug delivery by ear-drops and by aerosol spray. Clinical Otolaryngology 17: 359–360
243. Parnes SM, Chuma AV (2000) Acute effects of antileukotrienes on sinonasal polyposis and sinusitis. Ear Nose Throat J 79: 18–20, 24–25
244. Bachmann G, Hommel G, Michel O (2000) Effect of irrigation of the nose with isotonic salt solution on adult patients with chronic paranasal sinus disease. Eur Arch Otorhinolaryngol 257: 537–541
245. Rundles W (1946) Prognosis in the neurologic manifestations of pernicious anemia. Blood 1: 209–219
246. Tanaka O, Mukaino Y (1999) The effect of auricular acupuncture on olfactory acuity. Am J Chin Med 27: 19–24
247. Stevenson DD *et al* (1996) Aspirin desensitization treatment of aspirin-sensitive patients with rhinosinusitis-asthma: long-term outcomes. J Allergy Clin Immunol 98: 751–758
248. Moran DT *et al* (1992) Ultrastructural histopathology of human olfactory dysfunction. Microsc Res Tech 23: 103–110
249. Yamagishi M, Fujiwara M, Nakamura H (1994) Olfactory mucosal findings and clinical course in patients with olfactory disorders following upper respiratory viral infection. Rhinology 32: 113–118
250. Henkin RI *et al* (1976) A double-blind study of the effects of zinc sulfate on taste and smell dysfunction. Am J Med Sci 272: 285–299
251. Yee KK, Rawson NE (2000) Retinoic acid enhances the rate of olfactory recovery after olfactory nerve transection. Brain Res Dev Brain Res 124: 129–132
252. Hendriks APJ (1988) Olfactory dysfunction. Rhinology 26: 229–251
253. Nordin S *et al* (1996) Prevalence and assessment of qualitative olfactory dysfunction in different age groups. Laryngoscope 106: 739–744
254. Leopold D (1995) Distorted olfactory perception. Handbook of olfaction and gustation. In: Doty RL (ed) Marcel Dekker Inc. New York, p 441–454
255. Jafek BW, Murrow B, Linschoten M (2000) Evaluation and treatment of anosmia. Curr Opin Otol Head Neck Surg 8: 63–67
256. Reljanovic M *et al* (1999) Treatment of diabetic polyneuropathy with the antioxidant thioctic acid (alpha-lipoic acid): a two year multicenter randomized double-blind placebo-controlled trial (ALADIN II). Alpha Lipoic Acid in Diabetic Neuropathy. Free Radic Res 31: 171–179
257. Packer L, Kraemer K, Rimbach G (2001) Molecular aspects of lipoic acid in the prevention of diabetes complications. Nutrition 17: 888–895
258. Hounsom L *et al* (1998) A lipoic acid-gamma linolenic acid conjugate is effective against multiple indices of experimental diabetic neuropathy. Diabetologia 41: 839–843
259. Hounsom L *et al* (2001) Oxidative stress participates in the breakdown of neuronal phenotype in experimental diabetic neuropathy. Diabetologia 44: 424–428

260. Garrett NE *et al* (1997) alpha-Lipoic acid corrects neuropeptide deficits in diabetic rats via induction of trophic support. Neurosci Lett 222: 191–194
261. Coppey LJ *et al* (2001) Effect of antioxidant treatment of streptozotocin-induced diabetic rats on endoneurial blood flow, motor nerve conduction velocity, and vascular reactivity of epineurial arterioles of the sciatic nerve. Diabetes 50: 1927–1937
262. van Dam PS *et al* (2001) Glutathione and alpha-lipoate in diabetic rats: nerve function, blood flow and oxidative state. Eur J Clin Invest 31: 417–424
263. Lynch MA (2001) Lipoic acid confers protection against oxidative injury in non-neuronal and neuronal tissue. Nutr Neurosci 4: 419–438
264. Seiden AM (1997) The intial assessment of patients with taste and smell disorders. Taste and smell disorders. In: Seiden AM (ed) Thieme, New York, p 4–19
265. Garrett-Laster M, Russell RM, Jacques PF (1984) Impairment of taste and olfaction in patients with cirrhosis: the role of vitamin A. Hum Nutr Clin Nutr 38: 203–214
266. Henkin RI (1994) Drug-induced taste and smell disorders. Incidence, mechanisms and management related primarily to treatment of sensory receptor dysfunction. Drug Saf 11: 318–377
267. Ackerman BH, Kasbekar N (1997) Disturbances of taste and smell induced by drugs. Pharmacotherapy 17: 482–496
268. Hudry J, Perrin F, Ryvlin P, Mauguière F, Royet J-P (2003) Olfactory short-term memory and related amygdala recordings in patients with temporal lobe epilepsy. Brain 126: 1851–1863

Offprint from
Advances and Technical Standards in Neurosurgery, Vol. 30
Edited by J.D. Pickard

Cranial Venous Outflow Obstruction and Pseudotumor Cerebri Syndrome

B. K. OWLER[1,2], G. PARKER[3], G. M. HALMAGYI[1], I. H. JOHNSTON[1], M. BESSER[1,2], J. D. PICKARD[4], and J. N. HIGGINS[5]

[1] T. Y. Nelson Departments of Neurosurgery and Neurology, Institute of Clinical Neurosciences, Royal Prince Alfred Hospital, Sydney, Australia
[2] Department of Surgery, University of Sydney, Sydney, Australia
[3] Department of Radiology, Royal Prince Alfred Hospital, Sydney, Australia
[4] Department of Neurosurgery, Academic Neurosurgical Unit, University of Cambridge, Addenbrookes Hospital, Cambridge, UK
[5] Department of Neuroradiology, Addenbrookes Hospital, Cambridge, UK

With 14 Figures

Contents

Abbreviations

ACCS	Average Combined Venous Conduit Score
AVM	Arteriovenous Malformation
BMI	Body Mass Index
CBF	Cerebral Blood Flow
CBV	Cerebral Blood Volume
cmCSF	centimetres of CSF
cmH_2O	centimetres of water
$CMRO_2$	Cerebral Metabolic Rate of Oxygen
CPP	Cerebral Perfusion Pressure
CSF	Cerebrospinal Fluid
CT	Computed Tomography
CVT	Cerebral Venous Sinus Thrombosis
DRCV	Direct Retrograde Cerebral Venography
ICP	Intracranial Pressure
kg/m^2	kilograms per square metre
mmHg	millimetres of Mercury
mmHg/ml/min	millimetres of Mercury per millilitre per minute
MR	Magnetic Resonance
MRI	Magnetic Resonance Imaging
MRV	Magnetic Resonance Venography
PTS	Pseudotumor cerebri syndrome
R_{csf}	Resistance to CSF absorption
SSS	Superior Sagittal Sinus
SVC	Superior Vena Cava

Abstract

The pathophysiology of PTS including idiopathic intracranial hypertension or 'BIH', remains controversial. The older literature frequently referred to pathology in the cerebral venous drainage but more modern imaging techniques (CT and early MR) failed to reveal gross venous pathology. The role of impaired cranial venous outflow has recently been re-examined in the light of new methods of investigation (advanced MR venography and direct microcatheter venography with manometry) and of treatment (venous sinus stenting).

Venous sinus obstruction in PTS is a more common factor in the pathogenesis of the condition than previously recognised. Venous obstruction may be primary, that is, it is the underlying aetiological factor in PTS. Venous sinus obstruction may also be secondary to raised CSF pressure which may exacerbate problems with intracranial compliance and raised CSF pressure. Early experience with venous stenting suggests that it may be a helpful treatment for patients with PTS but more experience and longer follow-up is required to define the subgroups of patients for whom it is most appropriate.

Keywords: Pseudotumor cerebri syndrome; idiopathic intracranial hypertension; benign intracranial hypertension; papilloedema; intracranial pressure; cerebral venous drainage; venous stent, MR venography.

Introduction

There has been a recent revival of interest in the role of impaired cranial venous outflow in the causation of the PTS, even to the point of proposing that this may be the one underlying factor in all cases [78, 86]. For a general review of PTS please see reference 169a. Although the evidence is insufficient to establish such a unifying hypothesis on causation, nonetheless there is enough to suggest that some form of venous outflow pathology is present in a significant number of cases, the presence of which is likely to have been overlooked in the period when CT scanning was the basis of diagnosis [74]. Two recent developments make this an issue of significance. First, the means are now available as never before to detect the presence of cranial venous outflow tract pathology – advanced MR imaging methods [43, 64] and microcatheter venography with manometry. Second, there are now techniques available for direct treatment of intracranial venous sinus obstruction – endovascular clot lysis [88] and venous sinus stenting [63, 65, 128, 130]. Such developments, in their importance for the diagnosis and treatment of PTS, make a review of the subject of cranial venous outflow obstruction in this condition both pertinent and timely.

Historical Perspective

The association between cranial venous outflow obstruction and PTS dates back to the earliest descriptions of the latter condition, indeed even prior to the reports of Quincke [138, 139] and Nonne [123] which are generally taken to represent the first recognition of a specific condition initially called either meningitis serosa (Quincke, 1893) or pseudotumor cerebri (Nonne, 1904). Thus Taylor wrote in the 1890 edition of *The Practice of Medicine*: "It is important to remember what has now been verified in numerous cases that in mastoid suppuration there is often double optic neuritis with an entire absence of meningitis or abscesses proved by post-mortem examination, or by recovery after simple trephining of the mastoid cells." The link with ear disease was through involvement of the lateral sinuses by the disease itself, or through internal jugular vein ligation as a means of treatment. For example, Newton Pitt, also in 1890 [121], described three patients, all of whom had papilloedema but no other CNS signs, and who recovered, one having had a lateral sinus explored with clot removal, and the internal jugular vein ligated.

In the first three decades of the 20th century there were numerous reports of PTS occurring in conjunction with chronic suppurative otitis media and mastoiditis to the extent that Symonds [170] in 1931 felt able to define a condition of otitic hydrocephalus in which CSF excess due to over-production or impaired absorption followed cranial venous sinus, particularly but not exclusively transverse sinus obstruction, occurring secondary to chronic middle ear or other infection. In addition, Liedler in 1928 [102] was probably the first to describe PTS after ligation of one or both internal jugular veins in the treatment of chronic ear disease. Symonds' concept did not, however, survive the neuroradiological developments of the 1930s when the newly introduced techniques of encephalography and ventriculography showed that there was no demonstrable increase in the volume of fluid in the intracranial CSF-containing spaces in these cases. This has, of course, been an enduring difficulty in establishing a disturbance of CSF hydrodynamics as primary in PTS. Whatever the precise mechanism, the association of PTS with chronic ear disease and venous sinus pathology remained important. Thus in other significant reports from the 1930s, six of the fifteen cases reported by Davidoff & Dyke [33] had chronic suppurative otitis media or other infection whilst all of Gardner's [48] cases had chronic ear disease, the majority also having lateral sinus occlusion.

In parallel with the literature on so-called 'otitic hydrocephalus' a series of cases was published describing "traumatic hydrocephalus" [109]. This misnomer was used to describe cases of PTS occurring after closed head injury, either with or without skull fractures. The association between closed head injury and venous sinus obstruction was first proposed by Ecker in 1946 [40] but was not confirmed using venography until later by several other groups [11, 84, 109]. These cases were distinct from those in which there was a depressed fracture fragment or a penetrating injury. Most cases involved either a non-displaced linear occipital skull fracture crossing the sinus or a closed head injury without a fracture.

By the 1950s the relation between PTS and cranial venous sinus pathology, particularly secondary to chronic ear infection, was well-established, although hydrocephalus had been excluded. In an important paper in 1955, Foley [45] collected from the literature 46 cases of benign intracranial hypertension (a term which he introduced) which he labeled as otitic and described 60 cases of his own of which 13 were otitic or cerebral venous in origin. In 25 cases with right-sided otitis media mastoidectomy was performed in 21. Thrombosis of the lateral sinus was proven at surgery in 14 cases whilst another 6 had some type of sinus involvement such as perisinus abscess. Of the 13 cases with left-sided otitis media, mastoidectomy was performed in 9 with lateral sinus thrombosis in 3 and other involvement of the lateral sinus in 2 more. Five cases had bilateral ear disease and of these 4 went on to mastoidectomy. In 2 there was left-sided lateral sinus thrombosis and in 2 the sinuses were apparently normal. Foley proposed that lateral sinus thrombosis, presumably of the dominant sinus, rather than thrombosis of the superior sagittal sinus (SSS), was the primary pathology in otitic PTS.

Four other aspects of the possible link between PTS and cranial venous outflow obstruction were also documented by this time. First, Evans [42] gave further evidence of PTS following internal jugular vein ligation although in reviewing other reports as well as his own cases he found that none of the 7 patients having bilateral ligation for non-otitic problems developed papilloedema whereas the 3 cases who had the procedure in relation to ear disease did. Of 6 cases who had unilateral ligation for bilateral ear disease only one developed papilloedema. Second, there were several reports linking PTS with chronic respiratory disease and cardiac failure with the presumption of increased venous pressure [6, 21, 155]. Third, there were the first reports linking PTS with blood disorders which could be presumed in some instances to act through venous obstruction [38, 103, 172]. Fourth, and most significant, was the study by Ray and Dunbar [141] who used the technique of direct sinography in which a catheter was introduced via a burr-hole into the anterior segment of the SSS. They studied 4 patients, 2 of whom were identified as having PTS without any antecedent factors. Both had failed to respond satisfactorily to subtemporal decompression and both showed evidence of obstruction in the posterior segment of the SSS with elevation of intra-sinus pressure, one going on to clot removal with apparent benefit. A third case, described as a typical case of 'otitic hydrocephalus', was found to have complete obstruction of the right transverse sinus and a small left transverse sinus. There was measured elevation of SSS pressure [320 mmH_2O]. Their remaining case would not qualify for the diagnosis of PTS. The authors recommended sinography be introduced for the investigation of patients with PTS. They also proposed that the formation of collaterals after venous sinus obstruction was the mechanism for the spontaneous resolution of symptoms. Foley [45] also performed venography on a wider population of patients with PTS. Sixty cases were reviewed; 11 of which were classified as otitic hydrocephalus. Angiography was performed in 13 cases, however, only the patency of the SSS drew direct comment.

In the period between 1960 and 1980 the connection between PTS and cranial venous outflow obstruction fell out of prominence for a number of reasons. Among these may be numbered the introduction of effective antibiotic treatment which sharply reduced the incidence of chronic middle ear infection, the considerable number of other putative aetiological factors in

PTS which would not be thought to act through venous sinus obstruction, and the introduction of CT scanning which reduced the likelihood of recognizing cranial venous outflow tract pathology. The report of Janny *et al.* [71] in 1981 might, however, be taken as signalling the return of focus on the role of cranial venous outflow in PTS, a focus which has sharpened over the last two decades with the technical advances referred to in the introduction. Their study, and those that have followed, will be considered in detail in what follows.

Prevalence of Cranial Venous Outflow Obstruction in Pseudotumor Cerebri Syndrome

The study of Janny *et al.* [71] was a landmark in the study of PTS. The authors studied 16 patients with primary PTS using a combination of ventricular CSF pressure monitoring, SSS pressure monitoring and direct antegrade venography via a midline frontal burr-hole. They demonstrated venous sinus obstruction in 5 of 16 patients. In 4 of these patients the obstruction was at the level of the transverse sinus, being bilateral, or single in a functionally predominant sinus. Although issues of lesion morphology or aetiology were not addressed it suggested that the prevalence of venous sinus obstruction in primary PTS may be higher than previously appreciated.

Unfortunately, the issues raised by Janny *et al.* received little attention in the literature with only sporadic reports on the topic of PTS and venous sinus pathology. One example is that of Bortoluzzi *et al.* [17] who reported a case of PTS with severe radiculopathy in whom obstruction of a dominant right transverse sinus was clearly demonstrated with venography. Cremer *et al.* [27] also reported a single case of bilateral transverse/sigmoid sinus stenosis with pressure gradients on manometry, probably due to giant arachnoid granulations, in a patient with PTS.

In the mid-1990's two important papers detailed the findings of direct retrograde cerebral venography (DRCV) with manometry in the study of PTS patients. The first was that of King *et al.* [86] who studied 11 patients. All had undergone CT, static MRI and conventional angiography. One patient was suspected of harbouring a SSS thrombosis on MRI but this was not confirmed on other investigations. Venous manometry demonstrated elevated SSS pressures in 9 patients; the remaining 2 patients were those in whom minocycline had been implicated in the aetiology. Of the patients with elevated SSS pressures, there were focal pressure gradients (of at least 10 mmHg) at the junction of the middle and distal thirds of the transverse sinuses bilaterally in all patients. Venography demonstrated morphological abnormalities in these regions ranging from mild to severe focal narrowing. In some cases there appeared to be intraluminal filling defects and in others the sinus appeared smoothly tapered. These morphological characteristics were much more easily appreciated using venography than on conventional angiography. Shortly after a report from Karahalios *et al.* [78] suggested that venous sinus hypertension was the universal mechanism of PTS. In their study of 10 patients with PTS, 5 patients were found to have focal venous sinus obstruction. Later, in King *et al.'s* [85] second paper, a total of 21 patients with PTS for which no obvious cause was found (for example minocycline) were examined using venography and manometry. With the exception of 2 patients, SSS and CSF pressures followed each other closely and there were transverse sinus obstructions with significant pressure gradients.

The findings of Karahalios [78] and of King *et al.* [85, 86] of venous obstruction in PTC using DRCV were at odds with the conventional opinion regarding the pathophysiology of PTS. This probably reflects the fact that investigations of the venous sinuses in PTS were usually performed on the basis of four incorrect assumptions. First, the cause of venous sinus obstruction was thrombosis rather than stenosis or some other lesion. Second, the site of the obstruction was usually in the SSS. Third, static CT or MRI had sufficient sensitivity to detect the obstructing lesion. Fourth, when MRV was performed and was focused on the transverse sinuses, absence of flow in a transverse sinus was interpreted as a normal variant due to sinus hypoplasia or an artefact such as inflow turbulence.

The importance of the focus of the investigations and index of suspicion was evident from the report of Johnston *et al.* [74] who retrospectively reviewed 188 patients with PTS who had presented between 1968 and 1999. The group included 29 children. The overall incidence of venous sinus obstruction, they termed cranial venous outflow obstruction, was 19.7%. Of these 37 cases, an underlying cause of the obstruction could be identified in 20. Presumed aetiologies included thrombophilia (7 cases), trauma (2 cases), tumour (2 cases), and congenital jugular foraminal narrowing and infective internal jugular vein thrombosis with retrograde thrombosis. The remaining cases of idiopathic cranial venous outflow obstructions were all female patients. As cases had been accumulated over a 30 year period, investigations for PTS varied considerably as did the index of suspicion for venous sinus obstruction. In the first decade the incidence was only 4.2%, compared to 15% in the second and 31% in the third. In the final decade patients were likely to be investigated with MRI/MRV. While the patients in the first decade often underwent cerebral angiography rarely was the investigation focused on the venous sinuses which probably explains the low incidence of venous obstruction in that group. These authors also found the transverse and sigmoid sinuses to be the most common sites of obstruction (20/37 cases). In 11 cases obstruction was bilateral. Of the 9 patients with unilateral obstruction, the transverse sinus was definitely dominant in 6 cases and probably dominant in 2. In one case it was considered to be the non-dominant sinus although DRCV with manometry was not performed. Interestingly, only when the SSS became involved did obtundation or venous infarction become evident.

Two prospective case-control studies designed to investigate the prevalence of venous sinus obstruction in PTS have recently been published. Farb *et al.* [43] used a 3-D gadolinium enhanced MRV to examine the venous sinuses of 29 patients (age 37.2 years) with PTS and 59 control patients (age 60.3 years). The control group consisted of cancer patients that were undergoing MRI of the brain as a screening test for cerebral metastases. Patients with intracranial pathology were excluded from the study. The MRV of each patient was examined by 3 blinded radiologists. For each patient, an average combined venous conduit score (ACCS) of 2–8 was produced. For each side the patency of the transverse sinus was scored from 1–4 (1 = hypoplasia or severe stenosis and 4 = normal and patent). There was very high inter-observer reliability. With the exception of 2 patients, all had an ACCS of less than 5 (93.1%). Four of 59 controls (6.8%) had an ACCS of less than 5. Thus, an ACCS of less than 5 had a 93% specificity and sensitivity for PTS. For patients, there was no correlation between CSF pressure and ACCS. The morphology of the obstructing lesions amongst the PTS patients appeared extraluminal in 45 and intraluminal in 13 patients.

MRV was used by Higgins *et al.* [64] to examine 20 patients with PTS and 40 controls subjects. The control group consisted of patients who were recruited from patients presenting for MRI of another body region. Control patients were screened for headache and neurological disorders. Those with symptoms apart from very occasional minor headache were excluded. The PTS patients and asymptomatic controls were matched for sex and age. The MRVs were assessed for the existence of flow gaps in the transverse sinuses. No flow gaps could be seen in the transverse sinuses of any control patient on either side. In the PTS group, there were bilateral transverse flow gaps in 13 patients (65%). In only one patient were the sinuses normal bilaterally.

Interaction Between Venous Sinus Hypertension and CSF Pressure

The aforementioned recent observations demonstrating a much greater prevalence of venous sinus obstruction in patients with PTS raise the question of what role venous sinus obstruction occupies in the aetiology of PTS. Clearly, in cases of cerebral venous sinus thrombosis the role has been defined. However, the nature of venous sinus obstruction in PTS is different. Both intrinsic and extrinsic lesions have been identified and both characteristically are found in the region of the junction of the middle and distal thirds of the transverse sinus close to the asterion of the skull. Before attempting to answer the question of whether the lesions are the cause or effect of raised intracranial pressure, the effects of primarily raising venous pressure on CSF pressure, and the reverse, will be examined in both the experimental and clinical settings.

Effects of Raised Venous Pressure in Adults and Children

Experimentally, early studies aimed not to produce PTS but rather hydrocephalus by increasing venous sinus pressure. Attempts to produce sustained increases in venous sinus pressure, particularly by occluding large venous conduits, usually failed [8, 32]. Dixon & Halliburton [37] increased venous sinus (torcular) pressure acutely in the dog and found a small increase in CSF pressure (approximately 25% of the increase in venous pressure). The difficulties in these earlier experiments probably related to the difficulties in isolating the venous circulation in most species of laboratory animals, the propensity for the development of venous collaterals and the existence of alternative routes of CSF drainage by pathways such as the cribriform plate [135].

The first study in which adequate occlusion of cranial venous outflow was achieved was that of Bering & Salibi [15] in dogs. The external and internal jugular veins were ligated in the neck proximal to the facial vein. The condyloid foramen was also occluded. After one week, a neck dissection was performed and any collateral venous drainage was ligated. Of the 21 dogs subject to this procedure 13 developed hydrocephalus. In almost all animals both the CSF and SSS pressures were elevated. In dogs that developed hydrocephalus, CSF pressure fell after a few days and remained below the SSS pressure. However SSS pressure also fell with time and was associated with the development of collaterals as demonstrated using sinography. Of the 8 animals that did not develop hydrocephalus, ligation was incomplete in one while another was killed the day of completion of the surgery. In the remaining six animals both the CSF pressure and SSS venous pressures increased. Also examined were the pulse pressures of the CSF and SSS. In the eight animals so examined, 3 developed hydrocephalus and these animals had higher pulse pressures than those that did not. The results were taken as evidence of a long suspected link between hydrocephalus and venous sinus obstruction. However later studies did not confirm Bering and Salibi's findings. Guthrie *et al.* [57] obstructed the torcular and transverse sinuses of 10 adult dogs using cotton pledgelets in an attempt to produce hydrocephalus. Over a period of up to 29 weeks SSS pressure increased significantly as did CSF pressure. However both pressures fell towards 5 weeks and was associated with the development of venous collaterals around the torcular. There was no difference in ventricular size at post-mortem. The result of the experiment, at least initially, was thus not hydrocephalus but PTS. It has also been suggested that the extent of dissection required to isolate the venous circulation in the animals of Bering and Salibi's study was so great that alterative routes of CSF drainage including the lymphatics were also compromised.

Clinically, venous sinus hypertension due to obstruction is known to result in PTS. The most common clinical example of this is the venous obstruction of the cerebral sinuses that occurs in cerebral venous sinus thrombosis (CVT). Cerebral venous thrombosis is the most well recognized cause of venous sinus hypertension and when venous sinus thrombosis is limited to the lumen of the sinus and does not involve cortical veins the clinical picture may be identical to PTS [18, 147]. The acknowledgement of venous sinus thrombosis as a cause of PTS syndrome is evident in the need to exclude venous sinus thrombosis in cases of PTS [98]. This distinction between CVT and PTS is justified by the differences in management and prognosis of the two conditions [99]. These issues aside CVT does demonstrate the clinical effects of venous sinus obstruction on CSF and intracranial pressures.

Venous sinus obstruction may also occur from non-thrombotic venous sinus obstruction. There are a few small series and a large number of case reports documenting PTS due to mass lesion both intrinsic and extrinsic to the venous sinuses that result in PTS. A non-exhaustive list of these published cases are presented in Table 1. Furthermore relief of the obstruction by removal of the offending lesion usually results in a reduction in CSF pressure and relief of clinical symptoms.

Venous sinus obstruction causes an increase in venous sinus pressure proximally. The effects of this raised cranial venous outflow pressure on the brain and CSF do not occur in isolation and so need to be considered together. Elevated venous sinus pressure affects both CSF absorption and production. The main site of CSF absorption is thought to be the arachnoid villi of the lateral lacunae and SSS's. The absorption process is a pressure-dependent process. Davson [34] demonstrated that the absorption of CSF depends on a pressure gradient between the subarachnoid space and the venous sinus of approximately 3 mmHg in health. Thus when venous sinus pressure is raised, CSF pressure must also rise in order for CSF absorption to continue. This explains why CSF pressure is usually a few millimetres of mercury higher than venous sinus pressure in cases where both pressures are monitored simultaneously. In addition to causing an increase in CSF pressure, venous sinus hypertension may also effect production. CSF production is for the most part a pressure-independent process in respect to CSF pressure. However, venous sinus hypertension will affect venous outflow from the site of CSF production, that is, the choroid plexus. As part of the mechanism of CSF production is filtration of plasma

Table 1. *Case Reports and Series of Patients with a Pseudotumor Syndrome Secondary to Venous Sinus Obstruction of Various Aetiologies*

Reference	Pathology	Number of cases
[89] Kollar et al. 1999	AVM deep venous system – post embolisation	1 (child)
[97] Lee et al. 2001	Torcular epidermoid	1 (adult)
[92] Lam et al. 2001	Torcular epidermoid	1 (adult)
[91] Lam et al. 1992	Radical neck dissection/sigmoid sinus ligation	3 (adults)
[91] Lam et al. 1992	CVC thrombosis	2 (adults)
[53] Goldsmith et al. 1991	Ca prostate metastasis: SSS compression	1 (adult)
[132] Plant et al. 1991	Plasmacytoma & Ewing's sarcoma	2 (adults)
[80] Keiper et al. 1999	Suboccipital/translabyrinthine craniectomy	5 (adults)
[17] Bortoluzzi et al. 1982	Bilateral lateral sinus obstruction ?cause	1 (adult)
[132] Plant et al. 1991	Occipital skull tumours	2 (adults)
[82] Kim et al. 2000	Metastatic prostate cancer	1 (adult)
[133] Powers et al. 1986	Cholesteatoma	1 (adult)
[27] Cremer et al. 1996	Small meningioma & thrombosis	1 (adult)
[93] Lamas et al. 1977	Dural posterior fossa AVM	1 (adult)
[46] Ford et al. 1939	Occlusion left lateral sinus	1 (adult)
[48] Gardner 1939	Unilateral sinus occlusion	3 (adults)
[56] Greer 1962	Lateral sinus thrombosis – mastoiditis	3 (adults)
[117] Medlock et al. 1992	Depressed Skull Fracture	1 (adult)
[90] Kuker et al. 1997	Epidermoid	1 (adult)
[3] Angeli et al. 1994	Glomus Jugulare	1 (adult)
[9] Beck et al. 1979	Glomus Jugulare	1 (adult)
[141] Ray et al. 1951	SSS thrombosis	3 (adults)
[72] Jicha et al. 2003	Cardiac septal defect – L-R shunt	1 (adult)

across the choroid, the increase in venous sinus pressure, if the deep system is affected, might increase the hydrostatic pressure in the capillaries of the choroids plexus and increase CSF production. Kollar *et al.* [89] reported a case of PTS in a 5 year-old boy who had undergone embolisation of a deep temporal lobe AVM that drained via large venous varix into the vein of Galen. After embolisation a cerebral angiogram demonstrated that the vein of Galen did not fill and only sluggish flow in the straight sinus. The other dural sinuses were patent. The authors speculated that the venous outflow of the deep venous system would increase transcapillary CSF production in the choroid plexus. If this production was in excess of absorptive capacity, as may be the case if the system was underdeveloped, then PTS might result.

The other effect of raising venous sinus pressure is on venous outflow from the brain itself. Apart from perhaps the lumbar subarachnoid space, the cerebral venous system contributes most to the compliance of the intracranial space. Therefore, an increase in venous sinus and cerebral venous pressure increases the volume of the venous system proximal to the obstruction and reduces the compliance of the craniospinal axis. When venous sinus obstruction occurs, the high compliance cerebral venous system should increase in size and should be reflected in the observation of increased cerebral blood volume (CBV). In fact, Dandy [31] hypothesised that PTS was a result of increased CBV. Mathew *et al.* [113] calculated cerebral blood flow (CBF) and CBV before and after treatment using carotid injections of Xe^{133} and Tc^{99m}. Both patients demonstrated increased CBV and this decreased towards normal when CSF pressure had been reduced. CBF was also slightly reduced in both cases prior to treatment and increased after CSF pressure reduction. Mathew *et al.* [113] stated that the cases provide evidence of venous engorgement. Raichle *et al.* [140] studied CBF, cerebral metabolic rate oxygen ($CMRO_2$) and CBV using carotid injection of ^{15}O-labelled water, oxyhaemoglobin and carboxyhaemoglobin. Compared to normal values there was a small but significant reduction in CBF of 18.5% ($n = 9$) and an increase in CBV of 33% ($n = 8$). In 3 patients, the studies were repeated after CSF was removed to lower ICP. CBF remained unchanged but there was a 10% reduction in CBV. Most patients had undergone cerebral angiography and no evidence of venous outflow obstruction was reported. In contrast, Brooks *et al.* [19] used positron emission tomography and steady-state inhalation of $C^{15}O_2$, $^{15}O_2$ and ^{11}CO to study regional CBF, $CMRO_2$ and CBV. No difference in any of these variables could be demonstrated between the 5 patients and 15 controls. In one patient, the study was repeated after lumbo-peritoneal shunting. CBF and $CMRO_2$ appeared improved, at least in white matter. There was no change in CBV. Thus there is at least tentative evidence in a limited number of studies for an increase in CBV in PTS. Modern imaging techniques have yet to be applied to the study of CBV in PTS.

An increase in cerebral venous pressure will alter the Starling equation across the capillary bed as the venous outflow pressure and therefore capillary hydrostatic pressure is increased. This would normally result in an increase in ultrafiltrate and vasogenic oedema. There is little direct evidence for brain oedema in PTS. Although Sahs, Hyndman and Joynt [149, 150] provided histological evidence of brain oedema at post-mortem, their findings have been questioned on the basis of tissue preparation and artefact. Wall [176] could find no evidence of brain oedema in 2 patients with non-active PTS at post-mortem. However it should be clear that although there is no histological evidence supporting the finding of brain oedema there is no good evidence to refute such a claim.

More information has become available using MR imaging which has the ability to detect increased brain water. Early studies using low strength magnets without the benefits of diffusion weighted scans were contradictory. Connolly *et al.* [24] using qualitative examination of images obtained on a 0.15 Tesla magnet reported no signal change in 7 children with PTS. The same finding was reported by Silbergleit *et al.* [154] in 6 patients with PTS using a 0.35 Tesla magnet. Benefiting from improved technology, Moser *et al.* [120] used a heavily-weighted T2 MR technique (1.5 Tesla) to investigate the brain water content in 10 patients with PTS. They found an increase in the signal white matter free water content as reflected in prolongation of the T2 relaxation time. The authors concluded that this represents a diffuse low level of oedema. In addition, a triple-echo sodium MR technique was used to study 5 patients. Three demonstrated no change in their sodium signal. However, two patients who were clinically the most severely affected demonstrated increases in their sodium signal. As most sodium is extracellular, the authors concluded that the increase in brain water was likely to represent a vasogenic oedema. Sorenson *et al.* [161, 162] using diffusion sensitive sequences at 1.5 Tesla found that self-diffusion of white matter was increased. In some cases this was restricted to the periventricular region while in others it was distributed throughout the brain.

Gideon *et al.* [50] investigated a group of patients with PTS using diffusion-weighted MR imaging. They applied a diffusion gradient in one direction only and found that diffusion was increased in 10 patients with PTS. These studies indicate that there is a small but significant amount of brain oedema in PTS. In contrast, a more recent study by Bastin *et al.* [5] in 10 patients could demonstrate no evidence of brain oedema in PTS. These authors also used diffusion-weighted imaging (1.5 Tesla) but obtained their images using diffusion gradients in three orthogonal directions and used echo planar imaging. This allows very short acquisition times and minimises the effects of bulk brain motion. Using echo-planar imaging and diffusion tensor imaging at 3 Tesla we performed a regional analysis of the brains of 5 patients with PTS and 6 normal healthy controls [129]. Apart from small focal decreases in trace (diffusivity) in some grey matter regions, there were no differences in trace or the anisotropy of white matter regions between the two groups.

The effects of an increase in venous sinus pressure are therefore several although competing. Obstruction to cranial venous outflow appears to produce a balance between an increase in CSF pressure, capillary hydrostatic pressure, intraparenchymal pressure and CBV. This is brought about by these components being enclosed within the rigid cranium (the Munro-Kellie hypothesis). Although there is possibly a small increase in CBV, the pressure exerted on the parenchyma by an increased CSF pressure increases intraparenchymal pressure and so negates the ability of the increased capillary hydrostatic pressure to produce brain oedema. Thus the morphology of the brain does not appear to change although the intracranial pressure is increased and the compliance significantly reduced. A disturbance of this balance may be the reason for an increase tendency towards slit ventricle syndrome in PTS after ventricular shunting and the formation of the acquired Chiari syndrome after lumboperitoneal shunting in PTS. That is to say, reducing CSF pressure in the presence of continued raised venous pressure will leave the increased capillary hydrostatic pressure unbalanced resulting in brain oedema and an increase in cerebral volume.

Effects of Raised Venous Pressure in Infants

In neonates and infants the effects of raised venous sinus pressure are different from those in adults and children. Haar and Miller [60] as well as Rosman & Shands [146] catalogued a handful of cases of venous sinus obstruction with CSF circulation disorders. In patients under 18 months of age hydrocephalus developed while in those over 3 years of age PTS resulted. Both groups concluded that whether venous sinus obstruction results in hydrocephalus or PTS depends on the state of the cranial sutures [60, 146]. This difference in clinical expression of venous sinus hypertension was demonstrated experimentally by Olivero & Asner [126]. Occlusion of the posterior sagittal sinus in 10 craniectomised rabbits caused a moderate but significant increase in ventricular size compared to 5 animals that had undergone sinus occlusion but not craniectomy (distance from head of caudate to junction of septum pellucidum and corpus callosum: $7.2 +/- 0.7$ mm compared to $4.6 +/- 0.5$ mm).

Clinically, like in adults and older children, there are a large number of case reports in which various problems have caused venous hypertension (Fig. 1). However, instead of PTS, these cases develop hydrocephalus. These case series and reports are presented in Table 2. Although in some reports or series the age of the patients is greater than 18 months almost all cases had evidence of increasing head circumference from in the neonatal period. In some cases relief of the obstruction may produce relief of hydrocephalus and/or megalencephaly.

Due to these observations, there has been a gradual recognition of the role of venous sinus hypertension in the aetiology of some forms of infantile hydrocephalus and megalencephaly that have previously been thought to occur through other mechanisms. This is particularly true in achondroplasia [182] and patients with myelomeningoceles. More recently, venous outflow obstruction has been implicated in the aetiology of ventriculomegaly commonly seen in various forms of osteopetrosis [29].

Achondroplasia is frequently associated with hydrocephalus and megalencephaly. Growth in head circumference is most prominent in the first few months of life and is followed by a period of stabilisation between the 4th and 24th months. Pierre-Kahn *et al.* [131] found clinical evidence of increased prominence of venous collateral circulation in 17 of 18 patients with achondroplasia. One patient underwent angiography that depicted

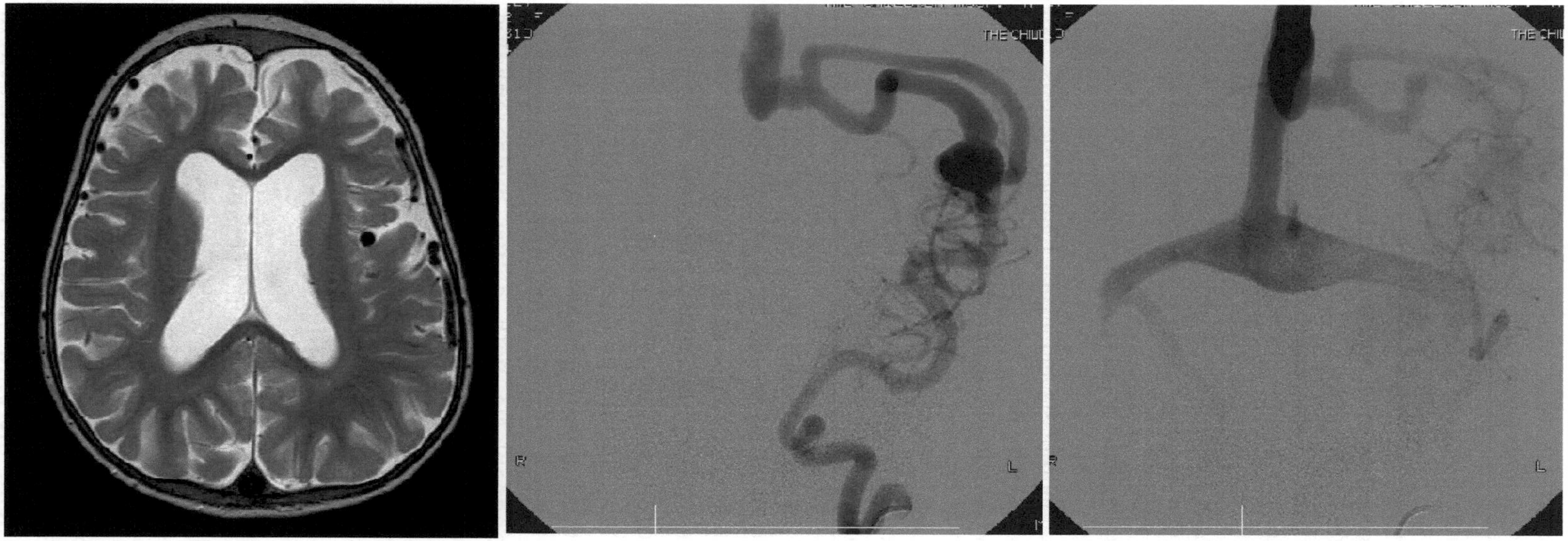

Fig. 1. Example of a 23 month old girl who presented with increasing head circumference and hydrocephalus. She was found to have an AV fistula fed by the middle cerebral arteries and which drained into the SSS

Table 2. *Case Series and Reports of Hydrocephalus Resulting from Venous Sinus Obstruction of Varying Aetiology*

Reference	Pathology	Number of cases
[58] Guttierrez et al. 1975	Agenesis of arachnoid granulations	2 (children)
[146] Rosman et al. 1978	CCF/Glenn procedure	1 (infant)
[41] Emery et al. 1956	Congenital abnormality SSS	2 (children)
[116] McLaughlin et al. 1997	SVC syndrome	3 (infants)
[79] Katznelson 1978	Cystic Fibrosis	3 (infants)
[83] Kinal 1966	Posterior fossa tumour compressing lateral sinus	4 (children)
[151] Sainte-Rose et al. 1984	Craniostenosis and achondroplasia	4 (children)
[182] Yamada 1981	Achondroplasia and jugular foramen stenosis	10 (children)
[60] Haar et al. 1975	SVC syndrome	1 (infant)
[67] Hooper 1961	SVC syndrome (thymic hyperplasia)	1 (infant)
[35] de Lange et al. 1970	AVM draining to both lateral sinuses	1 (16 years)
[28] Cronqvist et al. 1972	Cerebral AVM	2 (infants)
[49] Gibson et al. 1959	Cerebral AVM – vein of Galen	1 (infant)
[164] Stewart et al. 1975	Jugular vein thrombosis – TPN	4 (infants)

severe bilateral sigmoid sinus stenosis at the level of the jugular foramen. Friedman & Mickle [47] also reported a case with bilateral venous outflow obstruction at the level of the jugular bulb. Steinbok *et al.* [163] studied four achondroplastic children with active hydrocephalus using retrograde venography and documented significant venous hypertension in at least 2 patients associated with jugular vein stenosis and superior vena caval obstruction at the level of thoracic outlet. Furthermore, Lundar *et al.* [105] reported a case of achondroplasia with active hydrocephalus. Digital subtraction angiography demonstrated severe bilateral venous outflow obstruction at the foramen magnum. Operative decompression of the right sigmoid sinus and its junction with the jugular vein at the foramen was undertaken. A bony spur was found to be kinking the vein. Improved venous outflow was confirmed radiologically. Head circumference decreased and growth normalised.

Venous sinus obstruction is also common in craniosynostosis. Rollins *et al.* [143] using MRV studied 17 patients with craniosynostosis and a mean age of 7.3 years (4 months–34 years). The authors concentrated on the patency of the sigmoid sinus and jugular veins. No comments were made on the transverse sinuses. In 12 patients the MRV was abnormal. In 9 patients there was venous outflow obstruction at the sigmoid sinus and/or jugular bulb while in 3 patients there was jugular vein obstruction. Of these patients 9 had hydrocephalus. Venous sinus obstruction was associated with enlargement of collateral venous drainage particularly via the posterior condylar veins. Two of 11 patients with hydrocephalus had a normal MRV. However, the results of this study are difficult to interpret given the age of the patients and the time since the initial surgery in most cases. Taylor *et al.* [171] studied 23 such patients with digital subtraction angiography. ICP monitoring confirmed raised pressure in 21 cases and in 2 cases plain X-ray suggested the presence of large transcalvarial venous collateral drainage. In a total of 24 angiograms there was a greater than 50% stenosis or no flow in the sigmoid-jugular venous complex in 18 patients; in 7 unilateral and 11 bilateral. Of these 18 angiograms, florid transcalvarial collateral venous flow via a large stylomastoid emissary vein was observed in 11 cases. The severity of the stenosis did not correlate with ICP but did appear age-related. The mean age of patients with bilateral stenosis was 20.4 months, unilateral stenosis 25 months and 54 months in those with mild or no stenosis. The authors conclude that patients with more severe venous outflow obstruction tend to present with raised ICP earlier.

The importance of the state of the cranial sutures for the clinical manifestation of venous sinus obstruction is exemplified by the condition of craniosynostosis. If ICP is raised, this may manifest itself in several ways. Most commonly, raised ICP is noted after investigation for behavioural alteration or papilloedema. In the presence of normal or small sized ventricles, raised ICP is frequently attributed to inadequate intracranial volume and craniostenosis. However hydrocephalus is also recognised as occurring in association with craniosynostosis, particularly in syndromic cases where multiple sutures and the skull base are involved [44]. Hydrocephalus may occur with or without head enlargement [52]. Hydrocephalus with head enlargement may occur in the presence of sufficient uninvolved sutures or after the surgical suture release or cranial vault remodelling.

Cinalli *et al.* [23] reviewed 1727 cases of craniosynostosis. Of the 1447 cases of non-syndromic craniosynostosis, the prevalence of hydrocephalus at presentation requiring shunt insertion was just 0.28%; similar to the normal population. Two of these cases with complex craniosynostosis exhibited bilateral jugular foraminal narrowing. In comparison, syndromic cases of craniosynostosis (280) had a prevalence of hydrocephalus requiring a shunt of 12.1%. Non-progressive ventriculomegaly was seen in 15.7%. Hydrocephalus was most common in patients with Crouzon's disease (54%) and all of these patients had either their coronal, sagittal or both sutures open at the time of ventricular enlargement. In the other patients hydrocephalus occurred after surgical correction and all of these had had more severe craniosynostosis with fusion of both the coronal and sagittal sutures. Angiography demonstrated bilateral jugular vein stenosis in 13/16 patients (81.3%) examined. Of the patients with Apert's syndrome, progressive hydrocephalus requiring shunting was less common (6.5%) although 18 (23.7%) had non-progressive ventriculomegaly at presentation and another 12 developed ventriculomegaly not requiring shunting after surgery. Thirteen patients underwent angiography and 7 demonstrated jugular venous obstruction (53.8%).

To summarise, venous sinus hypertension produces an increase in CSF pressure. Whether hydrocephalus or PTS results depends on the state of the cranial sutures. Where the cranial sutures are fused, such as in adults, older children and infants with severe forms of craniosynostosis, raised venous pressure causes a PTS syndrome. There is an increase in CSF pressure and a reduction in intracranial compliance. Available evidence suggests that brain oedema does not occur and that there may be either a small or no increase in CBV. If the cranial sutures are patent, such as in the infant, hydrocephalus results as the raised CSF pressure is allowed to act on the cerebral mantle and cranium.

Effects of Raised CSF Pressure

In conditions that affect the CSF circulation and raise CSF pressure there are secondary effects on the cerebral veins and the dural venous sinuses. The cerebral veins must cross the subarachnoid space in order to reach the sinuses or lateral lacunae. At this point they are prone to compression by raised CSF pressure. The lateral lacunae, because of their wide surface area, are even more prone to compression and may assist in maintaining the patency of the cerebral venous outflow. It is also proposed that in health, these structures act as Starling resistors regulating CSF absorption according to changes in CSF pressure. That is when CSF pressure is high, the lacunae collapse, decreasing venous flow, dropping sinus pressure, increasing the gradient across the arachnoid villi and increasing CSF absorption.

The dural sinuses are enclosed between the two layers of the dura. The SSS and the transverse sinuses are triangular in cross section with their base on the dura lining the skull. At the apex of the triangle the other dural leaves fuse such that the walls of these sinuses are held open by the falx cerebri and tentorium cerebelli. The sigmoid sinus also appears protected as it usually runs in a deep groove to the jugular foramen. The assumption usually made is that the sinuses are not compressible due to the structures that maintain their shape. However, the sinuses, particularly the transverse sinuses, may be compressed due to significantly raised intracranial pressure as demonstrated by a number of experimental and clinical studies (*vide infra*).

Cushing [30] studied the effects of raised CSF pressure on the SSS of the dog and reported that increased CSF pressure resulted in SSS collapse. Wright [181] later repeated the study, again in the dog, but found that this collapse only occurred when the dura surrounding the sinus had been incised. He did however demonstrate that pressures within the SSS were affected by CSF pressure in the subarachnoid space. Both Becht [7] and Weed & Flexner [180] reported that venous sinus pressures were not influenced by changes in CSF pressure. In contrast Dixon & Halliburton [37] reported that increases in CSF pressure were accompanied by increases in torcular venous pressure. However, Wright [181] and Bedford [10] noted that increases in CSF pressure caused a small decrease in venous sinus pressure. This effect was usually seen as CSF pressure was initially being increased and most likely represented a compressive effect on the cerebral veins in the subarachnoid space and a decrease in venous return to the SSS. In terms of the relationship between the venous sinuses and CSF pressure, the arrangement of venous sinuses in the dog is different from that in humans in that the torcular and lateral sinuses are encased in bone and thus protected from any compressive effects. The applicability of these early studies to human physiology is therefore questionable.

Langfitt *et al.* [94] reported the effects of increasing ICP using a subdural balloon in the rhesus monkey. Initially, SSS pressure decreased, increased or was unchanged. When ICP approached 30 mmHg SSS pressure began to rise. The transverse and sigmoid sinus pressures were influenced far less by changes in ICP. The authors reported that a gradient was demonstrated in some instances between the SSS and distal transverse sinus. In an extension of that study, Shapiro *et al.* [152] described the morphological changes of the cerebral venous system that took place in the rhesus monkey during fatal increases in ICP due to brain oedema. These animals demonstrated collapse of the SSS and straight sinuses presumably secondarily to compression. However, as ICP approached arterial pressure in these animals the implications of the results are not clear. Johnston & Rowan's [77] study of the effects of raised intracranial pressure on cerebral venous blood flow in baboons also demonstrated that the sinuses may collapse with increasing ICP. Using saline infusion into the cisterna magna of 6 baboons to raise ICP, cortical venous pressure was noted to rise linearly with ICP and remained above subarachnoid CSF pressure at all times. Animals demonstrated two distinct patterns of SSS pressure response. In 3 animals SSS pressure remained less than 20 mmHg while in the other 3 animals SSS pressure rose linearly once ICP reached 40 mmHg and remained just below cortical venous pressure.

In man, Greenfield & Tindall [55] performed cerebral angiography on 3 patients at normal and raised CSF pressures and found compression of the cerebral veins in the subarachnoid space or of the venous sinuses. Kinal [83] performed sinography on 4 patients (age 7 months–15 years) who presented with evidence of a posterior fossa lesion. In all patients the lesions had resulted in obstructive hydrocephalus with raised pressure. Sinography revealed bilateral transverse/sigmoid sinus stenosis with development of venous collateral circulation. Removal of the posterior fossa lesions resulted in improvement in venous sinus flow demonstrated by opacification of the transverse/sigmoid sinuses and the disappearance of venous collaterals on sinography.

Osterholm [127] performed antegrade cerebral venography in patients with subdural (4 patients) or extradural (1 patient) haematomas. SSS pressure was measured and venography was performed immediately prior to operative decompression. Venous sinus pressure was 21–46 cm saline and venography demonstrated bilateral transverse sinus stenosis with opacification of venous collaterals. Cerebral decompression resulted in a fall in venous pressure to 0–4 cm saline and venography showed normally filling transverse sinuses without opacification of venous collaterals. To further examine the secondary collapse of the venous sinuses as a result of increasing ICP, Osterholm [127] also performed experiments on 3 fresh human cadavers. Into the anterior SSS was perfused 600 ml/min of saline. Distally the SVC was open. Ventricular and cisterna magna CSF pressures were monitored. Infusion of normal saline into the cisterna magna allowed ICP to be raised in 10 mmHg increments. No change in SSS flow was appreciable below 20 mmHg. At 50 mmHg there was a 30% flow reduction, at 70 mmHg there was a 60% SSS flow reduction and at 200 mmHg SSS flow was arrested.

Martins *et al.* [110] monitored CSF pressure and SSS pressure simultaneously in 12 patients undergoing ventriculography who were subsequently found to have cerebral tumours. In 9 patients SSS pressure was not related to CSF pressure and remained below 14 mmHg in the presence of spontaneous or artificial increases in CSF pressure up to 75 mmHg. On venography performed in 2 of these patients there was no change in CSF pressure while in the third patient there was a partial collapse of the transverse sinus at 40 mmHg despite there being no change in SSS pressure when CSF pressure was higher. In 3 patients, SSS pressure changed with ICP. In 2 patients it increased but to a lesser extent than CSF pressure while in another patient CSF pressure and SSS pressure remained closely related throughout. In one patient who demonstrated an increase in SSS pressure with increased CSF pressure there was partial collapse of the sagittal and transverse sinuses at 40 mmHg. In addition Martins *et al.* [110] noted that 3 patients had pressure waves during recording. In 2 patients the duration was less than 1 minute. These patients had no change in SSS pressure during the pressure wave. In contrast, the third patient experienced a pressure wave of 5 minutes and this was associated with an increase in SSS pressure.

The ability of increased CSF pressure to cause venous sinus obstruction is also seen in infants with hydrocephalus. Shulman and Ranshoff [153] measured CSF and SSS pressures of 15 such cases of varying causes and included both communicating and non-communicating forms of hydrocephalus. They found a close relationship between the SSS pressure and CSF pressure and the ratio of the former to latter was 1.08. By plotting the SSS pressure versus the CSF pressure they found a regression co-efficient of 0.95 with a standard error of 0.08 indicating that SSS pressure and CSF pressure were closely related in these patients. In some patients CSF was allowed to drain while CSF pressure and SSS pressure were recorded simultaneously. The regression slop of this curve was 1 and bisected the SSS pressure axis at approximately 30 mmH_2O (probably reflecting the venous outflow pressure). The authors postulated that the increase in SSS pressure was secondary to increased CSF pressure collapsing the sinus near the point of outflow from the skull. In 3 infants antegrade venography was performed. In 2 infants the sinuses appeared normal while in a third the lateral sinuses appeared to taper on each side and end in the jugular foramen. Collateral drainage was provided by enlarged parietal and mastoid emissary veins.

Norrell *et al.* [124] studied SSS and CSF pressures in 30 infants with hydrocephalus. Eleven patients had myelomeningoceles. CSF pressure exceeded SSS pressure in 12 patients, SSS pressure exceeded CSF pressure in 8 cases and the pressures were equal in 10 cases. SSS pressure was elevated more frequently in patients with myelomeningoceles (9/11) compared to those without (9/19). Of the non myelomeningocele group none of the patients with aqueduct stenosis had elevated sinus pressures compared to 9 of the 14 patients with communicating hydrocephalus. Adequate venography was performed in 28 patients. In 18 cases both lateral sinuses were opacified whereas only one sinus opacified in 10 cases. In the 17 patients without myleomeningoceles, the anatomical position and arrangement was considered normal. In one case there was stenosis of the lateral sinus. In 11 patients with myelomeningoceles, anatomical arrangement of the venous system was abnormal and consisted of a low lying torcular with the transverse/sigmoid sinuses running directly forward around the foramen magnum to the jugular bulb. Four of these patients had venography performed at CSF pressures of 10 and 60 cmCSF. This increase in ICP caused a partial or complete collapse of the lateral sinuses, an increase in opacification of emissary collateral veins, delayed sagittal sinus emptying time and a parallel rise in SSS pressure in all cases. Lateral sinus obstruction was not seen when the CSF pressure was elevated in hydrocephalic infants without myelomeningoceles and there was no change in emptying time.

These findings demonstrate that the falx cerebri and tentorium cerebelli do not afford protection of the cerebral venous sinuses from increased CSF pressure. There also appears to be considerable differences between individuals as to when secondary collapse of the sinuses occurs. The most common site of secondary compression of the venous sinuses is the distal transverse sinus; the same site that venous sinus obstruction is observed in PTS. It is possible that secondary venous sinus compression may be important in maintaining the patency of more proximal venous channels and or have some other important role. In addition, it appears that in some conditions such as myelomeningocele with Chiari malformations, abnormally positioned venous sinuses may be more exposed to CSF pressure elevations.

Venous Sinus Obstruction in Pseudotumor Cerebri Syndrome: Cause or Effect?

Whether venous sinus obstruction in PTS is cause or effect remains unresolved. The experimental and clinical evidence discussed above indicates that there is potential for primary venous sinus pathology to cause PTS and there is also potential for secondary venous sinus obstruction due to raised intracranial pressure. We discuss below the available evidence and also present a unifying theory based on the establishment of a disordered positive feedback cycle. The discussion will centre on the site of venous obstruction, the morphology of venous sinus obstruction and the effects of removing venous sinus obstruction.

Venous sinus obstruction in PTS is most commonly, although not exclusively, seen in the transverse sinus. This is very commonly at the junction of the distal and middle thirds of the sinus. It co-incides with the area of the skull at which a number of bony sutures meet called the asterion. Other regions that may demonstrate stenosis are the other portions of the transverse sinus, posterior third of the SSS, sigmoid sinuses and jugular bulb. This point appears to be the same site at which compression of the venous sinuses takes place when CSF pressure is raised in experimental and clinical studies. If the venous sinuses are considered as a series of collapsible tubes then collapse of the tubes in such a system tends to occur at its distal end. However, it is also the most common site for the location of large arachnoid granulations that are known to cause obstruction of the transverse sinuses.

The symmetrical bilateral nature of the obstructions is also a consistent finding. This is important because, although the right transverse sinus is dominant in most cases, there is usually sufficient communication through the torcular Herophili to overcome a unilateral obstruction. The exception occurs in cases where one sinus is atretic or hypoplastic; usually the left. In these cases unilateral venous sinus obstruction may be sufficient to compromise venous outflow sufficiently to produce symptoms.

Morphologically, there are four basic types of lesions that obstruct the venous sinuses: extrinsic compression and intrinsic lesions of three main forms (Fig. 2). Extrinsic compression may be implied from the appearance of a smooth tapering of the sinus. This appearance tends to indicate that increased CSF pressure may be secondarily collapsing the venous sinus. Intrinsic lesions are of three main types: 1) broad-based lesions with an undu-

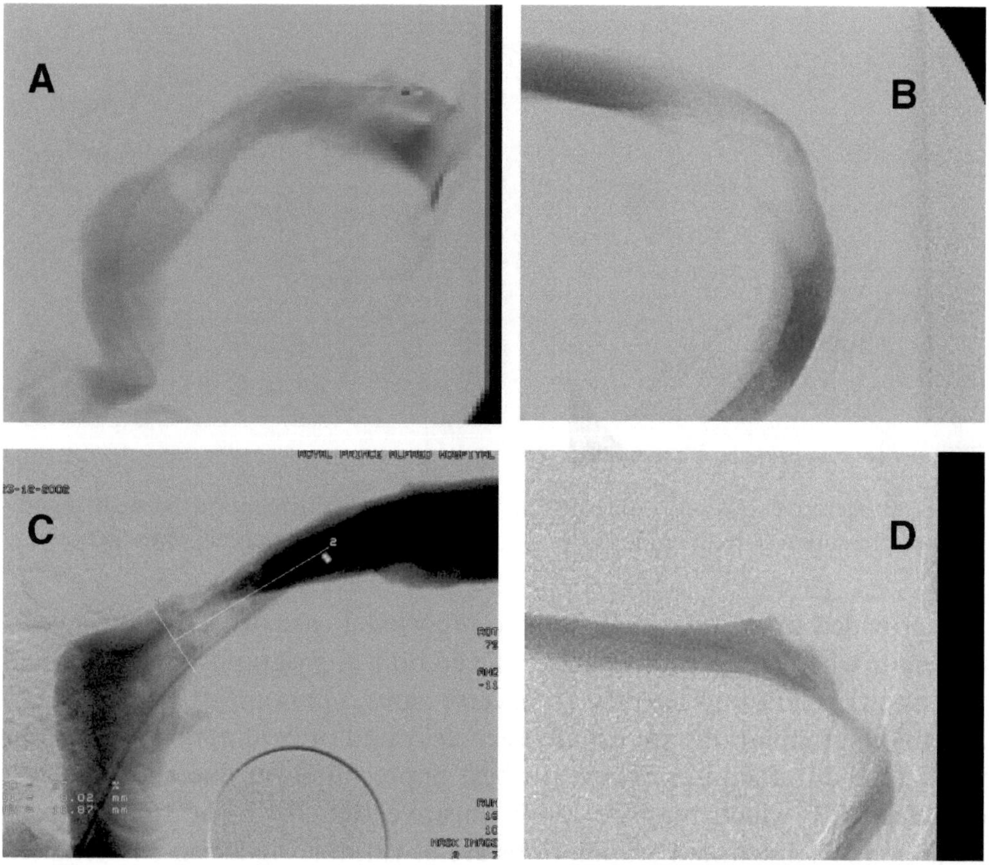

Fig. 2. Various forms of focal venous sinus obstruction in the transverse sinuses of 4 patients. Intrinsic filling defects are obvious in the top two radiographs being an arachnoid granulation (A) and a broad based undulating lesion (B). Of the lower two, the radiograph on the left (C) could be a focal stricture or extrinsic compression while the radiograph on the right (D) appears to indicate secondary venous sinus compression or irregularity due to old thrombus

lating surface that are of unknown aetiology. These may be difficult to distinguish from extrinsic compression but may require significant pressure application during stent deployment indicating the presence of a focal stenotic lesion (Fig. 3); 2) Abnormally large arachnoid granulations forming a round well defined filling defect in the venous sinus (Fig. 4), and; 3) Irregular lesions suggesting old thrombus.

Arachnoid granulations are most frequently reported as incidental findings on angiography, contrast enhanced CT or MR imaging [22, 142]. The true incidence and range of variation of 'normal' arachnoid granulations in a large series of asymptomatic individuals is yet to be established. On angiographic studies, arachnoid granulations of small to moderate size are frequently seen as filling defects of the sagittal and transverse sinuses. An example of a likely arachnoid granulation causing venous sinus obstruction

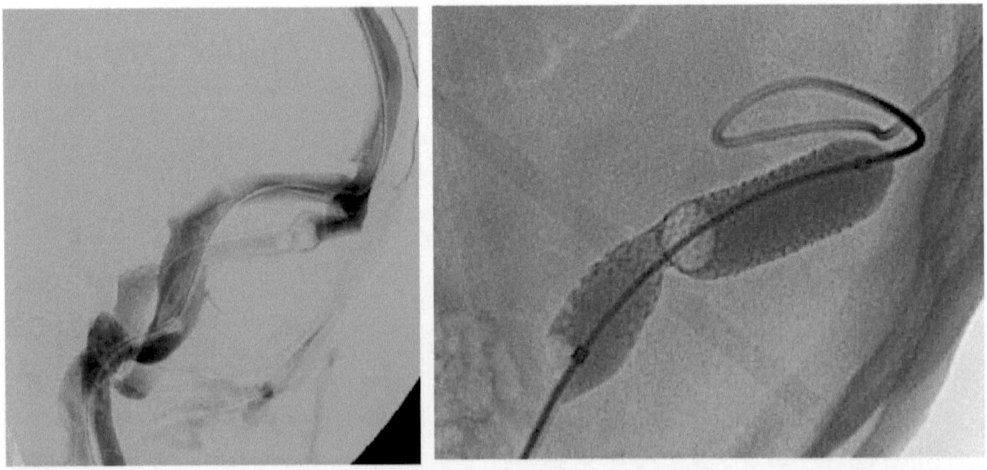

Fig. 3. Bilateral venous sinus obstruction. During the deployment of the stent, significant pressure was required in order to overcome this focal venous sinus stenosis

was provided by Arjona *et al.* [4] who reported a case of PTS in a 51 year-old man where the venous phase of cerebral angiography demonstrated the lesions protruding into the transverse sinus. On contrast enhanced CT arachnoid granulations are usually seen as round or oval hypodense lesions within the dural sinuses. They are best appreciated on fine slice contrast enhanced CT which may be more sensitive than MR for small lesions [96]. On MR, arachnoid granulations are of variable signal on T1-weighted images and hyperintense on T2 weighted images. Compared to CSF, arachnoid granulations are usually isointense to CSF on T1 weighted, T2 weighted and FLAIR MR imaging but may also have signal characteristics suggesting fat content. Their appearance is variable on proton density images [70] and may be altered by the presence or absence of calcification [142]. Oblique views on MR venography may give the impression of elongated lesion that may be mistaken for thrombus [142].

Roche & Warner [142] reported 41 arachnoid granulations in 32 patients (17 males and 15 females) on either CT or MR imaging in a 5 year period. Thirty-five (85.4%) of the arachnoid granulations were found in the distal or middle thirds of the transverse sinus. One or more vessels were closely associated with the granulation in 16 cases and appeared to enter 4 granulations. There were 2 arachnoid granulations located at the torcular and 4 in the sigmoid sinuses. Leach *et al.* [96] found 168 arachnoid granulations in 138 patients (24%) on reviewing 573 contrast enhanced CT scans; 92% of the granulations were found in the transverse sinus, especially in the middle and lateral parts. A vein entered the sinus adjacent to the granulation in 62% of cases and there was a tendency to increased incidence with age. There was no difference in the male to female distribution. On reviewing 100 MR scans there were 14 granulations identified in 13

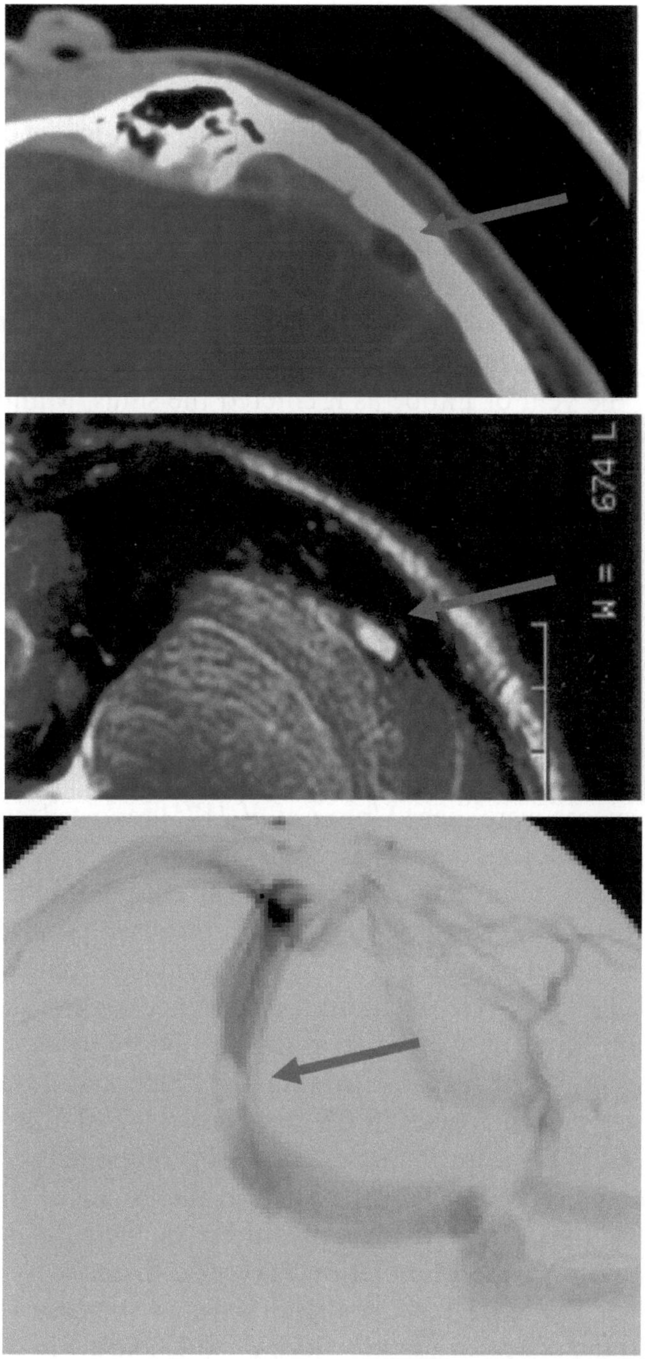

Fig. 4. A Patient with PTS with an intrinsic lesion on venography with a pressure gradient. On review of the MR images focal lesions were demonstrated in the transverse sinuses of both sides. These may represent either large arachnoid granulations or fat within the sinuses

patients; 85% were closely associated with a vein draining into the sinus. Ikushima *et al.* [70] reviewed static MR images of 1118 patients. Arachnoid granulations were present in 8.3% of males and 12.2% of females. A total of 134 arachnoid granulations were found with an overall incidence of 10%. In 14 patients there were was more than one granulation. The most common site of the granulations was the transverse sinus (85.8%).

In Browder *et al.'s.* [20] report arachnoid granulations were described as benign tumours of the cerebral venous sinuses. In an anatomical study of 295 sinuses, 25 arachnoid granulations were identified; all but 2 were in the transverse sinuses. Of those in the transverse sinuses they were almost always associated with the vein of Labbe as it entered the sinus. Mamourian & Towfighi [107] studied 10 patients without known venous sinus disease, 2 patients were found to have giant arachnoid granulations of the distal transverse sinus. In one patient there were bilateral lesions. Leach *et al.* [96] reported on the inspection of the sinuses of 29 cadavers for the presence of focal intraluminal protuberances. Ninety-one protuberances were observed in 19 cases (66%) and ranged in size from less than 1 mm to 8 mm diameter. Ninety-five percent were located in the transverse sinus – predominantly the left. In Rosenberg *et al.'s* [145] report of 4 cases of giant arachnoid granulations presenting as osteolytic skull lesions, histological examination demonstrated loss of the normal stromal organisation. Instead large CSF filled cystic spaces were seen. Upton *et al.* [175] reported on the structure of arachnoid granulations obtained from 23 autopsies. There was a tendency for the granulations to become larger and more complex with age.

Given the typical site and the correspondence with intrinsic lesions seen on MRV or DRCV, arachnoid granulations may represent a proportion of the obstructing lesions in PTS. While the natural occurrence suggests a primary role for these lesions, they may increase in size secondary to increases in CSF pressure for a number of reasons. In animals, arachnoid granulations may increase in size with increases in CSF pressure [54, 100]. It may also be that increased CSF pressure increases the CSF component of the arachnoid granulation which becomes incarcerated in the lumen of the sinus further exacerbating the increased CSF pressure. An increase in the size of the collagenous core of the granulation may also increase its size. Chronic inflammation of this core may result in such an increase [175]. Hayes *et al.* [62] studied the effects of dietary vitamin A deficiency in calves and rats and confirmed Eaton's [39] finding that it resulted in increased CSF pressure. Histological examination of the arachnoid granulations in rats and calves revealed the granulations to be larger in the vitamin A deficient animals compared to controls. This was particularly so along the transverse sinuses of the calves. The increase in size of the granulations was associated with an increase in collagen with stimulation of fibroblasts with an increase and extreme dilation of the golgi endoplasmic reticulum of these cells. Hernias of brain tissue into the core of arachnoid granulations may also result in enlargement of the granulation. Kollar *et al.* [87] reported a case of where an abnormally large arachnoid granulation was removed and the histological sections published clearly demonstrate a hernia of brain into the granulation. The authors at the time described this as 'ectopic' brain tissue but the true nature of the lesion is evident. Small hernias of brain tissue are frequently seen in the bases of arachnoid granulations at craniotomy (*M Besser; personal communication*).

There are various other pathological entities that may be represented in the spectrum of obstructive intrinsic venous sinus lesions. Fat deposits in the dural sinuses on CT was reported by Tokiguchi *et al.* [173]. These authors reported 8 cases in which macroscopic, non-obstructing fat deposits were demonstrated in the walls of the sinuses. In 5 cases fat was located at the torcular while in 3 it was located in the SSS. Anatomical proof that these lesions did indeed consist of fat was supplied in a later report where 2 of these patients had later undergone autopsy [174]. Finally, nodules of cavernous tissue have been identified in the sinuses, especially at the junction of the straight sinus and vein of Galen [13, 95]. The nodules, distinct from arachnoid granulations, were common at autopsy and were composed of endothelium lined sinusoids resembling erectile tissue.

Cerebral venous sinus thrombosis (CVT) may also contribute a proportion of the cases of PTS. Acute CVT should be distinguished from chronic CVT. In the former, provided the thrombus does not involve the cortical veins, the condition may certainly cause a pseudotumor syndrome [14, 99, 159]. The approach to management of these patients is quite different. The prevention of thrombus propagation and dissolution of the thrombus are important in the management [12]. Thus anticoagulation either systemically or locally via the endovascular route are used. Persistence of thrombosis with reorganisation and recanalisation may result in improved cerebral venous outflow. In such cases intrinsic venous sinus obstructive lesions may represent old thrombus. However, more frequently such chronic CVT cases appear to have less focal stenosis and may be difficult to treat.

There is also significant circumstantial evidence for a role of occult CVT in PTS from the association of a large number of prothrombotic states. These associations include the thrombophilias [81, 101, 112, 119] as well as other prothrombotic states such as essential [118] and iron deficiency anaemia [69]. This association between prothrombotic conditions and PTS was reviewed by Sussman *et al.* [169]. A mixed group of 38 retrospectively and prospectively accumulated patients. Eighteen patients were subject to angiography and three patients were found to have venous sinus thrombosis although the location and other details were not given. Each of these patients had a prothrombotic disorder.

In order to test whether the venous obstruction is primary or secondary one may either relieve the obstruction and observe the effects on CSF pressure or, reduce CSF pressure and observe the effects on venous obstruction. However even these measures may not produce a conclusive result due to the nature of positive feedback loops which can be interrupted without producing conclusive information on causation. For example, stopping chickens breeding does not determine whether the chicken or the egg came first.

King *et al.* [85] reduced CSF pressure in a total of 21 patients with PTS for which no obvious cause was found (for example minocycline) were examined using venography and manometry. With the exception of 2 patients, SSS and CSF pressures followed each other closely. In these patients there were transverse sinus stenoses with significant pressure gradients. Of these patients 8 underwent C1-2 puncture with removal of 20–25 mls CSF. Manometry was then repeated. The procedure was also performed on 3 patients with so-called non-idiopathic PTS. The drop in CSF pressure produced by C1-2 puncture was measured in only 3/11 patients. The reductions were 40, 23 and 10 cm CSF. All patients had a reduction in the proximal venous sinus pressures. In 5/8 idiopathic PTS patients, the pressure gradient in the transverse sinus disappeared. In 2 other patients in this group the gradient remained although it was reduced. One patient had no transverse sinus obstruction or gradient before C1-2 puncture. Of the 3 patients with so-called non-idiopathic PTS who were examined after C1-2 puncture, two patients had no venous pressure gradient to begin prior C1-2 puncture. The results of the patient with a high CSF pressure and a transverse sinus pressure gradient prior to C1-2 puncture are difficult to interpret as although proximal venous pressure fell to 10 mmHg, the distal sinus pressure was not recorded.

Given the reduction in venous sinus pressures and associated pressure gradients after C1-2 puncture and the CSF pressure reduction, the authors [85] and others [25] concluded that transverse sinus obstruction in PTS was a result of raised CSF pressure but not the cause. However, such a conclusion is difficult to justify on the basis of these results. First, although no normal healthy controls were examined, venography and manometry were performed on 10 patients with diseases other than idiopathic PTS. CSF pressure was raised in 7 patients (20–50 cmCSF) but venography and manometry demonstrated transverse sinus gradients in only 3 (43%) of these patients. In the other four there was no evidence of venous sinus hypertension. In comparison, of the 21 patients with idiopathic PTS venous sinus hypertension with transverse sinus pressure gradients were found in 19 (90%). Although the average CSF pressure was higher in the idiopathic PTS group, the higher incidence of venous sinus hypertension indicates that it may be an aetiological factor. Second, the results indicate that venous sinus pressure fell almost universally after C1-2 puncture. This is an expected outcome even in the presence of a fixed venous sinus stenosis. In at least 2 patients, the venous sinus gradient remained although reduced. Finally, the authors avoided the issue of morphological change in the stenoses after C1-2 puncture. Although they allude to the presence of either tapering or intraluminal filling defects on pre-C1-2 puncture venograms they were unable to state whether these lesions were present or absent after CSF pressure reduction. The heterogeneous nature of the stenosing or obstructing lesions means that the conclusions of the authors may only be valid for one subtype of obstructing lesion, in particular extrinsic compression.

In support of King *et al.* is a case report by McGonigal *et al.* [115] who document a case of bilateral transverse sinus obstruction in a 19 year old with quite severe PTS. The obstructions had the appearance of a smooth tapering on CT venogram. After insertion of a lumboperitoneal shunt symptoms resolved and there was marked improvement in the degree of narrowing bilaterally. Higgins and Pickard [66] also reported resolution of venous sinus obstruction in a very similar case after lumboperitoneal shunting. We have also observed this phenomenon after ventriculoperiteoneal shunting in one case. In contrast, morphological and functional venous sinus obstruction in the presence of a functioning shunt has been observed in 2 of 8 cases from our series and in 2 patients in the series of Higgins *et al.* [63].

Sainte-Rose *et al.* [151] studied the relationship between CSF and venous pressures in 31 infants (age 1–23 months). These patients consisted of 6 cases of communicating hydrocephalus, 6 cases of hydrocephalus associated with a myelomeningocele, 14 cases of craniostenosis, 3 cases of achondroplasia and a case each of aqueduct stenosis and subdural haematoma. In the first part of the study, consisting of a group of 11 infants mainly with craniostenosis, intraventricular CSF and SSS venous pressures were recorded simultaneously. In all patients the difference between CSF and SSS venous pressures were small (< 3 mmHg). CSF pressure was elevated (15–25 mmHg) in 8 patients but there was no relationship between underlying pathology and pressure recordings. In the second part of the study, a second ventricular catheter was also introduced to allow CSF pressure reduction via CSF withdrawal. The jugular venous pressure was also monitored. In 16 of 20 patients, after withdrawal of CSF to reduce CSF pressure to zero, SSS venous pressure also fell to the jugular venous pressure. Re-injection of the same volume of CSF usually restored CSF pressure to the same or a slightly higher CSF pressure than at baseline. SSS venous pressure also increased to baseline. In these patients there was no evidence of sigmoid sinus compression on sinography. In the remaining 4 patients; two with achondroplasia and two with craniostenosis; SSS did not decrease to the jugular venous pressure when CSF pressure was reduced to zero. Instead illustrated traces demonstrate a modest fall in SSS pressure.

Also suggestive of a fixed obstruction was the finding that reinjection of CSF produced a rise in SSS pressure above baseline. Fixed venous sinus obstruction of the transverse or sigmoid sinus was confirmed on sinography.

The effects of relieving the venous sinus obstruction in PTS have also been studied. The results of these studies will be considered fully in the section on treatment (*vide infra*). However, it is clear that reduction of venous sinus hypertension in PTS may result in rapid clinical resolution and a reduction in CSF pressure. Of the cases reported to date, significant clinical improvement was apparent in 4 of 4 cases in our series [130] and in 8 of 12 cases in the series from Cambridge [63]. These results, along with the observation that some obstructions are intrinsic, indicate that venous sinus obstruction in PTS may have a primary aetiological role, possibly exacerbated by raised ICP due to disordered positive biofeedback.

It is clear that arguments exist for primary and secondary aetiological roles for venous sinus obstruction in PTS. Certainly there are cases in which either may exist. Failure of treatment of venous sinus obstruction in a handful of cases indicates that it may be important to differentiate between the two. However, there may be a role for treatment of the obstruction whether the obstruction is primary or secondary. The argument for treatment of primary lesions is intuitive; however secondary venous sinus obstruction may exacerbate any underlying CSF circulation disorder; an argument supported by King [85] and Quattrrone *et al.* [137]. When CSF pressure becomes increased and venous sinus obstruction ensues, the compliance of the craniospinal axis is reduced because of engorgement of the cerebral venous compartment. Therefore small increase in CSF or blood volume will cause rapid increases in CSF pressure. More importantly though, if CSF pressure is increased sufficiently to overcome venous pressure and collapse the sinuses, venous pressure must increase in order to overcome the obstruction and maintain adequate cranial venous outflow. Venous sinus hypertension therefore becomes increased. If the primary problem is one of CSF absorption and raised R_{csf}, then in order for CSF absorption to continue, the pressure gradient between the subarachnoid CSF must be even higher than the normal gradient of 3 mmHg. CSF pressure must rise further and thus a vicious circle of rising CSF pressure, venous sinus obstruction and rising venous sinus pressure is established. Treatment of the venous sinus obstruction may interrupt this positive feedback loop and restore the normal compliance of the cerebral venous compartment.

We therefore propose the following unifying hypothesis. Normally, increased CSF pressure leads to increased CSF drainage and restoration of normal CSF pressure, representing classical physiological biofeedback control where the response of the system has a negative effect on the stimulus (Fig. 5) [59]. In PTC, with venous abnormalities, the presence of the lesion in the venous outflow system creates a possibility for an abnormal

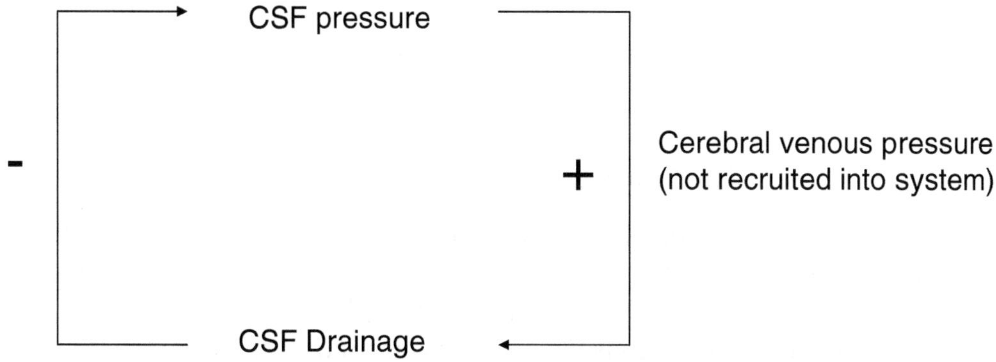

Fig. 5. Normal negative feedback. Increases in CSF pressure are controlled by an increase in the rate of CSF absorption as it is a pressure dependent process

positive biofeedback to develop as increases in CSF pressure can worsen the degree of venous compromise leading to a further increase in CSF pressure, a further increase in venous obstruction, and so on (Fig. 6). The degree to which cerebral venous pressure is recruited into the control system could determine the extent to which CSF pressure rises until other negative feedback mechanisms (for example, CSF absorption through other routes) become significant and re-establish control of CSF pressure at a higher level. Recruitment of cerebral venous pressure into this control system is variable, and due to a variety of factors, potentially explaining the variability of PTC symptoms over time. The prolonged benefit produced by a single CSF tap in PTC patients may be understood in this way if one postulates that removal of CSF reduces secondary venous compression and improves CSF drainage due to uncoupling of cerebral venous pressure from the control system for a period much greater than that required to re-

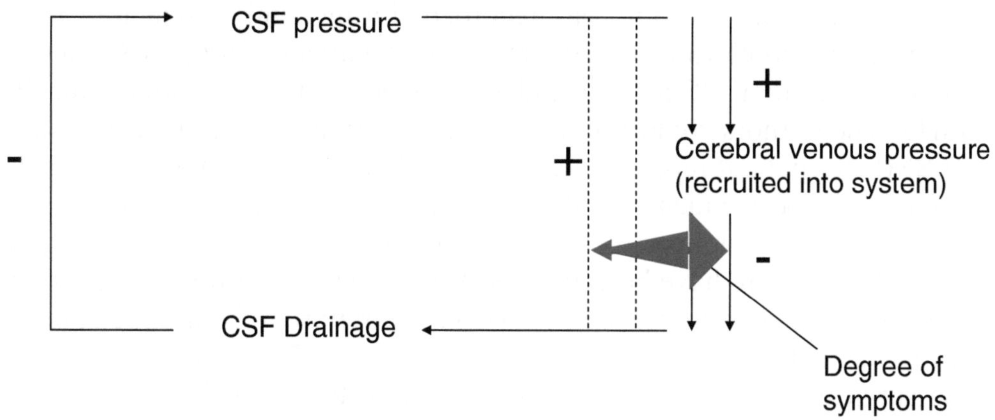

Fig. 6. Disordered positive feedback. Recruitment of the cerebral venous sinuses into the feedback loop due to venous sinus collapse, secondary to increased CSF pressure, causes venous sinus pressure (particularly SSS pressure) to increase. This inhibits CSF drainage and results in further increases in CSF pressure and so on

place the volume of CSF removed. Permanent decoupling of cerebral venous pressure by stenting an obstructive cause may be therapeutic, regardless of the initial cause.

Non Obstructive Venous Hypertension

Venous sinus hypertension also occurs in the absence of venous sinus obstruction. Karahalios *et al.* [78] drew attention to systemic venous hypertension in their series of patients studied with venography. There are several case reports in which cardiac lesions have been associated with PTS [72] or hydrocephalus [134]. However, the most well-studied scenario is that of morbid obesity and its association with venous sinus hypertension and PTS.

Obesity is strongly associated with development of PTS in both men and women [36, 51, 178]. Johnston & Paterson [75] found that of 110 patients with PTS, 35 were moderately or grossly obese. These patients were all female and 27 of them had no recognisable aetiology. Foley [45] found that only 1 of 46 cases with 'otitic hydrocephalus' compared to 20 of 60 cases with 'toxic hydrocephalus' were obese. In a prospective study of 50 patients with PTS by Wall & George [177] 94% were noted to be overweight.

Corbett & Mehta [26] investigated CSF pressure in 116 acute PTS patients, 8 chronic PTS patients, 41 normal obese volunteers and 15 normal non-obese volunteers. CSF pressure was only slightly higher in the normal obese subjects compared with normal non-obese subjects. There was no correlation with the degree of obesity and CSF pressure in this study. Patients with acute PTS all had CSF pressures markedly higher than both patients with chronic papilloedema and normal subjects. In a study of 19 patients (mean BMI 39.3 kg/m^2) randomly selected from an obesity clinic CSF pressure was recorded during lumbar puncture [61]. CSF pressure was elevated (> 20 mmHg) in 15 patients (79%); in 8 patients it was greater than 25 mmHg and in 2 patients it was greater than 30 mmHg. The authors could not find any correlation between CSF pressure and BMI. No patient reported headaches and no patient had papilloedema.

There has been much speculation on the relationship between obesity and PTS. Numerous hormonal and metabolic links between obesity and raised CSF pressure have been proposed. In some cases a weight reduction can result in clinical improvement. There are reports of dramatic clinical improvement in patients with extreme obesity and PTS who have undergone surgically induced weight loss. Noggle & Rodning [122] reported a case of PTS in a morbidly obese patient who was successfully treated (weight reduction from 150 to 86 kg; resolution of clinical PTS) with gastric exclusion surgery. Furthermore symptoms recurred 3 years later with failure of the gastroplastic stapling line and recurrent weight gain. After revision of the gastric surgery, significant weight loss and clinical PTS resolved. At around the same time Amaral *et al.* [2] reported a similar case (138 kg) in whom surgically induced weight loss resulted in clinical resolution of PTS and a reduction in CSF pressure.

In 1995 Sugerman *et al.* [167] reported resolution of symptoms of PTS in morbidly obese patients (BMI 49 +/− 3 kg/m^2) following surgically induced weight loss. At mean follow-up of 34 months average CSF pressure had reduced from 35.3 +/− 3.5 to 16.8 +/− 1.2 cmH_2O. These initial results were confirmed in a second study of 24 severely obese patients (mean BMI 47 +/− 6 kg/m^2) previously diagnosed with PTS [168]. Mean CSF pressure was 32.4 +/− 8.3 cmH_2O. Twenty three patients underwent gastric bypass and one had laproscopic gastric banding. Follow-up was 62 +/− 52 months. At the time of the report, one year follow-up was available in 19 patients who had lost an average of 71 +/− 18% of their excess weight (45 +/− 12 kg). Headache and pulsatile tinnitus resolved within 4 months of surgery with the exception of one patient. Papilloedema and cranial nerve dysfunction improved in all patients. Interestingly, 2 patients who had initially lost weight and experienced resolution of their symptoms developed recurrent PTS on regaining weight. As a non-surgical solution, Sugerman *et al.* [165] designed a counter-traction mechanism to reduce intra-abdominal pressure and central venous pressures. Improvement in headache and pulsatile tinnitus was reported with the nocturnal application of this external negative abdominal pressure device to the abdomen of 5 patients with severe obesity and PTS.

Sugerman *et al.* [166] studied 6 obese patients (mean BMI 45 +/− 3 kg/m^2) with PTS undergoing gastric banding. Mean ICP was 29.3 +/− 8.0 cm H_2O. Intra-abdominal pressure (22 +/− 3 cmH_2O), central venous pressure (20 +/− 6 mmHg; $n = 5$) and transoesopohageal pleural pressure (15 +/ − 10 mmHg; $n = 3$) were all elevated in these patients. The authors concluded that central obesity raises intra-abdominal, pleural and cardiac filling pressures. The later impedes cranial venous outflow and causes PTS. Gastric bypass or laproscopic gastric banding in these patients resulted in significant weight loss. At the time of the report 5 of the 6 patients had resolution of their PTS symptoms, including pulsatile tinnitus. One patient had only recently undergone their surgery. Although the right atrial pressures demonstrated using venography and manometry in the study of Karahalios *et al.* [78], the mechanism that these authors propose may still be plausible. However, as Sugerman and colleagues themselves noted, there remains no satisfactory explanation for why some obese individuals develop PTS but most do not. In addition, why are women more commonly affected than males given that the proposed mechanism is increased intra-abdominal pressure and males tend to have more central obesity compared to females.

Experimentally, Luce *et al.* [104] demonstrated that in anaesthetized dogs, an increase in pleural pressure increases lumbar and intracranial CSF pressure. This increase in CSF pressure was secondary to elevation of venous pressure in the superior vena cava. In the swine, Bloomfield *et al.* [16] demonstrated that elevation of intraabdominal pressure 25 mmHg above baseline caused an increase in central venous and intracranial pressures (7.6 +/− 1.2 to 21.4 +/− 1.0 mmHg). In addition there was a reduction in cardiac index and CPP decreased. Expansion of intravascular volume returned cardiac index and CPP to normal and also resulted in a further increase in ICP (27.8 +/− 1.0 mmHg). Decompression of the abdomen returned ICP to normal. The effects on central venous pressure and ICP were negated by sternotomy and pleuropericardotomy [16]. Even when ICP was already artificially elevated (mean 25.8 mmHg) increases of 15–25 mmHg in intra-abdominal pressure resulted in significant increases in intra-thoracic pressure and ICP (25.8 to 33.8 and 39.0 mmHg, respectively) [148].

There exists, therefore, a co-hort of patients with systemically increased venous sinus pressure without focal obstruction. These patients appear to be those with morbid obesity. It should be stressed that the degree of obesity in the patients that Sugerman *et al.* have dealt with surgically is much greater than that in the average overweight patient with PTS. In addition, it should also be noted that venous sinus obstruction, from both intrinsic lesions and extrinsic compression does also occur in obese patients. Non-obstructive venous sinus hypertension are a particularly difficult group to diagnose as there are no static CT or MR examinations that will demonstrate this aetiology. Instead cerebral venography with manometry including right atrial pressures, preferably in the awake patient, should be performed. These patients are also difficult to distinguish from a co-hort of patients with normal veins, normal venous pressures and increased CSF pressure possibly due to disordered function of the arachnoid granulations.

Cerebrospinal Fluid Dynamics in Pseudotumor Cerebri Syndrome

Johnston & Patterson [76] proposed that PTS resulted from either a problem of CSF absorption at the level of the arachnoid villi or cranial venous outflow obstruction. To determine whether there is an obstruction to CSF absorption at the level of the arachnoid villi, the resistance to CSF absorption (R_{csf}) is calculated using the CSF infusion study. In Martin's [111] study, 4 patients with PTS were studied and had a similar R_{csf} value to 2 patients with venous outflow obstruction. Sklar *et al.* [158] reported their findings in 10 patients with PTS who underwent a total of 17 investigations using the constant-pressure variable infusion rate method at various stages of their disease. The results of their study are difficult to interpret as at least half of the patients had normal baseline CSF pressures and several patients were being treated with diuretics or steroids. While the authors interpreted their findings as demonstrating evidence of an CSF absorption deficit in PTS, 6 of the studies in 2 patients were normal. Ropper & Marmarou [144] presented a case of PTS secondary to Guillain-Barre syndrome in whom serial measurements of R_{csf} were made. R_{csf} was elevated at presentation and fell to normal with clinical improvement. However, the authors calculated the contribution of this raised R_{csf} to the elevated CSF pressure and concluded that the R_{csf} recorded was insufficient to produce this extent of pressure elevation. They proposed that elevation of the venous sinus pressure must also contribute to the elevation in CSF pressure in their patient. Lamas *et al.* [93] reported a case of a dural AVM with PTS and raised SSS pressure. A constant-infusion lumbar CSF infusion study had to be terminated at a pressure of 50 mmHg before reaching equilibrium. Calculating R_{csf} on the basis of this result gives a value of at least 36 mmHg/ml/min. There was no evidence of any other CSF circulation disorder in that patient.

In most series the R_{csf} is measured using a perfusion or infusion technique and the average R_{csf} measured in patients with PTS is raised. However, patients are usually analysed as a group. One of the most valuable studies therefore was that of Janny *et al.* [71] who, using a constant intraventricular CSF infusion technique, measured the R_{csf} and differentiate patients with venous sinus obstruction from those without. In the patients without venous sinus obstruction mean R_{csf} was 46.6 mmHg/ml/min while in the patients with venous sinus obstruction it was 14.5 mmHg/ml/min. In addition, Janny *et al.* [71] demonstrated a reversal of the normal pressure gradient between the SSS and CSF in patients with venous sinus obstruction causing a PTS syndrome. The mean pressure was 3.16 mmHg higher in the SSS. In contrast, their patients without venous sinus obstruction ($n = 12$), the mean pressure was 9.5 mmHg higher in the CSF. The difference suggests that in patients without venous sinus obstruction the underlying problem may be that of CSF absorption.

The results of CSF infusion studies in PTS are difficult to interpret. Studies of CSF dynamics are performed using the potentially invalid assumption that the venous sinus pressure remains stable during mock CSF infusion. Using the constant infusion CSF study, baseline CSF pressure is monitored and then mock CSF is infused at a constant rate until a new equilibrium CSF pressure is reached. The R_{csf} is calculated on difference in CSF pressure at equilibrium and baseline divided by the infusion rate. However, as noted in the earlier discussion regarding the effects of raised CSF pressure on the venous system, elevations of CSF pressure may cause venous sinus collapse, secondary venous obstruction and an elevation of venous sinus pressure. Therefore, as CSF absorption depends upon a pres-

Table 3. *Classification of Venous Sinus Pathology in PTS*

Classification of PTC based on Venous Sinus Pathology
A. Raised Venous Sinus Pressure
i. Obstructive
Intrinsic Vs Extrinsic
Primary Vs Secondary
ii. Non-obstructive
B. Normal Venous Sinus Pressure

sure gradient between the subarachnoid space and venous sinus, CSF pressure must rise further. Therefore, the increase in CSF pressure required to reach equilibrium may be falsely high. Likewise in the presence of a primary venous sinus obstruction, venous sinus pressure will be high. If during increases in CSF pressure the venous sinus pressure remained at the same level then the increase in R_{csf} should be normal. However, due to the low compliance of the venous compartment, elevation of CSF pressure may produce increases in venous sinus pressure and therefore may result in falsely high calculated R_{csf} values. It should be noted that measurement of R_{csf} in situations where CSF pressure is very high prior to testing makes such studies technically difficult.

Despite these problems there is sufficient evidence to conclude that most cases of PTS are related to either problems of venous sinus obstruction or problems of increased R_{csf} at the level of the arachnoid villi as proposed by Johnston and Patterson [73, 76, 77], and we have seen patients with PTC with normal veins and normal venous pressures and elevated CSF pressure.

It is clear therefore from the above discussion that there is no single cause of PTC and that a multitude of pathological states may cause the symptom complex. We have found it useful to divide patients into venogenic and non-venogenic groups based on the results of DRCV and manometry studies which provides guidance for therapeutic options Table 3.

Investigation of Venous Aetiology in Pseudotumor Cerebri Syndrome

The high proportion of patients with venous sinus obstruction coupled with the ability to treat such obstruction should be an indication to investigate all patients with PTS for the existence of venous sinus obstruction. In performing these investigations one should be cognisant of the location and morphology of the venous obstruction that is sought.

The most important distinction in the investigation of venous sinus obstruction in PTS is that between thrombosis and stenosis. Until recently, almost all papers reviewing and recommending management strategies for PTS concentrate on the exclusion of cerebral venous thrombosis rather

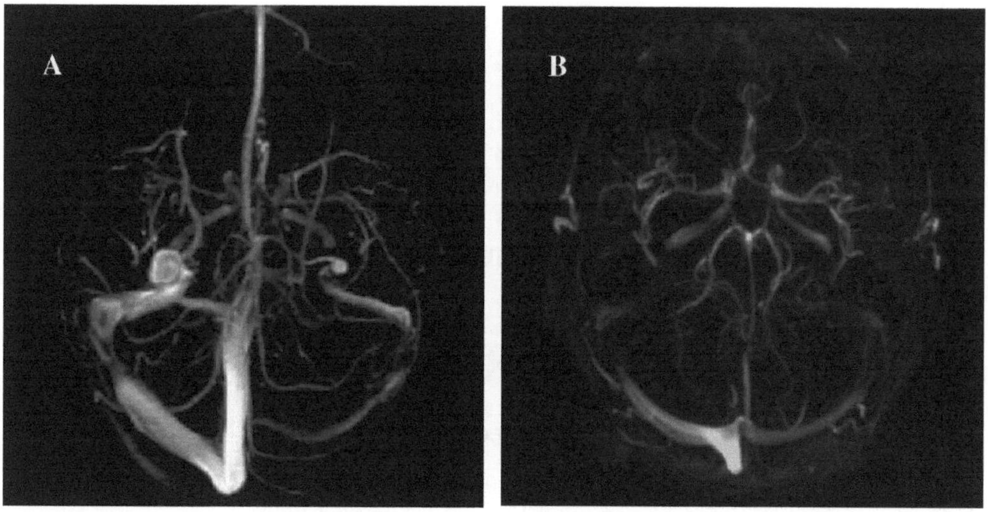

Fig. 7. Axial MRVs of two different patients. (A) There is a dominant large right dominant transverse sinus with absence of flow in the distal third. This patient later went on to have a venous sinus stent. (B) Another patient with a dominant right transverse sinus. Both patients also have obstructions in the small left transverse sinuses confirmed using DRCV and manometry

than stenosis [98, 99, 160]. The possibility of the other causes of obstruction is rarely mentioned. Strategies that aim to exclude thrombosis often use CT and/or static MRI with most attention given to the SSS, for example, the empty delta sign. CT, even with contrast enhancement, should not be relied upon for the diagnosis of venous sinus obstruction in PTS. This is exemplified by Leker *et al.'s* [98] report of 46 cases of PTS with normal CT results. When conventional angiography or MRI/MRV was performed, 12 patients (26%) had evidence of venous sinus thrombosis.

We recommend that as a minimum static MR as well as MR venography with full coverage of the cerebral venous system should be used to investigate patients with PTS for venous sinus disease (Fig. 7). The combination of these exams should be able to identify most cases of thrombosis or venous sinus stenosis. MR also has the advantages of identifying intraluminal lesions such as giant arachnoid granulations. Absence of flow in the distal transverse sinuses in particular should be treated as a real finding and not attributed to artefact.

Conventional angiography may detect venous sinus obstruction. However, the sensitivity for venous sinus obstruction is less than for DRCV. King *et al.* [86] as well as Karahalois *et al.* [78] found that even with benefit of hindsight, venous sinus obstruction was more difficult to identify on conventional angiography than on DRCV. In addition, there is a risk of embolic stroke or arterial dissection with cerebral angiography that is not

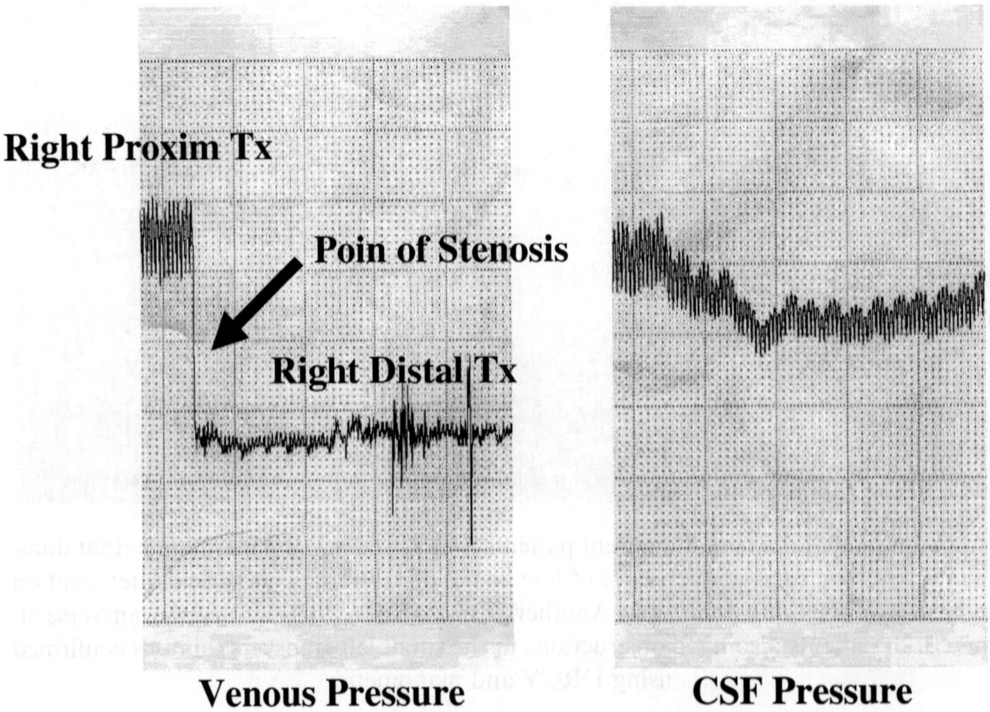

Fig. 8. Venous sinus manometry and CSF pressure recordings. The venous catheter has been pulled back across the point of stenosis during pressure recording demonstrating the venous sinus pressure gradient

present with DRCV. However, conventional angiography may be useful when there is a suspicion of a dural AV fistula.

Direct retrograde cerebral venography (DRCV) combined with manometry is the investigation of choice for venous sinus obstruction in PTS. It is probably the most sensitive investigation for detecting venous sinus obstruction. However, because of its more invasive nature, in cases where venous sinus obstruction has been demonstrated and medical therapy is to be trialled, presuming there is no immediate threat to vision, it is reasonable to keep DRCV with manometry in reserve until treatment of the stenosis is contemplated.

The advantages of DRCV with manometry are that it provides the most accurate information regarding the nature of venous sinus flow and allows measurement of venous sinus pressures to be obtained. Unlike either MRV or conventional angiography, manometry affords information regarding the functional significance of the obstruction. It allows determination of whether venous sinus pressure is raised and whether any pressure gradients exist across morphological obstructions (Fig. 8). If there is doubt regarding the presence or absence of a lesion due to inflow or contrast streaming, manometry will provide clarification. King *et al.* [86] commented that in some cases the filling defects were not impressive on venography, or when initially examined, these changes were not appreciated on the venous phase of the carotid angiogram or were attributed to streaming of the contrast medium. In addition central venous pressures can also be measured using manometry and allows patients with systemic venous hypertension to be identified.

The advantages of venography and manometry over conventional angiography were exemplified in a case of PTS reported by Cremer *et al.* [27]. A small (1 cm diameter) falcine meningioma caused partial obstruction of the posterior one-third of the SSS. Conventional angiography had been revealed a normal arterial and venous systems. Venography however revealed the obstruction of both transverse and sigmoid sinuses. Manometry demonstrated venous pressures to be 40 mmHg through the entire SSS and transverse sinuses and dropped to 3–5 mmHg in the distal sigmoid sinuses. There was no pressure gradient at the meningioma itself.

Manometry should be performed with patient awake unless the patient is unable to co-operate fully. DRCV and manometry are usually well tolerated however the sinuses are sensitive to stimuli, particularly stretch and intermittent discomfort from the catheters may be reported. Anaesthetic agents and positive pressure ventilation may interfere with accurate measurement of venous pressures by changing both the ICP and the intrathoracic pressures. Therefore for diagnostic purposes the procedure should be performed, if feasible, in the awake patient.

Venous pressures should be measured for all segments of the venous sinuses including the SSS, torcular, transverse and sigmoid sinuses, jugular bulb, internal jugular vein and right atrium. Where a morphological obstruction has been demonstrated, the pressures proximal to, at and immediately distal to the stenosis are recorded. By pulling the catheter back across the stenosis under radiological control and recording the pressure simultaneously, a sudden fall in venous pressures is often recorded demonstrating the functional significance of the stenosis. By measuring right atrial pressures non obstructive venous hypertension can be demonstrated and is particularly relevant in the grossly obese patients.

Venography and manometry are invasive procedures compared to MR imaging. The risks however are less than those of conventional cerebral angiography, for example, embolic stroke. There is however a risk of perforation of a vein or sinus, for instance, if the guidewire inadvertently enters a fragile draining cerebral vein. Another potential risk is the formation of a thrombosis around the catheter which might case pulmonary embolus. Overall however the risks of DRCV and manometry appear lower than those of conventional angiography.

We have performed DRCV with manometry in 22 patients diagnosed with PTS. In 11 (50%) there were either bilateral venous sinus obstructions (7 cases) or a unilateral venous sinus obstruction in a dominant venous sinus (4 cases). These obstructions were all associated with pressure gradients across the point of stenosis. Pressure gradients ranged from 7–41 mmHg (mean 18.3 +/− 10.6 mmHg). The resulting SSS pressure in this group ranged from 17–43 mmHg (mean 27.9 +/− 9.3 mmHg). In comparison, the other 11 patients had gradients of less than 5 mmHg. The mean SSS pressure in these patients was 15.0 +/− 6.8 mmHg. However, 5 of these 11 patients had venous sinus pressures of 15 mmHg or above. The mean of the right atrial pressures in the 11 non obstructed patients was 8.1 +/ − 5.3 mmHg compared to 4.7 +/− 4.4 mmHg in those with venous sinus obstructions. Therefore some of the patients in the group without morphological obstruction or focal pressure gradients still had significantly raised venous sinus pressures that may be contributing to their PTS.

The location of obstructing lesions was always within the distal two-thirds of the transverse sinus. The morphology of the obstructions varied as described previously. In some cases the obstructions appeared to be intrinsic and were well defined rounded filling defects. Other cases appeared exhibit the features of a tight primary stenosis. In some cases the obstruction was less well defined and may have been consistent with extrinsic compression.

Treatment of Venous Sinus Obstruction

In most cases of PTS spontaneous resolution of the condition will occur. In these cases, removal of any offending agent such as tetracyclines, conventional medical therapy including acetazolamide and/or steroids, and intermittent lumbar punctures will enable sufficient control of the condition until spontaneous resolution occurs. However, consideration should be considered to more aggressive intervention when the condition is refractory to medical therapy, it does not undergo spontaneous resolution within a reasonable period, vision is threatened or in the rare case where PTS exhibits a fulminant course.

The options for surgical treatment that currently exist are CSF shunting, usually either lumbo-peritoneal or ventriculo-peritoneal, optic nerve sheath fenestration or bilateral subtemporal decompressions. These procedures all have their own limitations and certainly are not always effective. Ventriculoperitoneal shunting, even with use of stereotaxis, may be difficult and the shunt prone to blockage in small ventricles. Lumbo-peritoneal shunts have the added problem of the acquired Chiari malformation. Optic nerve sheath fenestration may help with vision but do not always effectively reduce ICP and headaches. Finally subtemporal decompressions provide symptomatic relief by increasing the effective intracranial compliance but do not address the underlying aetiology. As an alternative to these treatments, where a venous sinus obstruction has been identified, consideration should be given to treatment of the venous sinus obstruction itself.

The options for treatment include direct surgical treatment of lesion and/or surgical venous bypass or endovascular therapy, the most useful of which appears to be venous sinus stenting.

Direct Surgical Treatment

There is limited experience with direct surgical treatment of venous sinus obstruction in PTS. From a historical perspective, Ray & Dunbar [141] reported a 49 year old woman with PTS of 14 months duration who was found to have an obstruction at the junction of the posterior and middle thirds of the SSS. At surgery a sterile thrombus was removed from the sinus but a partial obstruction remained on the post-operative venogram although pressure was slightly reduced and she was clinically improved.

Venous sinus bypass surgery has been performed by Sindou and Auque [156, 157] in 5 cases of PTS. The underlying aetiology was post-otitic sinus thrombosis, surgical internal jugular vein ligation, a torcular meningioma and dural AV fistula excision with sinus thrombosis in two cases. In 4 of 5 cases internal saphenous vein was used and the authors describe excellent clinical outcome with post-operative patency documented on angiography. In one case, a Gore-Tex graft was used but this became occluded. The authors emphasize the importance of post-operative anticoagulation.

Sainte-Rose *et al.* [151] reported 3 infants with craniostenosis and fixed venous sinus outflow obstructions related to bony stenosis who were treated with a saphenous vein bypass. In the first patient (8 months of age) the graft progressively dilated over a 6 month period with a gradual reduction in ICP. The authors emphasized the problems related to the size of the saphenous vein grafts in infants (about 1 mm diameter). In the other two patients, both with craniostenosis and fixed venous sinus obstruction, saphenous vein grafts did not reduce ICP after surgery and to protect vision a VP shunt was required. However, in both patients the grafts dilated and cranial remodelling could proceed without jeopardizing the venous collaterals in the scalp.

Direct treatment of venous sinus obstruction, as yet, has not been undertaken in PTS. However, the role of surgery, especially in removing intrinsic venous sinus obstructing lesions should be considered. Although, there are obvious risks of haemorrhage and thrombosis, there is the potential benefit of avoiding placement of a permanent venous sinus stent.

Endovascular Treatment

Endovascular therapy for venous sinus obstruction in PTS may be divided into those treatments aimed at thrombolysis and those aimed at mechanical relief of the obstruction. The later includes venous sinus angioplasty and venous sinus stenting.

Thrombolysis is effective in cases of venous sinus thrombosis. The first reported application of anticoagulation in patients with PTS was that of Ray & Dunbar [141] which was successful in 2 patients. However, apart from cases of acute thrombosis thrombolysis is usually ineffective in cases of PTS. For example, thrombolysis was also attempted in 2 patients by Karahalios *et al.* [78] using urokinase but without success. King *et al.* [86] reported their unsuccessful attempt at clot dissolution using urokinase in one patient with transverse sinus narrowing and PTS. The authors speculate that as the lesion had been present for some months dissolution is unlikely as a mural thrombus would probably have organised and become epithelialised. In contrast, Kollar *et al.* [88] attempted thrombolysis using urokinase in 2 patients with mixed success. In one patient with symptoms of 3 weeks duration, thrombotic occlusion of the transverse sinus was effectively treated with urokinase and systemic anticoagulation. A second patient, also with a short duration of symptoms and an apparent transverse sinus thrombosis underwent thrombolysis with urokinase followed by systemic anticoagulation. Recurrent thrombosis was problematic and raised CSF pressures responded to acetazolamide. The later two cases are really cases of cerebral venous sinus thrombosis and the role of thrombolysis in such acute cases is well established. However, most cases of PTS with venous sinus obstruction will not have thrombosis as their primary pathology. In these cases of course there will be no effect of thrombolytic therapy.

Venous Sinus Angioplasty

Balloon angioplasty of venous sinus obstructions would be the ideal treatment for patients with venous sinus obstruction as the need for implanting a stent would be avoided. However, the results of venous sinus angioplasty have been disappointing due to a high rate of recurrence. Karahalios *et al.* [78] attempted angioplasty in 2 patients with sigmoid sinus stenosis and 1 patient with jugular bulb stenosis. While in 2 patients the stenosis initially improved, one patient developed restenosis in one year and another experienced no resolution despite a good hemodynamic result. In their report of PTS syndrome secondary to venous sinus stenosis after suboccipital craniotomy or translabyrinthine craniectomy, Keiper *et al.* [80] described a case of post-operative stenosis in a dominant sinus with a pressure gradient of 24 mmHg on manometry. Repeated balloon angioplasty reduced that gradient to 14 mmHg although the patient remained symptomatic. An attempt to place a stent across the stenosis was unsuccessful although the details of why this was were not provided. Kollar *et al.* [88] utilized angioplasty for bilateral transverse sinus lesions in a patient who had been symptomatic for 3 years. One lesion was consistent with a large arachnoid granulation. Symptomatic improvement was found post-procedure but she experienced a recurrence of symptoms after 3 months and underwent surgical therapy. There are two reasons why recurrence of stenosis might occur after venous sinus angioplasty. First, the situation is not akin to performing angioplasty of stenosing arterial plaques instead the obstructing lesions, whether enlarged arachnoid granulations or venous sinus strictures, appear to have elastic properties and reconstitute their shape. Second, if the stenosis is secondary to raised CSF pressure then the cause of the obstruction has not been addressed. Due to these reasons, we do not recommend venous sinus angioplasty in the treatment of venous sinus obstruction secondary to PTS.

Venous Sinus Stenting

The first report of deployment of stents in the transverse/sigmoid sinuses is attributed to Marks *et al.* [108]. The first of two cases was a 25 year-old woman with disabling right sided tinnitus but no evidence of PTS. There were bilateral transverse sinus stenoses and the right was dominant. There was a 20 mmHg pressure gradient across the stenosis which was abolished, along with the tinnitus, after deployment of a stent. The second case was an 8 year-old boy with episodic ischemic symptoms and absence of a deep venous drainage who also demonstrated bilateral sigmoid sinus stenoses with a 14 mmHg pressure gradient. The right sided stenosis, which was resistant to angioplasty, was successfully stented. However his episodic symptoms remained.

Hunt *et al.* [68] reported two cases of papilloedema secondary to what was reported to be venous sinus thrombosis. In both cases patients developed right sided venous sinus thrombosis. The first patient had systemic lupus erythematosis and the second had undergone a right radical neck dissection. In both cases left sided focal venous obstruction was also demonstrated; one patient in the sigmoid sinus and one in the transverse sinus. Thombolysis was not followed by clinical improvement in either case. Therefore the left sided focal stenosis was stented in both cases with good effect and resolution of papilloedema. Detail regarding the anatomical arrangement of the cerebral venous system in these cases was not provided.

The first reported case of venous sinus stenting for PTS was that of Higgins *et al.* from Cambridge [65]. This 30 year-old obese woman (30.1 kg/m^2) with a 22 month history of typical PTS refractory to medical therapy was shown to have bilateral transverse sinus stenoses on MRV and venography that were associated with 18 mmHg pressure gradients on manometry. A stent was deployed across the right sided stenosis with dramatic clinical improvement. CSF pressure fell from 20.6 to 13.7 mmHg while intracranial compliance was normalised from 3.6 mL to 16.7 mL. She remained well although with persisting mild residual headache at 1 year.

After that initial report two other single case reports were also published. The Sydney Group reported the case of a woman with a previous diagnosis of PTS who had presented with CSF rhinorrhoea and striking radiological evidence of raised pressure [128]. Following a craniotomy and repair of the anterior cranial fossa defect, a large subgaleal collection developed that required shunting. Venography and manometry demonstrated a filling defect in the right transverse sinus that was associated with a pressure gradient of 14 mmHg and a SSS pressure of 22 mmHg. The left transverse sinus was hypoplastic. Deployment of a stent reduced the gradient to 1 mmHg and subsequently the LP shunt was removed. Ogungbo *et al.* [125] reported a case of typical PTS in a 37 year old woman. CSF pressure was 40 cm CSF and an MRV showed a focal obstruction in the right transverse sinus which was dominant; the left transverse sinus was hypoplastic. There was a pressure gradient of 25 mmHg with a proximal venous pressure of 40 mmHg. Deployment of a stent across the lesion re-established normal venous flow, complete resolution of clinical symptoms with a reduction in CSF pressure to 26 cm CSF.

The Sydney Group reported 4 patients with PTS treated with stent placement of a series of 9 consecutive patients investigated with DRCV and manometry [130]. One of those stented was the subject of the earlier case report [128]. Of the other 3 patients who underwent venous sinus stenting, clinical improvement was seen in all. The first patient, an obese 17 year old girl had bilateral transverse sinus obstruction on DRCV with associated pressure gradients of 23 and 25 mmHg. Stenting of the left side resulted in a dramatic reduction in CSF pressure (35 to 11 mmHg) and resolution of headache. Unfortunately despite these changes vision acuity did not improve due to optic atrophy. A 27 year old thin male had typical PTS with bilateral papilloedema. Again there were bilateral transverse sinus obstructions that appeared intrinsic in nature. A review of the MR scan revealed symmetrical lesions in the transverse sinuses that at the time we suggested might represent fat but they may also represent large arachnoid granulations. Stenting resulted in a reduction in CSF pressure and resolution of papilloedema and headache. After a period of 18 months there was some recurrence of headache although less than previously but vision and fundi remained normal. The third patient was another obese young woman (27 years). Bilateral transverse sinus obstructions with 25 mmHg pressure gradients were demonstrated. These obstructions had the appearance of large arachnoid granulations. The right lesion was stented with restoration of normal CSF pressure and clinical resolution. Apart from a short period of mild left-sided headaches which resolved she remains well. Of the other 5 patients, one had bilateral transverse sinus obstructions with moderate gradients; in this case we opted for a ventriculo-peritoneal shunt. The other four, all of which had chronic PTS and had undergone numerous other

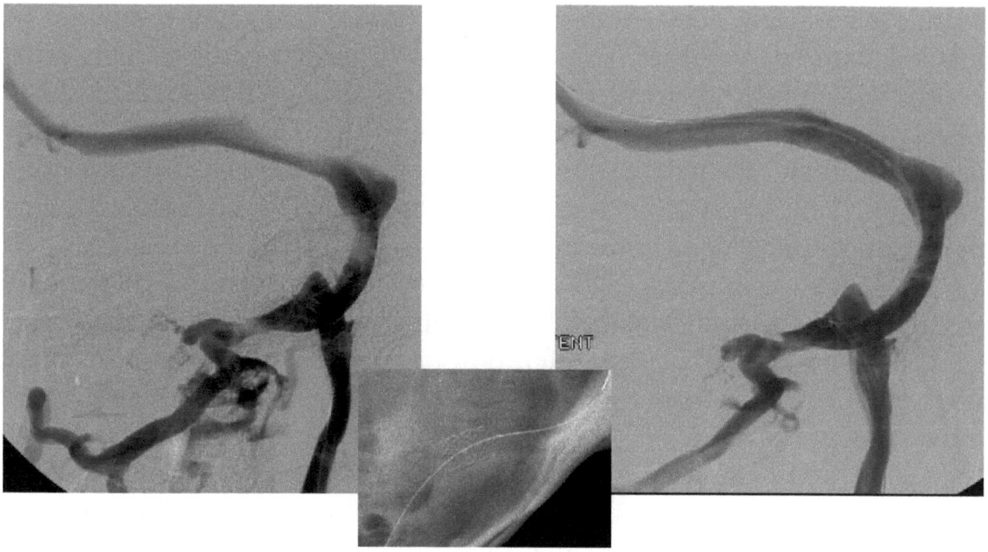

Fig. 9. Venous sinus obstruction treated with a venous sinus stent

treatments did not demonstrate morphological or functional transverse sinus obstructions.

More recently, a series of 12 patients with PTS treated with venous sinus stenting was reported from Cambridge by Higgins *et al.* [63]. All had headache and some form of visual disturbance prior to stenting. The duration of symptoms ranged from several months to 12 years and 5 patients had undergone surgical intervention (at least 10 months prior) for their PTS. All 12 patients had significant pressure gradients across obstructions of the transverse sinuses. After stenting of the transverse sinuses 5 patients were asymptomatic, 2 were improved but with residual headache and 5 were unchanged. There were apparently no predictors of clinical improvement after stenting in these patients. Papilloedema was present in 8 cases prior to stenting and follow-up was available in 7 cases; it resolved in 4, improved in another and in the 2 cases where there was no change papilloedema had been chronic.

The Sydney Group has have now had experience in 8 cases of stenting in eight cases of PTS (Fig. 9). In 7 cases there is long term follow-up available. Headache has improved in all cases. Pain over the region of the stent has been reported in all but one cases and generally resolves over a period of days to weeks with simple analgesia. In 2 cases there was return of a mild headache. This was slightly difference in nature in certainly much less severe than their previous PTS headache. In one case, a second contralateral venous sinus stent was deployed (*vide infra*) with some improvement. Vision improved in 6 of 7 patients. In one patient, optic atrophy developed despite a reduction in CSF pressure. In some cases improvement of vision was dramatic changing in one case from light perception only to almost normal vision within weeks. One patient also had resolution of bilateral sixth nerve palsies. Papilloedema was present in all patients prior to stenting and resolved in all. There was one case of recurrence after stenting in which the first procedure produced immediate and dramatic clinical improvement and the patient returned a few months later with her previous symptoms. Venous sinus obstruction was again noted and treated with dramatic clinical improvement once more (*vide infra*).

The longest published follow-up of the Cambridge series of patients was 26 months with a mean of 14.1 months. In no series has stenosis of the stent due to endothelial proliferation or venous sinus thrombosis been observed during long term follow-up. Of the patients published in the original series from Sydney, the longest follow-up is 30 months with a mean of 22 +/− 8.5 months. However, we are cognisant of the longer follow-up that is required. We inform all patients and their families of the unknown long-term performance of venous sinus stents. It is obviously important that patients treated in the Sydney and Cambridge groups to be followed over an extended period of time in order to assess these issues.

In cases with bilateral transverse venous sinus stenosis where stenting of one side has either not resulted in clinical improvement or the patient is improved but not asymptomatic presents the therapeutic dilemma of whether to stent the contralateral side. In the Cambridge series [63] 2 patients received bilateral transverse sinus stents. The first patient, who improved after an initial stent, improved further after a contralateral transverse sinus stent but was still not asymptomatic. In the second patient, no improvement was demonstrated after the initial stent and a partial but non-sustained improvement was evident after a contralateral stent. The 27 year-old male of the previous series from Sydney [130] also underwent bilateral transverse sinus stenting (Fig. 10). Initially he had bilateral transverse sinus stensoses with pressure gradients of 13 mmHg across each. The right transverse sinus was successfully stented. Although there was resolution of papilloedema, improvement in headache and he had returned to work, some headache remained. CSF pressure had fallen to 11 mmHg on last lumbar puncture. Stenting of the contralateral sinus resulted in marginal improvement in headache and he remains well at 7 months. It is difficult to argue that stenting of the contralateral venous sinus will produce further clinical improvement if there is free communication of the transverse sinuses at the torcular. However, in cases where the transverse sinuses appear to drain the deep and superficial cerebral venous systems independently and venography with manometry demonstrates that there is a persistent stenosis with a pressure gradient, stenting the second transverse sinus may be useful.

Complications of venous sinus stenting have been few in the cases reported so far. Higgins *et al.* [63] performed venograms on some patients in the immediate post-stent period due to concern about stent patency.

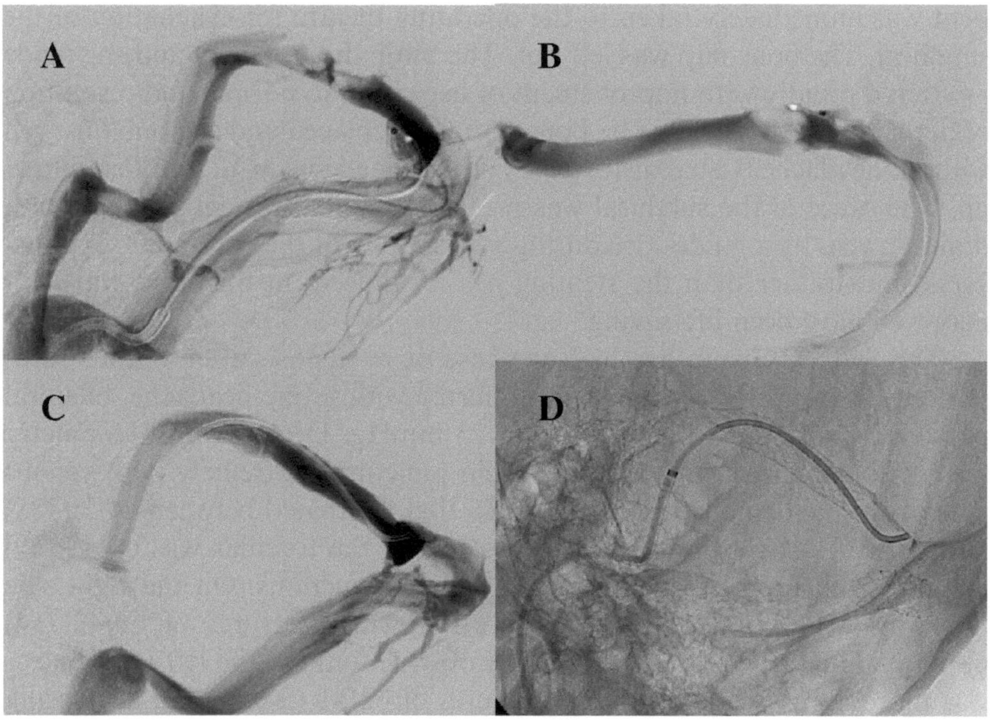

Fig. 10. A case of bilateral transverse sinus stenting. (A) Lateral DRCV demonstrating stented right transverse sinus with normal calibre and an obstruction of the left transverse sinus. (B) AP view of left transverse sinus with filling defect. (C) Lateral DRCV demonstrating that the left transverse sinus has also now been stented. (D) Plain film showing the bilateral transverse sinus stents in situ

Although all stents were patent, probable intraluminal thrombi were observed in 2 patients and were successfully treated with thrombolytic therapy. Transient hearing loss ipsilateral to the side of stenting was observed in both the patients from Sydney (2 patients) and Cambridge (2 patients). Hearing returned to normal within a few days. Higgins *et al.* [63] also reported one patient who complained of unsteadiness temporarily.

There was one life-threatening complication of venous sinus stenting – an acute subdural haematoma on a previously unpublished case from Sydney. This patient presented with severe global headache, gross papilloedema, bilateral sixth nerve palsies and marked visual loss with light perception only. CSF pressure was >64 cmCSF. She underwent optic nerve sheath fenestration and had an external ventricular drain inserted to control her CSF pressure. Venography and manometry were performed and demonstrated the bilateral transverse sinus stenoses in the typical location with large pressure gradients. During the venogram the CSF pressure was noted to rise and blood was noted in the external ventricular drain. The right transverse sinus was stented successfully and CT scan immediately after the procedure demonstrated a left acute subdural haematoma. The patient was immediately taken to the operating theatre for evacuation of the subdural. The bone flap was left out. The brain did not swell and she made recovered rapidly with improvement of her vision to normal and resolution of her sixth nerve palsies. The bone flap was replaced and she remains well without headache, visual disturbance or papilloedema at 12 months follow-up. The cause of the subdural was probably the inadvertent puncture of a draining vein by a guide-wire during venography of the contralateral transverse sinus rather than the stenting *per se*. Indeed, in this case, the stent may well have been life-saving.

The Sydney Group has had one case of re-stenosis after venous sinus stenting. A 23 year-old woman was referred with severe headache, bilateral papilloedema and a CSF pressure of 23 mmHg. DRCV with manometry was performed via a right femoral vein puncture. Extremely high venous pressures were demonstrated in the SSS that fluctuated between 45/30 (38) mmHg to 70/45 (56) mmHg. The pressure at the torcular was 69/41 (50) mmHg. Both transverse sinuses had tight obstructions. On the right side this was localized to its anterior portion with pressures of 66/42 (54) mmHg falling to 16/12/(13) below the obstruction. On the left the stenosis was more diffuse with proximal pressures of 62/41 (50) mmHg decreasing to 18/14 (12) mmHg. After a discussion with the patient, informed consent was obtained and the patient was returned to the angiography suite. Under general anaesthesia, a 10 × 30 mm and a 8 × 20 mm Wall stent were placed in the right transverse and sigmoid sinuses (Fig. 11). The pressure gradient was reduced but not completely obliterated. The pressure proximal to the stent was 41/26 (32) and was 20/17 (18) mmHg distal to the stent. The patient was recovered and returned to the neurosurgical high dependency unit on heparin and oral antiplatelet medications. Post-procedure the patient was improved. Her headache resolved and she reported that her vision had also improved. However she complained of muffled hearing. This resolved after several days. The LP was repeated under fluoroscopy and this demonstrated CSF pressure to be reduced 13 cm CSF (10 mmHg). She was discharged home under ophthalmological surveillance and on anti-platelet medication.

However, she experienced recurrence of her symptoms less than 2 months later with headache and visual disturbance. Papilloedema had returned. DRCV with manometry was performed under general anaesthesia. Mean venous pressure in the SSS was 41 mmHg. The stenosis in the left transverse sinus was again seen. On the right side, the stent appeared to be patent. However, there had been some minor collapse of the stent along with the development of a stenosis at the proximal edge of the stent. Mean venous pressure above the stent was 37 mmHg which fell to 24 mmHg inside and 12 mmHg distal to the stent. Another overlapping stent was therefore deployed across the stenosis. This abolished the pressure

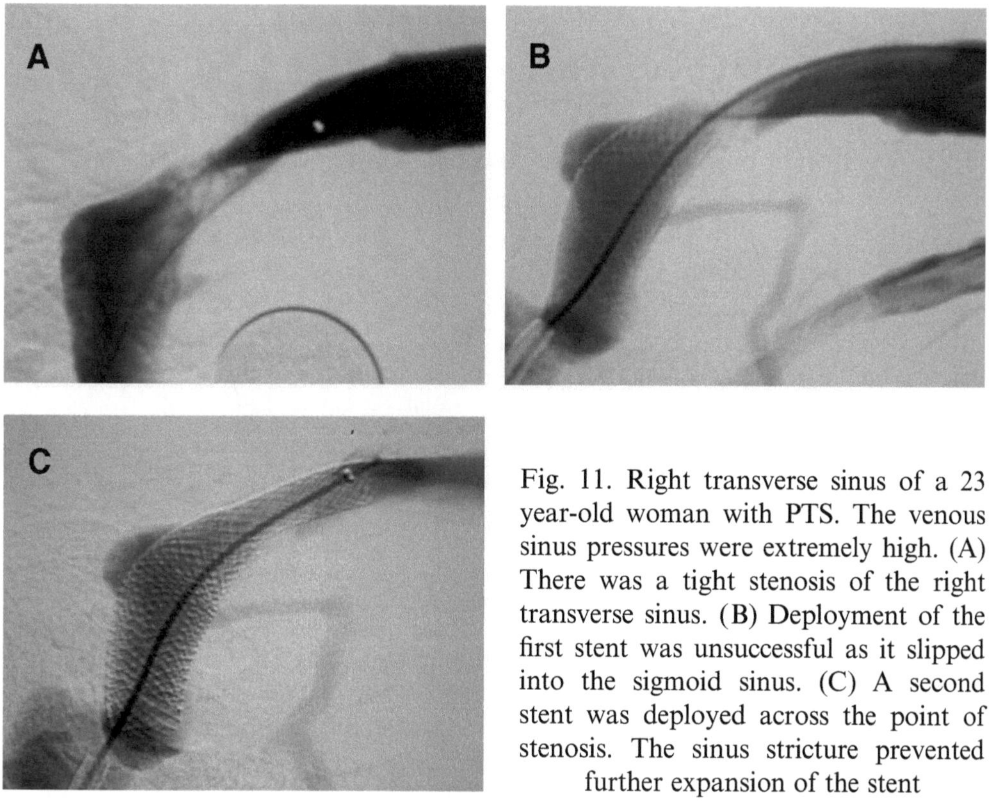

Fig. 11. Right transverse sinus of a 23 year-old woman with PTS. The venous sinus pressures were extremely high. (A) There was a tight stenosis of the right transverse sinus. (B) Deployment of the first stent was unsuccessful as it slipped into the sigmoid sinus. (C) A second stent was deployed across the point of stenosis. The sinus stricture prevented further expansion of the stent

gradient across the right transverse sinus. Mean venous pressure in the SSS was 17 mmHg compared to 15 mmHg in the right sigmoid sinus. Post-procedure, there was again remarkable clinical improvement. Her headache had resolved. There was some minor hearing disturbance which again resolved. Since then papilloedema had improved. She has now remained well for 12 months (Fig. 12). In the series of Higgins *et al.* [63], 2 patients required overlapping stents in a sinus to produce satisfactory reduction of the pressure gradient. We now recommend the use of balloon expandable stents rather than self expanding stents as these can be placed more precisely and with far less chance of stent migration and better initial expansion.

Paediatric PTS may also be related to venous sinus obstruction. In Sydney venography with manometry has been performed in 2 paediatric patients with PTS (both 9 years of age). In the first patient who had been treated first diagnosed at 3 years of age had undergone many suffered recurrent headaches and visual disturbance despite acetazolamide and had been controlled with repeated lumbar punctures. However there had been a slow worsening of symptoms and MRV demonstrated a dominant right transverse sinus with a hypoplastic left transverse sinus. There was a stenosis of the right transverse sinus at the junctions of the middle and distal thirds. Venography and manometry confirmed the stenosis and pressure gradient. A venous sinus stent was deployed across the stenosis with reso-

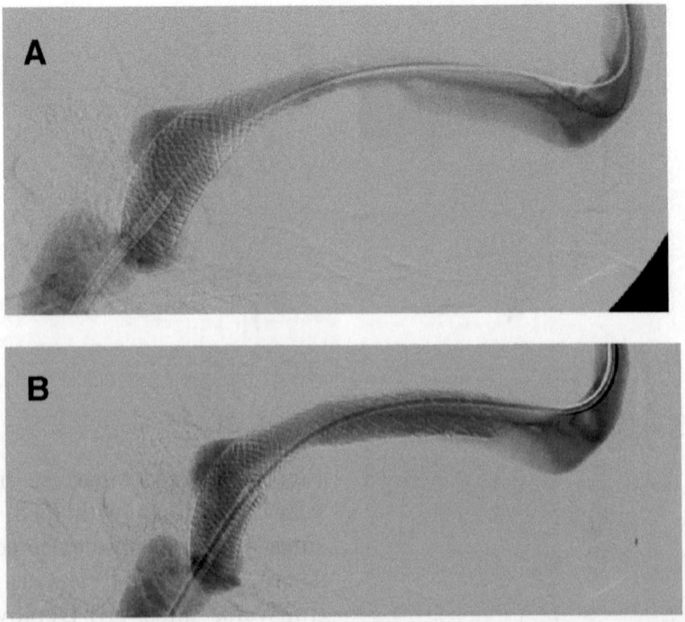

Fig. 12. Second procedure for recurrent PTS symptoms. (A) The stenosis of the right transverse sinus had recurred just proximal to the stents. (B) A third transverse sinus stent was deployed with dramatic resolution of symptoms and she remains well at 12 months

lution of the pressure gradient (Fig. 13). He remains well in early follow-up. A second patient did not have any pressure gradients although venous pressures were systemically high. He was subsequently treated with a lumbo-peritoneal shunt. In most cases of paediatric PTS there is spontaneous resolution and therefore medical treatment should always be trialled first. However there is a small group of patients that may require further treatment. In older children venous sinus stenting should be considered.

Technical Considerations

Once a patient is considered for venous sinus stenting, they are commenced on aspirin and clopidigrel several days before the procedure. Clopidigrel is omitted in children. The procedure is always performed under general anaesthesia. This is because the dura that constitutes the venous sinuses is sensitive to stretch and patients would not tolerated deployment of the stent if awake. A venogram is first performed to define the point of stenosis and the venous sinus pressures are checked. The diameter and the length of sinus to be stented are checked against a reference and appropriate stent is chosen. The stents used are uncovered stents and we prefer the slightly stiffer balloon expandable stents over self expanding stents for the reasons listed above.

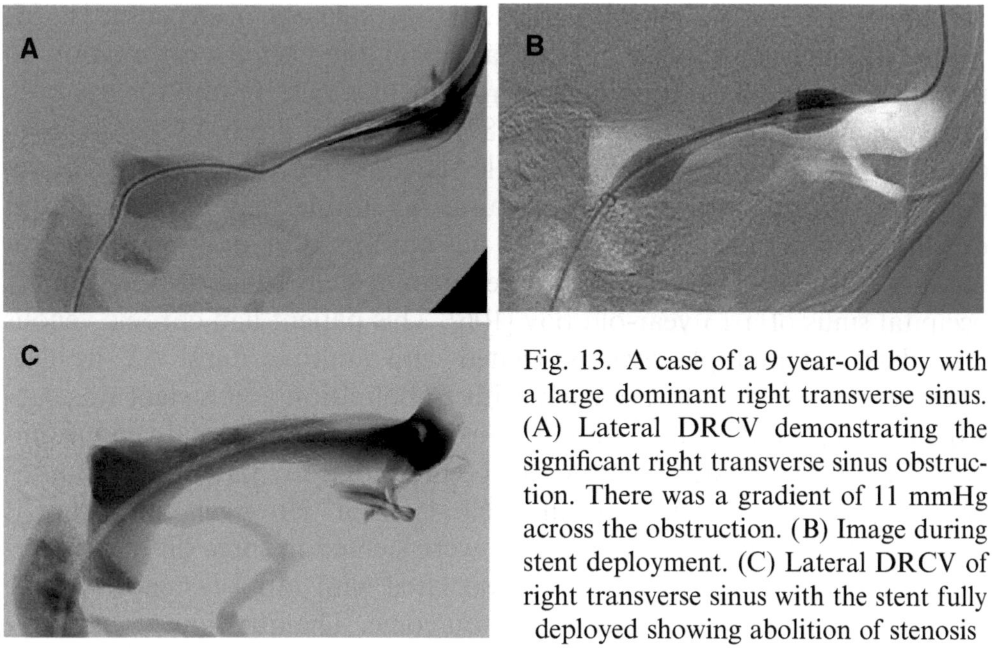

Fig. 13. A case of a 9 year-old boy with a large dominant right transverse sinus. (A) Lateral DRCV demonstrating the significant right transverse sinus obstruction. There was a gradient of 11 mmHg across the obstruction. (B) Image during stent deployment. (C) Lateral DRCV of right transverse sinus with the stent fully deployed showing abolition of stenosis

We prefer to deploy the stent via a sheath in the right femoral vein. It may be difficult to guide the stent through the jugular foramen and sigmoid sinus due to the tight curvature and stiffness of the catheter. Alternatively, the jugular vein may be used. Another group have deployed venous sinus stents into the transverse sinus via a frontal burr-hole over the SSS in order to overcome this problem (*Prof. Peter Reilly, Adelaide, Australia; personal communication*). Once the stent is situated over the point of stenosis the balloon is inflated and the stent is deployed. Several inflations of the balloon may be necessary in order to achieve the desired result. Care must be taken when withdrawing the delivery device as it may catch the edge of the stent, particularly with self expanding stents, and pull it along the sinus distally. Once the stent is deployed venography is performed to assess the position of the stent and resolution of the stenosis. Venous sinus pressures are also recorded to ensure that the pressure gradient has resolved.

At the present time it is our practice to recover the patient and have them observed in the neurosurgical high-dependency unit overnight. Heparin is either allowed to wear off or is continued for 24 hours. Aspirin is continued for at least six months. In adults, clopidigrel is given for one month.

Related Disorders

Dural AV Fistulas

The relationship between venous sinus stenosis and dural arteriovenous fistulae remains unresolved. There are several reports of dural AV fistulae developing in association with transverse or sigmoid obstruction [1, 87, 106]. Although most case reports attribute the venous obstruction to thrombosis, there have been no case reports in which thrombosis has been clearly demonstrated. Kollar *et al.*'s [87] report of a dural AV fistula associated with a heterotopic brain nodule suggests that the pathology of venous obstruction causing PTS and dural AV fistula may have significant overlap. Stenting of the venous sinus has been used in an attempt to treat dural AV fistulas. There is one case report of a stent being deployed in the occipital sinus of a 13 year-old boy [106]. This patient had chronic venous thromboses at multiple sites associated with multiple dural AV fistulae. These were treated endovascularly with embolisation and a stent was subsequently deployed in the occipital sinus to re-establish cranial venous outflow. The stent was demonstrated to be radiologically patent at 3 months and the patient remained clinically well at 1 year. However, the author is aware of three unpublished cases that were stented in three different international centres all of which were associated with rapid restenosis of the stent and in at least one case, a poor outcome. The question of aetiology in these cases remains unanswered however venous sinus stenting must be used with great caution in cases of dural AV fistulae. One suggestion is that restenosis may be a reaction of the sinuses to continued high pressure inflow from feeding arteries.

Other Headache Disorders

Recently attention has also been given to the condition of idiopathic intracranial hypertension without papilloedema. These patients are characterised by chronic headache. Fundoscopy is normal and there is an absence of visual symptoms. Mathew *et al.* [114] performed lumbar punctures on 85 patients with chronic daily headache. Twelve of these patients had raised CSF pressure and responded to treatment with acetazolamide and frusemide. Quattrone *et al.* [136] examined patients with chronic daily headaches of at least six months. No patient had papilloedema. Of 114 consecutive patients 9.6% of patients had venous sinus abnormalities which consisted of marked irregular or absent flow in the distal portion of one or both transverse sinuses. While the authors considered these abnormalities to represent venous sinus thrombosis, the true nature of these flow gaps was not ascertained. A control group of 28 subjects had no flow gaps in the transverse sinuses on MRV. In contrast, Wang *et al.* [179] reported on 25 consecutive patients diagnosed with idiopathic intracranial hypertension without papilloedema. Pulsatile tinnitus (odds ratio 13) and obesity (odds ratio 4.4) in patients with chronic daily headaches were significant predictors of idiopathic intracranial hypertension without papilloedema. However, of the 6 patients that underwent MR venography none demonstrated

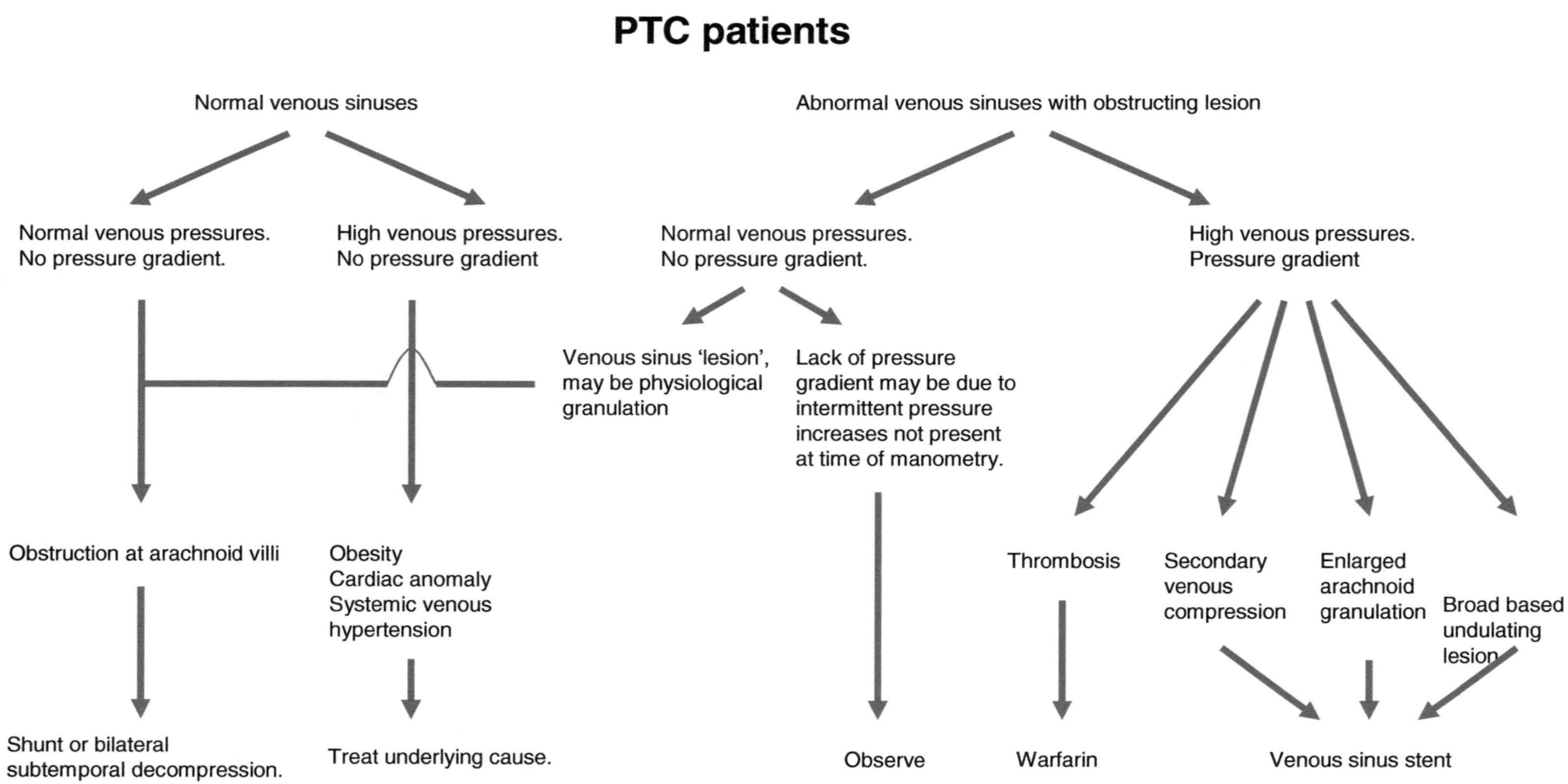

Fig. 14. Diagnostic and Therapeutic Scheme of patients with PTC based on DRCV and manometry

venous sinus occlusion. The more vexed question of prevalence of venous sinus obstruction in patients with chronic daily headache remains to be addressed.

Conclusions

Venous sinus obstruction in PTS is a more common factor in the pathogenesis of the condition than previously recognised. Venous sinus obstruction usually occurs around the junction of the middle and distal thirds of the transverse sinus and is often bilaterally symmetrical. Venous obstruction may be primary, that is, it is the underlying aetiological factor of PTS. Venous sinus obstruction may also be secondary to raised CSF pressure. The latter may exacerbate problems with intracranial compliance and raised CSF pressure. Venous sinus obstruction does not only occur in thin patients but occurs across the spectrum of PTS patients including young overweight females.

In the investigation of PTS, the index of suspicion for venous sinus obstruction should be high. Examinations should not only exclude thrombotic obstruction but should also focus on detecting venous sinus obstruction, especially in the region of the transverse and sigmoid sinuses. Static MR and contrast-enhanced MR venography are the most useful non-invasive investigations for this purpose and should be performed in all patients with PTS. Patients with PTS should preferably undergo DRCV with manometry. This should be performed in all patients who are considered for non-medical therapy whether or not an obstruction has been demonstrated on MR imaging. Venography with manometry will also diagnose systemic venous hypertension.

Treatment for cases of PTS with venous sinus obstruction should be medical initially. In cases where clinical PTS and raised CSF pressure persist or if vision is threatened, consideration should be given to other treatments. Figure 14 provides a useful scheme for the interpretation of venographic and manometric studied and may be used as a guide to therapy. Venous sinus stenting should be considered as a first-line option in cases of venous sinus obstruction with associated pressure gradients, especially where the obstruction appears to intrinsic. For patients with extrinsic compression, venous sinus stenting may still be effective, especially in the instance of disordered feedback loops and should be considered as a viable treatment alternative to other forms of surgical management including CSF shunting.

References

1. Alexander M, Rajaratanam S, Singh S, Korah IP, Gnanamuthu C, Seshadri MS (1999) Acquired dural fistulae in benign intracranial hypertension: a short case report. Acta Neurol Scand 99: 318–321

2. Amaral JF, Tsiaris W, Morgan T, Thompson WR (1987) Reversal of benign intracranial hypertension by surgically induced weight loss. Arch Surg 122: 946–949
3. Angeli SI, Sato Y, Gantz BJ (1994) Glomus jugulare tumors masquerading as benign intracranial hypertension. Arch Otolaryngol Head Neck Surg 120: 1277–1280
4. Arjona A, Delgado F, Fernandez-Romero E (2003) Neurological Picture: Intracranial hypertension secondary to giant arachnoid granulation. J Neurol Neurosurg Psychiatry 74: 418
5. Bastin M, Sinha S, Farrall A, Wardlaw J, Whittle I (2003) Diffuse brain oedema in idiopathic intracranial hypertension: a quantitative magnetic resonance imaging study. J Neurol Neurosurg Psychiatry 74: 1693–1696
6. Beaumont G, Hearn J (1948) A case of reversible papilloedema due to heart failure. BMJ 1: 50
7. Becht F (1920) Studies on the cerebrospinal fluid. Am J Physiol 51: 1
8. Beck D, Russell D (1946) Experiments on thrombosis of the superior longitudinal sinus. J Neurosurg 3: 337–347
9. Beck DW, Kassell NF, Drake CG (1979) Glomus jugulare tumor presenting with increased intracranial pressure. Case report. J Neurosurg 50: 823–825
10. Bedford T (1942) The effect of variations in the subarachnoid space pressure in the superior longitudinal sinus and in the torcular of the dog. J Physiol 101: 362–368
11. Beller A (1964) Benign post-traumatic inttracranial hypertension. J Neurol Neurosurg Psychiatry 27: 149–152
12. Benveniste R, Patel A, Post K (2004) Management of cerebral venous sinus thrombosis. Neurosurg Q 14: 27–35
13. Bergquist E, Willen (1974) Cavernous nodules in the dural sinuses. J Neurosurg 40: 330–335
14. Bergui M, Bradac G (2003) Clinical picture of patients with cerebral venous thrombosis and patterns of dural sinus involvement. Cerebrovasc Dis 16: 211–216
15. Bering E, Salibi B (1959) Production of hydrocephalus by increased cephalic-venous pressure. Arch Neurol Psychiatry 81: 693–698
16. Bloomfield G, Ridings P, Blocher C, Sugerman H (1997) A proposed relationship between increased intra-abdominal, intrathoracic, and intracranial pressure. Crit Care Med 25: 496–503
17. Bortoluzzi M, Di Lauro L, Marini G (1982) Benign intracranial hypertension with spinal and radicular pain. Case report. J Neurosurg 57: 833–836
18. Bousser M, Russell R (1997) Cerebral venous thrombosis. Major problems in neurology, vol 33. In: Warlow C, Van Gijn J (eds) W. B. Saunders, London
19. Brooks DJ, Beaney RP, Leenders KL, Marshall J, Thomas DJ, Jones T (1985) Regional cerebral oxygen utilization, blood flow, and blood volume in benign intracranial hypertension studied by positron emission tomography. Neurology 35: 1030–1034

20. Browder J, Kaplan H, Howard E (1973) Hyperplasia of Pacchionian granulations. Arch Path Lab Med 95: 315–316
21. Cameron A (1933) Marked papilloedema in pulmonary emphysema. Brit J Ophthalmol 17: 167–169
22. Chin S, Chen C, Lee C *et al* (1998) Giant arachnoid granulation mimicking dural sinus thrombosis in a boy with headache: MRI. Neuroradiol 40: 181–183
23. Cinalli G, Sainte-Rose C, Kollar E *et al* (1998) Hydrocephalus and craniosynostosis. J Neurosurg 88: 209–214
24. Connolly MB, Farrell K, Hill A, Flodmark O (1992) Magnetic resonance imaging in pseudotumor cerebri. Dev Med Child Neurol 34: 1091–1094
25. Corbett J, Digre K (2002) Editorial: Idiopathic intracranial hypertension: an answer to, "the chicken or the egg?". Neurology 58: 9–10
26. Corbett JJ, Mehta MP (1983) Cerebrospinal fluid pressure in normal obese subjects and patients with pseudotumor cerebri. Neurology 33: 1386–1388
27. Cremer PD, Thompson EO, Johnston IH, Halmagyi GM (1996) Pseudotumor cerebri and cerebral venous hypertension. Neurology 47: 1602–1603
28. Cronqvist S, Granholm L, Lundström N (1972) Hydrocephalus and congestive heart failure caused by intracranial arteriovenous malformation in infants. J Neurosurg 36: 249–254
29. Cure J, Key L, Goltra D, VanTassel P (2000) Cranial MR imaging of osteopetrosis. Am J Neuroradiol 21: 1110–1115
30. Cushing H (1902) Some experimental and clinical observations concerning states of increased intracranial tension. Am J Med Sci 124: 375
31. Dandy W (1937) Intracranial pressure without brain tumor, diagnosis and treatment. Ann Surg 106: 492–513
32. Dandy W, Blackfan K (1914) Internal hydrocephalus. An experimental and clinical and pathological study. Am J Diseases Childhood 8: 406–482
33. Davidoff L, Dyke C (1937) Hypertensive meningeal hydrops: syndrome frequently folling infection in the middle ear or elsewhere in the body. Am J Ophthalmol 20: 908–927
34. Davson H, Hollingsworth G, Segai M (1970) The mechanism of drainage of the cerebrospinal fluid. Brain 93: 665–678
35. de Lange S, de Vlieger M (1970) Hydrocephalus associated with raised venous pressure. Dev Med Child Neurol 12[Suppl] 22: 28–32
36. Digre KB, Corbett JJ (1988) Pseudotumor cerebri in men [published erratum appears in Arch Neurol 1989 Feb; 46(2): 172]. Arch Neurol 45: 866–872
37. Dixon W, Halliburton W (1914) The cerebrospinal fluid II. Cerebro-spinal pressure. J Physiol 48: 128–153
38. Drew J, Grant F (1945) Polycythemia as a neurosurgical problem. Arch Neurol 54: 25–36
39. Eaton H (1969) Chronic bovine hypo- and hypervitaminosis A and cerebrospinal fluid pressure. Am J Clin Nutrition 22: 1070–1080
40. Ecker A (1946) Linear skull fracture across the venous sinuses. N Y St J Med 46: 1120–1121

41. Emery J, Zachary R (1956) Hydrocephalus associated with obliteration of the longitudinal sinus. Arch Dis Childhood 31: 299–292
42. Evans M (1942) Bilateral jugular vein ligation following bilateral suppurative mastoiditis. Ann Otology, Rhinol Laryngol 51: 615–625
43. Farb R, Vanek I, Scott J *et al* (2003) Idiopathic intracranial hypertension. The prevalence and morphology of sinovenous stenosis. Neurology 60: 1418–1424
44. Fishman M, Hogan G, Dodge P (1971) The concurrence of hydrocephalus and craniosynostosis. J Neurosurg 34: 621–629
45. Foley J (1955) Benign forms of intracranial hypertension – "toxic" and "otitic" hydrocephalus. Brain 78: 1–48
46. Ford F, Murphy E (1939) Increased intracranial pressure. A clinical analysis of causes and characteristics of several different types. Bull Johns Jopkins Hospital 64: 369–398
47. Friedman WA, Mickle JP (1980) Hydrocephalus in achondroplasia: a possible mechanism. Neurosurgery 7: 150–153
48. Gardner W (1939) Otitic sinus thrombosis causing intracranial hypertension. Arch Otolaryngol 30: 253–268
49. Gibson J, Taylor A, Richardson A (1959) Congenital arteriovenous fistula with an aneurysm of the great cerebral vein and hydrocephalus treated surgically. J Neurol Neurosurg Psychiatry 22: 224–228
50. Gideon P, Sorensen PS, Thomsen C, Stahlberg F, Gjerris F, Henriksen O (1995) Increased brain water self-diffusion in patients with idiopathic intracranial hypertension. AJNR Am J Neuroradiol 16: 381–387
51. Giuseffi V, Wall M, Siegel PZ, Rojas PB (1991) Symptoms and disease associations in idiopathic intracranial hypertension (pseudotumor cerebri): a case-control study. Neurology 41: 239–244
52. Golabi M, Edwards M, Oussterhout D (1987) Craniosynostosis and hydrocephalus. Neurosurgery 21: 63–67
53. Goldsmith P, Burn D, Coulthard A, Jenkins A (1999) Extrinsic cerebral venous obstruction resulting in intracranial hypertension. J Neurol Neurosurg Psychiatry 1999: 550–551
54. Gomez D, Potts D, Deonarine V (1974) Arachnoid granulations of the sheep. Structural and ultrastructural changes with varying pressure differences. Arch Neurol 30: 169–175
55. Greenfield JJ, Tindall G (1965) Effect of acute increase in intracranial pressure on blood flow in the internal carotid artery of man. J Clin Investigation 44: 1343–1351
56. Greer M (1962) Benign intracranial hypertension. I. Mastoiditis and lateral sinus thrombosis. Neurology (Minneapolis) 12: 472–476
57. Guthrie T, Dunbar H, Karpell B (1970) Ventricular size and chronic increased intracranial venous pressure in the dog. J Neurosurg 33: 407–414
58. Guttierrez Y, Friede R, Kaliney W (1975) Agenesis of arachnoid granulations and its relationship to communicating hydrocephalus. J Neurosurg 43: 553–558

59. Guyton A (1991) Textbook of medical phsyiology, 8th edn. W. B. Saunders Company, Philidelphia pp 5–7
60. Haar F, Miller C (1975) Hydrocephalus resulting from superior vena cava thrombosis in an infant. Case report. J Neurosurg 42: 597–601
61. Hannerz J, Greitz D, Ericson K (1995) Is there a relationship between obesity and intracranial hypertension? [see comments]. Int J Obes Relat Metab Disord 19: 240–244
62. Hayes K, McCombs H, Faherty T (1971) The fine structure of vitamin A deficiency. II. Arachnoid granulations and CSF pressure. Brain 94: 213–224
63. Higgins J, Cousins C, Owler B, Sarkies N, Pickard J (2003) Idiopathic intracranial hypertension: 12 cases treated with venous sinus stenting. J Neurol Neurosurg Psychiatry 74: 1662–1666
64. Higgins J, Gillard G, Owler B, Harkness K, Pickard J (2004) MR venography in idiopathic intracranial hypertension: unappreciated and misunderstood. J Neurol Neurosurg Psychiatry 74
65. Higgins J, Owler B, Cousins C, Pickard J (2002) Venous sinus stenting for refractory benign intracranial hypertension. Lancet 359: 228–230
66. Higgins J, Pickard J (2004) Lateral sinus stenoses in idiopathic intracranial hypertension resolving after CSF diversion. Neurology 62: 1907–1908
67. Hooper R (1961) Hydrocephalus and obstruction of the superior vena cava in infancy. Clinical study of the relationship between cerebrospinal fluid pressure and venous pressure. Pediatrics 28: 792–799
68. Hunt M, Lee A, Kardon R, Lesley W, Chaloupka J (2001) Improvement in papilloedema and visual loss after endovascular stent placement in dural sinus thrombosis. Neuro-Ophthalmol 26: 85–92
69. Ikkala E, Laitinen L (1963) Papilloedema due to iron deficiency anaemia. Acta Haematoligica 29: 368–370
70. Ikushima I, Korogi Y, Makita O *et al* (1999) MRI of arachnoid granulations within the dural sinuses using a FLAIR pulse sequence. Brit J Radiol 72: 1046–1051
71. Janny P, Chazal J, Colnet G, Irthum B, Georget AM (1981) Benign intracranial hypertension and disorders of CSF absorption. Surg Neurol 15: 168–174
72. Jicha G, Suarez G (2003) Pseudotumor cerebri reversed by cardiac septal defect repair. Neurology 60: 2016–2017
73. Johnston I, Hawke S, Halmagyi M, Teo C (1991) The pseudotumor syndrome. Disorders of cerebrospinal fluid circulation causing intracranial hypertension without ventriculomegaly. Arch Neurol 48: 740–747
74. Johnston I, Kollar C, Dunkley S, Assaad N, Parker G (2002) Cranial venous outflow obstruction in the pseudotumor syndrome: incidence, nature and relevance. J Clin Neurosci
75. Johnston I, Paterson A (1974) Benign intracranial hypertension. I. Diagnosis and prognosis. Brain 97: 289–300
76. Johnston I, Paterson A (1974) Benign intracranial hypertension. II. CSF pressure and circulation. Brain 97: 301–312
77. Johnston I, Rowan J (1974) Raised intracranial pressure and cerebral blood flow. 3. Venous outflow tract pressures and vascular resistances in experi-

mental intracranial hypertension. J Neurol Neurosurg Psychiatry 37: 392–402
78. Karahalios DG, Rekate HL, Khayata MH, Apostolides PJ (1996) Elevated intracranial venous pressure as a universal mechanism in pseudotumor cerebri of varying etiologies. Neurology 46: 198–202
79. Katznelson D (1978) Increased intracranial pressure in cystic fibrosis. Acta Paediatr Scand 67: 607–609
80. Keiper GL, Jr, Sherman JD, Tomsick TA, Tew JM, Jr (1999) Dural sinus thrombosis and pseudotumor cerebri: unexpected complications of suboccipital craniotomy and translabyrinthine craniectomy. J Neurosurg 91: 192–197
81. Kesler A, Ellis M, Reshef T, Kott E, Gadoth N (2000) Idiopathic intracranial hypertension and anticardiolipin antibodies. J Neurol Neurosurg Psychiatry 68: 379–380
82. Kim AW, Trobe JD (2000) Syndrome simulating pseudotumor cerebri caused by partial transverse venous sinus obstruction in metastatic prostate cancer. Am J Ophthalmol 129: 254–256
83. Kinal M (1966) Infratentorial tumors and the dural venous sinuses. J Neurosurg 25: 395–401
84. Kinal M (1967) Traumatic thrombosis of dural venous sinuses in closed head injuries. J Neurosurg 27: 142–145
85. King J, Mitchell P, Thomsen K, Tress B (2002) Manometry combined with cervical puncture in idiopathic intracranial hypertension. Neurology 58: 26–30
86. King JO, Mitchell PJ, Thomson KR, Tress BM (1995) Cerebral venography and manometry in idiopathic intracranial hypertension. Neurology 45: 2224–2228
87. Kollar C, Johnston I, Parker G, Harper C (1998) Dural arteriovenous fistula in association with heterotopic brain nodule in transverse sinus. AJNR Am J Neuradiol 19: 1126–1128
88. Kollar C, Parker G, Johnston I (2001) The endovascular treatment of cranial venous sinus obstruction resulting in pseudotumor syndrome. Report of three cases. J Neurosurg 94: 646–651
89. Kollar CD, Johnston IH (1999) Pseudotumour after arteriovenous malformation embolisation [letter]. J Neurol Neurosurg Psychiatry 67: 249
90. Kuker W, Mull M, Mayfrank L, Weis J, Schiefer J, Thron A (1997) A cystic lesion within the dural sinuses: a rare cause of increased intracranial pressure. Neuroradiology 39: 132–135
91. Lam BL, Schatz NJ, Glaser JS, Bowen BC (1992) Pseudotumor cerebri from cranial venous obstruction. Ophthalmology 99: 706–712
92. Lam C, Solomon R, Brent Clark H, Casey S (2001) Reversal of increased intracranial pressure with removal of a torcular epidermoid: case report. Neurosurgery 48: 929–932
93. Lamas E, Lobato R, Esparza J, Escudero L (1977) Dural posterior fossa AVM producing raised sagittal sinus pressure. J Neurosurg 46: 804–810

94. Langfitt T, Weinstein J, Kassell N, Gagliardi L, Shapiro H (1966) Compression of the cerebral vessels by intracranial hypertension. I. Dural sinus pressures. Acta Neurochir (Wien) 15: 212–222
95. Le Gros Clark W (1920) On the pacchionian granulations. J Anatomy 55: 40–48
96. Leach J, Jones B, Tomsick T, Stewart C, Balko M (1996) Normal appearance of arachnoid granulations on contrast-enhanced CT and MR of the brain: differentiation from dural sinus disease. Am J Neuroradiol 17: 1523–1532
97. Lee G, Seex K, Scott G (2001) Pseudotumor cerebri due to a torcular epidermoid cyst. ANZ J Surg 71: 385–388
98. Leker RR, Steiner I (1999) Features of dural sinus thrombosis simulating pseudotumor cerebri. Eur J Neurol 6: 601–604
99. Leker RR, Steiner I (2000) Isolated intracranial hypertension as the only sign of cerebral venous thrombosis [letter; comment]. Neurology 54: 2030
100. Levine J, Povlishock J, Becker D (1982) The morphological correlates of primate cerebrospinal fluid absorption. Brain Res 241: 31–41
101. Levine S *et al* (1987) Cerebral venous thrombosis with lupus anticoagulants: report of 2 cases. Stroke 18: 801–804
102. Liedler, RabTA (1928) Otitic general septic infection with bilateral optic neuritis. J Laryngol Otology 43: 672–673
103. Loman J, Damashek W (1944) Increased intracranial venous and cerebrospinal fluid pressures in polcythemia. Trans Amer Neurol Assoc 70: 84–87
104. Luce J, Husebuy J, Kirk W, Butler J (1982) Mechanism by which positive end-expiratory pressure increases cerebrospinal fluid pressure in dogs. Am J Physiol
105. Lundar T, Blakke S, Nornes H (1990) Hydrocephalus in an achondroplastic child treated by venous decompression at the jugular foramen. J Neurosurg 73: 138–140
106. Malek A, Higashida R, Balousek P *et al* (1999) Endovascular recanalization with balloon angioplasty and stenting of an occludedd occiptal sinus for treatment of intracranial venous hypertension: technical case report. Neurosurgery 44: 896–901
107. Mamourian A, Towfighi J (1995) MR of giant arachnoid granulation, a normal variant presenting as a mass within the dural venous sinus. AJNR 16: 901–904
108. Marks M, Dake M, Steinberg G, Norbash A, Lane B (1994) Stent placement for arterial and venous cerebrovascular disease: preliminary experience. Neuroradiology 191: 441–446
109. Martin J (1955) Signs of obstruction of the superior longitudinal sinus following closed head injuries (traumatic hydrocephalus). Brit Med J 2: 467–470
110. Martins A, Kobrine A, Larsen D (1974) Pressure in the sagittal sinus during intracranial hypertension in man. J Neurosurg 40: 603–608
111. Martins AN (1973) Resistance to drainage of cerebrospinal fluid: clinical measurement and significance. J Neurol Neurosurg Psychiatry 36: 313–318

112. Massons J *et al* (1992) Cerebral venous thrombosis and hereditary protein C deficiency. Neurologia 7: 34–38
113. Mathew NT, Meyer JS, Ott EO (1975) Increased cerebral blood volume in benign intracranial hypertension. Neurology 25: 646–649
114. Mathew NT, Ravishankar K, Sanin LC (1996) Coexistence of migraine and idiopathic intracranial hypertension without papilledema [see comments]. Neurology 46: 1226–1230
115. McGonigal A, Bone I, Teasdale E (2004) Resolution of transverse sinus stenosis in idiopathic intracranial hypertension after L-P shunt. Neurology 62: 514–515
116. McLaughlin J, Loeser J, Roberts T (1997) Acquired hydrocephalus associated with superior vena cava syndrome in infants. Child's Nerv Syst 13: 59–63
117. Medlock MD, Olivero WC, Hanigan WC, Wright RM, Winek SJ (1992) Children with cerebral venous thrombosis diagnosed with magnetic resonance imaging and magnetic resonance angiography. Neurosurgery 31: 870–876; discussion 876
118. Mitchell D, Fisher J, Irving D, Gazzard BG, Guiloff RJ (1986) Lateral sinus thrombosis and intracranial hypertension in essential thrombocythaemia [letter]. J Neurol Neurosurg Psychiatry 49: 218–219
119. Mokri B, Jack CR, Jr, Petty GW (1993) Pseudotumor syndrome associated with cerebral venous sinus occlusion and antiphospholipid antibodies. Stroke 24: 469–472
120. Moser FG, Hilal SK, Abrams G, Bello JA, Schipper H, Silver AJ (1988) MR imaging of pseudotumor cerebri. AJR Am J Roentgenol 150: 903–909
121. Newton Pitt G (1890) An analysis of 57 fatal cases of ear disease and of the complication which led to death. BMJ 1: 643–647
122. Noggle JD, Rodning CB (1986) Rapidly advancing pseudotumor cerebri associated with morbid obesity: an indication for gastric exclusion. South Med J 79: 761–763
123. Nonne M (1904) Über Fälle vom Symptomkomplex "Tumor cerebri" mit Ausgang in Heilling (Pseudotumor cerebri): Über letal verlaufene Fälle von "Pseudotumor cerebri" mit Sektionsbefund. Dtsch Z Nervenheilk 27: 169–216
124. Norrell H, Wilson C, Howieson J *et al* (1969) Venous factors in infantile hydrocephalus. J Neurosurg 31: 561–569
125. Ogungdo B, Roy D, Gholkar A, Mendelow A (2003) Endovascular stenting of the transverse sinus in a patient presenting with benign intracranial hypertension. Br J Neurosurg 17: 565–568
126. Olivero W, Asner N (1992) Occlusion of the sagittal sinus in craniectomized rabbits. Child's Nerv Syst 8: 307–309
127. Osterholm J (1970) Reaction of the cerebral venous sinus system to acute intracranial hypertension. J Neurosurg 32: 654–659
128. Owler B, Allan R, Parker G, Besser M (2003) Pseudotumor cerebri, CSF Rhinorrhoea and the role of Venous Sinus Stenting in Treatment. Brit J Neurosurg 17: 79–83

129. Owler B, Pena A, Green H, Donovan T, Carpenter A, Pickard J (2001) A study of benign intracranial hypertension using diffusion tensor imaging. In World Congress of Neurological Surgeons, Sydney
130. Owler BK, Parker G, Halmagyi GM *et al* (2003) Pseudotumor Cerebri Syndrome: Venous sinus obstruction and treatment with venous sinus stenting. J Neurosurg 98: 1045–1055
131. Pierre-Kahn A, Hirsch J, Renier D, Metzger J, Maroteaux P (1980) Hydrocephalus and Achondroplasia. A study of 25 observations. Child's Brain 7: 205–219
132. Plant G, Donald J, Jackowski A, Vinnicombe S, Kendall B (1991) Partial, non-thrombotic, superior sagittal sinus occlusion due to occipital skull tumors. J Neurol Neurosurg Psychiatry 54: 520–523
133. Powers JM, Schnur JA, Baldree ME (1986) Pseudotumor cerebri due to partial obstruction of the sigmoid sinus by a cholesteatoma. Arch Neurol 43: 519–521
134. Pritz MB (1984) Monitoring cardiac function and intravascular volume in neurosurgical patients. Neurosurgery 15: 775–780
135. Purves M (1972) The physiology of the cerebral circulation. Cambridge University Press, Cambridge
136. Quattrone A, Bono F, Oliveri R *et al* (2001) Cerebral venous thrombosis and isolated intracranial hypertension without papilloedema in CDH. Neurology 57: 31–36
137. Quattrone A, Bono F, Pardatscher K (2002) Manometry combined with cervical puncture in idiopathic intracranial hypertension. Neurology 59: 963
138. Quincke H (1893) Meningitis serosa. Samml Klin Votr, Leipzig 67. Inn Med 23: 655
139. Quincke H (1897) Ueber meningitis serosa und verwandte zustände. Dtsch Z Nervenheilk 9: 149–168
140. Raichle ME, Grubb RL, Jr, Phelps ME, Gado MH, Caronna JJ (1978) Cerebral hemodynamics and metabolism in pseudotumor cerebri. Ann Neurol 4: 104–111
141. Ray B, Dunbar H (1951) Thrombosis of the dural venous sinuses as a cause of "pseudotumor cerebri". Ann Surg 134: 376–385
142. Roche J, Warner D (1996) Arachnoid granulations in the transverse and sigmoid sinuses: CT, MR, and MR angiographic appearances of a normal anatomic variation. Am J Neuroradiol 17: 677–683
143. Rollins N, Booth T, Shapiro K (2000) MR venography in children with complex craniosynostosis. Pediatr Neurosurg 32: 308–312
144. Ropper AH, Marmarou A (1984) Mechanism of pseudotumor in Guillain-Barre syndrome. Arch Neurol 41: 259–261
145. Rosenberg A, O'Connell J, Ojemann R, Palmer W (1993) Giant cystic arachnoid granulations: a rare cause of lytic skull lesions. Human Pathol 24: 438–441
146. Rosman NP, Shands KN (1978) Hydrocephalus caused by increased intracranial venous pressure: a clinicopathological study. Ann Neurol 3: 445–450

147. Röther J, Waggie K, van Bruggen N, de Crespgny A, Moseley M (1996) Experimental cerebral venous thrombosis: evaluation using magnetic resonance imaging. J Cereb Blood Flow Metabolism 16: 1353–1361
148. Saggi B, Bloomfield G, Sugerman H *et al* (1999) Treatment of intracranial hypertension using nonsurgical abdominal decompression. The Journal of Trauma: Injury, Infection and Critical Care 46: 646–651
149. Sahs A, Hyndman O (1939) Intracranial hypertension of unknown cause: cerebral oedema. Arch Surg 38: 429–434
150. Sahs A, Joynt R (1956) Brain swelling of unknown cause. Neurol Minneapolis 6: 791–803
151. Sainte-Rose C, LaCombe J, Pierre-Kahn A, Renier D, Hirsch JF (1984) Intracranial venous sinus hypertension: cause or consequence of hydrocephalus in infants? J Neurosurg 60: 727–736
152. Shapiro K, Langfitt T, Weinstein J (1966) Compression of the cerebral vessels by intracranial hypertension. II. Morphological evidence for collapse of vessels. Acta Neurochir (Wien) 15: 223–233
153. Shulman K, Ransohoff J (1964) Sagittal sinus venous pressure in hydrocephalus. 169–173
154. Silbergleit R, Junck L, Gebarski SS, Hatfield MK (1989) Idiopathic intracranial hypertension (pseudotumor cerebri): MR imaging. Radiology 170: 207–209
155. Simpson T (1948) Papilloedema in emphysema. BMJ II: 639–641
156. Sindou M, Auque J (2000) The intracranial venous system as a neurosurgeon's perspective. Adv Techn Stand Neurosurg 26: 131–216
157. Sindou M, Mercier P, Bokor J, Brunon J (1980) Bilateral thrombosis of the transverse sinuses: microsurgical revascularization with venous bypass. Surg Neurol 13: 215–220
158. Sklar FH, Beyer CW Jr, Ramanathan M, Cooper PR, Clark WK (1979) Cerebrospinal fluid dynamics in patients with pseudotumor cerebri. Neurosurgery 5: 208–216
159. Soleau S, Schmidt R, Stevens S *et al* (2003) Extensive experience with dural venous sinus thrombosis. Neurosurgery 52: 534–542
160. Soler D, Cox T, Bullock P, Calver DM, Robinson RO (1998) Diagnosis and management of benign intracranial hypertension. Arch Dis Child 78: 89–94
161. Sorensen PS, Thomsen C, Gjerris F, Henriksen O (1990) Brain water accumulation in pseudotumour cerebri demonstrated by MR-imaging of brain water self-diffusion. Acta Neurochir (Wien) [Suppl] 51: 363–365
162. Sorensen PS, Thomsen C, Gjerris F, Schmidt J, Kjaer L, Henriksen O (1989) Increased brain water content in pseudotumour cerebri measured by magnetic resonance imaging of brain water self diffusion. Neurol Res 11: 160–164
163. Steinbok P, Hall J, Flodmark O (1989) Hydrocephalus in achondroplasia: the role of intracranial venous hypertension. J Neurosurg 71: 42–48
164. Stewart D, Johnson D, Myers G (1975) Hydrocephalus as a complication of jugular catheterisation during total parental nutrition. J Pediatric Surg 10: 771–777

165. Sugerman H, Felton W, Sismanis A *et al* (1999) Effect of externally applied negative abdominal pressure device (ABSHELL) on headaches and pulsatile tinnitus in patients with pseudotumor cerebri. Neurology 52 6 [Suppl] 2: A34–35
166. Sugerman HJ, DeMaria EJ, Felton WL, 3rd, Nakatsuka M, Sismanis A (1997) Increased intra-abdominal pressure and cardiac filling pressures in obesity-associated pseudotumor cerebri. Neurology 49: 507–511
167. Sugerman HJ, Felton WL, 3rd, Salvant JB, Jr, Sismanis A, Kellum JM (1995) Effects of surgically induced weight loss on idiopathic intracranial hypertension in morbid obesity. Neurology 45: 1655–1659
168. Sugerman HJ, Felton WL 3rd, Sismanis A, Kellum JM, DeMaria EJ, Sugerman EL (1999) Gastric surgery for pseudotumor cerebri associated with severe obesity. Ann Surg 229: 634–640; discussion 640–642
169. Sussman J, Leach M, Greaves M, Malia R, Davies-Jones GA (1997) Potentially prothrombotic abnormalities of coagulation in benign intracranial hypertension. J Neurol Neurosurg Psychiatry 62: 229–233
169a. Sussman J, Sarkies N, Pickard JD (1998) Benign intracranial hypertension. Adv Tech Stand Neurosurg 24: 261–305
170. Symonds C (1931) Otitic Hydrocephalus. Brain 54: 55–71
171. Taylor W, Hayward R, Lasjaunias P *et al* (2001) Enigma of raised pressure in patients with complex craniosynostosis: the role of abnormal intracranial venous drainage. J Neurosurg 94: 377–385
172. Tinney W, Hall B, Giffin H (1943) CNS manifestations of polycythemia vera. Proc Mayo Clin 18: 300–303
173. Tokiguchi S, Ando K, Tsuchiya T, Ito J (1986) Fat in the dural sinus. Neuroradiol 28: 267–270
174. Tokiguchi S, Kurashima A, Ito J, Takahashi H, Shimbo Y (1988) Fat in the dural sinus – CT and anatomical correlations. Neuroradiol 30: 78–80
175. Upton M, Weller R (1985) The morphology of cerebrospinal sluid drainage pathways in human arachnoid granulations. J Neurosurg 63: 867–875
176. Wall M, Dollar JD, Sadun AA, Kardon R (1995) Idiopathic intracranial hypertension. Lack of histologic evidence for cerebral edema. Arch Neurol 52: 141–145
177. Wall M, George D (1991) Idiopathic intracranial hypertension. A prospective study of 50 patients. Brain 114: 155–180
178. Wall M, Giuseffi V, Rojas P (1989) Symptoms and disease associations in pseudotumor cerebri: a case-control study. Neurology [Suppl] 39: 210
179. Wang S, Silberstein S, Patterson S, Young W (1998) Idiopathic intracranial hypertension without papilloedema: a case-control study in a headache center. Neurology 51: 245–249
180. Weed L, Flexner L (1933) The relations of the intracranial pressures. Am J Physiol 105: 266–272
181. Wright R (1938) Experimental observations on increased intracranial pressure. ANZ J Surg 7–8: 215–235
182. Yamada H (1981) Neurological manifestations of paediatric achondroplasia. J Neurosurg 54: 49–57

Patient demographics

Case No.	Age (years)	Body mass index (kg/m^2)	Duration of symptoms (years)	Previous procedures	Symptoms		Papilloedema	CSF pressure (cms H_2O)
Cambridge Patients					Headache	Visual symptoms		
1	34	44	3	LPS × 6*	yes	blurring	absent	25
2	30	30	1.3	none	yes	sparkling	present	35
3	46	32	4	none	yes	decreased acuity	chronic	30
4	49	33	11	LPS	yes	obscurations, constricted fields	chronic	40
5	52	41	12	LPS, ONSF, BSTD, VPS*	yes	decreased acuity, constricted fields	absent	31 (before VPS)
6	32	45	5	LPS, VPS*	yes	poor vision	chronic	39 (before VPS)
7	33	30	0.4	none	yes	obscurations	present	46
8	24	31	0.7	none	yes	obscurations	present	30
9	21	29	1	none	yes	constricted fields	mild	30
10	19	43	3.5	LPS × 3*	yes	obscurations, constricted fields	present	refused LP

11	25	42	2.3	none	yes	blurring	absent	40
12	32	43	5	none	yes	blurring, constricted fields	absent	25
Sydney Patients								
1	17	38	3 months	Diamox, ONSF, Ext CSF drainage	yes	obscurations	present	48
2	27	25	1	Diamox, LP	yes	obscurations, pulsatile tinnitus	present	30
3	27	34	>5	Diamox, ONSF, LP shunt, subtemporal decompression	moderate	blurring	present	39
4	38	23	10	LP shunts, Anterior fossa repair	Yes, CSF Rhinorrhoea	blurring	minimal	N/A
Newcastle Patients								
1	37	–	5 months	–	yes	obscurations	present	40

* LPS lumboperitoneal shunt
* ONSF optic nerve sheath fenestration
* BSTD bilateral subtemporal decompressions
* VPS ventriculoperitoneal shunt

Summary of published cases of venous stenting for PTC

Venous pressures and clinical outcome								
Case	pressures prior to stenting (mm Hg)		pressures after stenting (mm Hg)		Pcsf after stenting cms H_2O (mmHg)	Clinical outcome		
	torcular	jugular bulb	torcular	jugular bulb		symptoms	papilloedema	follow-up (months)
Cambridge Patients								
1	45	8	23	8		no change	absent^{+}	26
2	29	8	13	7		improved	resolved	24
3	25	4	19	9		no change	no change	18
4	25	7	13	9		asymptomatic	improved	18
*5	23	11	16	12		no change	no change	14
*6	15	7	14	11		no change	absent^{+}	14
7	34	6	13	11		asymptomatic	resolved	12
8	29	7	12	8		asymptomatic	resolved	12
9	26	9	17	10		improved	resolved	14
10	24	13	17	11		asymptomatic	unknown	2
11	20	8	12	10		asymptomatic	absent^{+}	9
12	31	11	18	12		no change	absent^{+}	7

Sydney Patients								
1	35	11	–	–	15 (11)	improved	resolved	12
2	15	2	–	–	15 (11)	asymptomatic	resolved	11
3	36	9	–	–	16 (12)	modest headache	resolved	5
4	22	8	–	–	N/A	resolved	–	11
Newcastle Patients								
1	40	15	–	–	26 (19)	resolved	mild	6

+ papilloedema resolved prior to stenting
* ventriculoperitoneal shunt in situ

Technical Standards

Offprint from
Advances and Technical Standards in Neurosurgery, Vol. 30
Edited by J.D. Pickard

Sacral Neuromodulation in Lower Urinary Tract Dysfunction

J. R. VIGNES[1], M. DE SEZE[2], E. DOBREMEZ[3], P. A. JOSEPH[2], and J. GUÉRIN[1]

[1] Department of Neurosurgery, Medical School Hospital, Hôpital Pellegrin, Bordeaux, France
[2] Neurorehabilitation Unit, Medical School Hospital, Hôpital Pellegrin, Bordeaux, France
[3] Department of Pediatric Urology, Medical School Hospital, Hôpital Pellegrin, Bordeaux, France

With 4 Figures

Contents

Abstract

Vesico-urethral dysfunction is a major problem in daily medical practice due to its psychological disturbances, its social costs and its high impact on quality of life. Recently, sacral neuromodulation, namely the electrical stimulation of the sacral nerves, appears to have become an alternative for radical bladder surgery particularly in cases of idiopathic bladder overactivity. The mechanism of action is only partially understood but it seems to involve a modulation in the spinal cord due to stimulation of inhibitory interneurons.

Temporary sacral nerve stimulation is the first step. It comprises the temporary application of neuromodulation as a diagnostic test to determine the best location for the implant and to control the integrity of the sacral root. If test stimulation is successful, a permanent device is implanted. This procedure is safe in experienced hands.

So-called idiopathic bladder overactivity still the major indication for this technique. Patients not likely to benefit from the procedure were those with complete or almost complete spinal lesions, but incomplete spinal lesions seemed to be a potential indication. This technique is now also indicated in the case of idiopathic chronic retention and chronic pelvic pain syndrome.

When selection is performed, more than three-quarters of the patients showed a clinically significant response with 50% or more reduction in the frequency of incontinent episodes, but the results vary according to the author's mode of evaluation. From the economic point of view, the initial investment in the device is amortized in the mid-term by savings related to lower urinary tract dysfunction.

Finally, this technique requires an attentive follow-up and adjustments to the electric parameters so as to optimize the equilibrium between the neurological systems.

Keywords: Bladder neurogenic; electric stimulation therapy; voiding dysfunction; urinary urge incontinence; urinary retention.

Introduction

Vesico-urethral dysfunction is a major problem in daily medical practice due to its psychological disturbances, its social costs and its high impact on quality of life. A complex neuroanatomic network governs the relationships between the spinal, pons and supra-pons centers, and the vegetative and somatic systems. Despite this complexity, the consequence of these relationships is always the same, i.e. retention and emptying of the bladder to ensure continence and micturition and to protect the upper urinary tract. To restore its function, surgical techniques intervening in the peripheral or central nervous systems have always played an important role. Recently, sacral neuromodulation, namely the electrical stimulation of the sacral nerves, appears to have become an alternative for radical bladder surgery particularly in cases of idiopathic bladder overactivity. The mechanism of action is only partially understood but it seems to involve a modulation in the spinal cord due to stimulation of inhibitory interneurons. This technique is also indicated in the case of idiopathic chronic retention and chronic pelvic pain syndrome.

Anatomy and Physiology of the Lower Urinary Tract

The lower urinary tract has two main functions: storage and periodic elimination of urine. These two functions are regulated by a complex neural control system involving a central pathway located in the spinal cord, pons and brain and as well as the peripheral autonomic and somatic neural pathways. This control system functions like a switching circuit to maintain a reciprocal relationship between the bladder and outlet components of the lower urinary tract. Because of these complex neural regulations, the central nervous system control of the lower urinary tract is susceptible to a variety of neurologic disorders which, among a wide range of non-invasive therapeutic modalities, may be improved by sacral neuromodulation.

The storage and periodic elimination of urine are dependent on the reciprocal activity of two functional units in the lower urinary tract: a reservoir, the bladder and an outlet represented by the bladder neck and the smooth and striated sphincter muscles of the urethra. During urine storage, the bladder outlet is closed and the bladder smooth muscle is quiescent, allowing intravesical pressure to remain low over a wide range of bladder volumes. During voluntary voiding, the initial event is a relaxation of the pelvic floor and striated urethral muscles, followed by a detrusor muscle contraction and opening of the bladder neck. This activity is mediated by three sets of peripheral nerves: parasympathetic (pelvic), sympathetic (hypogastric) and somatic (pudendal) nerves (Fig. 1). These nerves also contain afferent axons terminating in the lower urinary tract which are involved in initiating micturition.

Spinal Levels

Efferent Pathway

The parasympathetic efferent pathway is the main excitatory input to the bladder. Parasympathetic preganglionic axons originate in the intermedio-

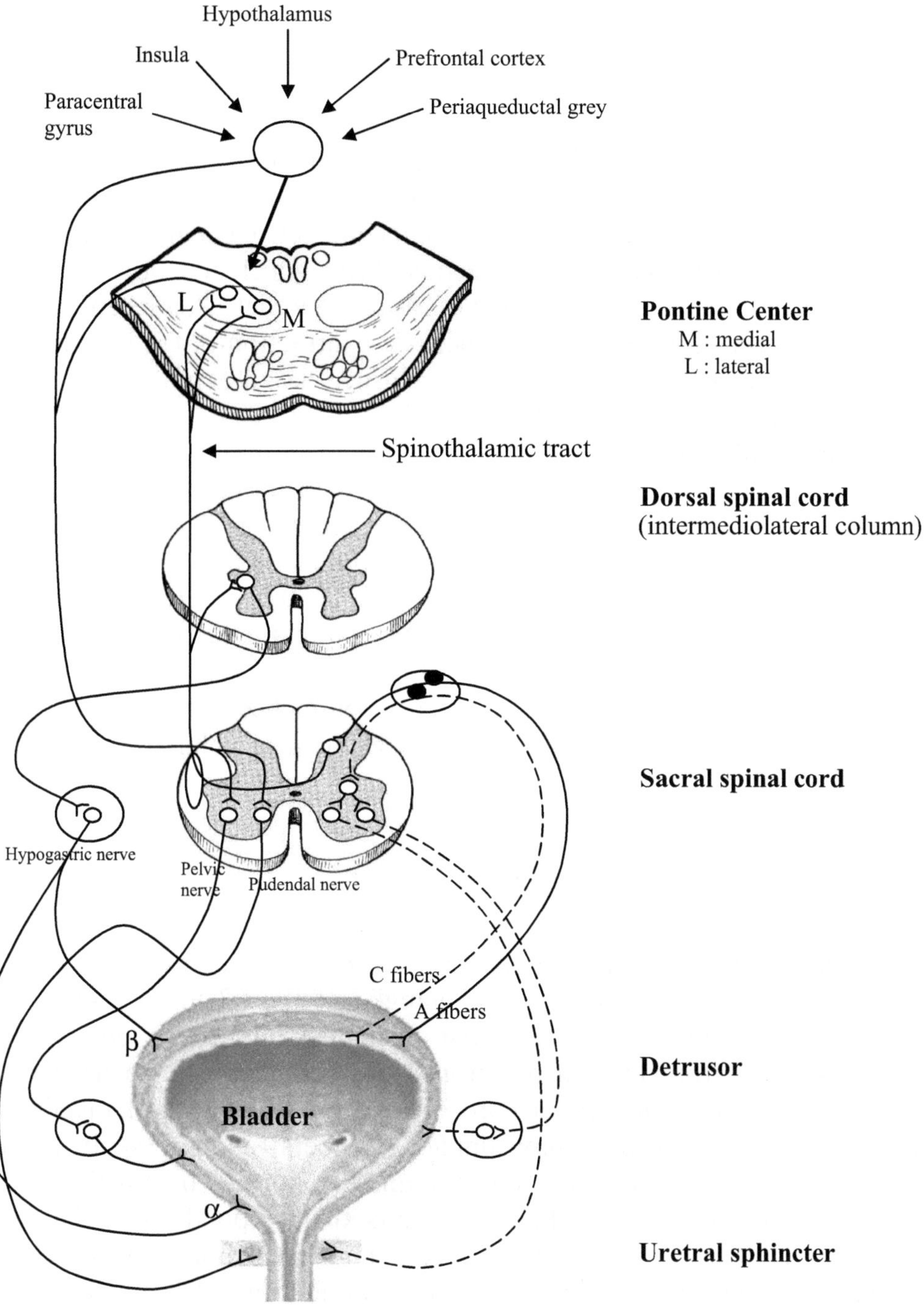

Fig. 1. Anatomy and physiology of the lower urinary tract

lateral column of the S2 to S4 spinal cord and terminate in the postganglionic neurons in the bladder wall and in the pelvic plexus [98]. The main neurotransmitter released by the parasympathetic postganglionic nerve terminals is acetylcholine. Acetylcholine can act on different subtypes of detrusor muscarinic receptors, among which the M3 are most important for mediating evoked smooth-muscle contractions neurally in the bladder [40].

The sympathetic preganglionic neurons are located within the intermediolateral cell column of the T11 to L2 spinal cord. They make synaptic connections with postganglionic neurons in the inferior mesenteric ganglionic neurons in the paravertebral ganglia and pelvic ganglia. Sympathetic postganglionic terminals release norepinephrine which acts on alpha-1 vesical and urethral receptors and beta-2 adrenergic detrusor receptors. The effect of norepinephrine on the former is a contraction of the bladder base and urethral smooth muscle. Norepinephrine, via an action of the Beta 2 receptors, can also relax the bladder body.

Somatic afferent pathways that originate from the motoneurons in the Onuf nucleus of the anterior horn of the S2 to S4 spinal cord innervate the external striated urethral sphincter muscle and the pelvic floor musculature. Somatic nerve terminals release acetylcholine, which acts on nicotinic receptors to induce a muscle contraction. The striated urethral sphincter also receives noradrenergic input from the sympathetic nerves. The combined activation of the sympathetic and somatic pathways elevates bladder outlet resistance and contributes to urinary continence. The striated sphincter (via the pudendal nerve) is the unique element of voluntary continence and micturition.

Afferent Pathway

Sensory information regarding bladder fullness is conveyed to the spinal cord via afferent axons in the pelvic and hypogastric nerves, which possess neuronal somata in the dorsal root ganglia at the S2 to S4 and T11 to L2 spinal segmental levels. Afferent fibers passing in the pelvic nerve carry impulses from tension receptors in the bladder wall to neurons in the dorsal horn of the spinal cord. These are mainly small myelinated (Aδ fibers) [49, 110] and unmyelinated (C fibers) axons [28]. In several mammalian species including the human, the normal micturition reflex is mainly mediated by Aδ fibers afferents that respond to bladder distension [91]. The C fibers, which have a high mechanical threshold, are usually unresponsive to bladder distension and are thus called silent C-fibers, but many of them do respond to chemical, noxious or cold stimuli [58, 74].

Spinal Centers

The sacral micturition center involves laminae VI, VII and X. The innterneurones participate in local control of elementary programs via parasympathetic and somatic pathways [73]. The C fibers project to the dorsal horn and via a polysynaptic reflex [48] with medullar interneurones [140] to form the «C reflex» of Bradley [28].

Pontine Centers

Among the sub-encephalic centers involved in micturitional control (Fig. 1), the most important are localized at the pontine level [10, 16]. This part of the tegmentum receives afferent pathways from collateral spino-thalamic (from dorsal horn, laminae I and IV) to form the spino-ponto-spinal reflex or the «A reflex» of Bradley [28]. Two pontine centers have been characterized in mammalians [98]. The first is localized in the medial part of the dorsolateral pontine tegmentum, and is thus called the M-region or Pontine Micturitional Center (PMC) [121]. The PMC projects to the sacral intermediolateral cell column, in which are localized the para-sympathetic center connected to the bladder motoneurons and to the sacral intermedioventral cell column. The PMC is involved in the voiding phase via both these projections. The excitatory PMC projection to bladder motoneurons is responsible for an increase in bladder pressure during micturition. The relaxation of the striated uretral sphincter during micturition is due to excitatory projection to inhibitory interneurones in the spinal dorsal gray commissure.

The second pontine center, located more ventrally and more laterally in the pontine tegmentum than the PMC, is involved in the storage of urine during continence. During the storage phase, this L-center or Pontine Storage Center (PSC) acts by direct excitatory projection to the urethral sphincter in the nucleus of Onuf [85].

Suprapontine Controls

Several other central structures located in the forebrain and the cerebral cortex have been thought to be involved in lower urinary tract control. At the mesencephalic level, the periaqueductal gray (PAG) is considered as the main center involved in micturitional control. The PAG is thought to act as a central sensorimotor integrative relay of the micturition reflex, via the reception of sensory information concerning bladder fullness and the direct projection to the PMC [15].

In the forebrain, the most documented structure is the pre-optic area of the hypothalamus, which is thought to play a role in the initiation of the voiding phase via direct projection to the PMC. In addition, the anterior cingulated gyrus, amygdala, bed nucleus of the stria terminalis and septal nuclei are susceptible, when excited, to elicit bladder contraction [16]. The superomedial part of the precentral gyrus and the superolateral part of the precentral gyrus seem to be involved in voluntary control on the pelvic floors and in abdominal straining, respectively. Finally, the exact role of the cerebellum is not fully understood, but both afferent and efferent contributions to the micturitional reflex have been proposed [122].

Reflex Mechanisms Controlling Micturition

Storage Reflexes

The bladder functions as a low pressure reservoir during urine storage due to the combined effect of the visco-elasticity of the bladder wall and the quiescence of the parasympathetic pathway to the bladder. Continence during bladder filling is reinforced by the activation of a sacral-to-thoracolumbar intersegmental spinal reflex pathway, initiated by afferent fibers linked to a bladder tension receptor, which triggers firing in sympathetic pathways to the bladder, thus mediating an inhibition of bladder activity and a contraction of the bladder neck and proximal urethra. Simultaneously, the activation of pudendal motoneurons during bladder filling induces a contraction of the striated sphincter muscle, which in turn contributes to urinary continence.

In addition to these spinal continence reflexes, a supraspinal urine storage center located in the dorsolateral pons is involved in continence via descending inputs activating the pudendal motoneurons to increase urethral resistance (Fig. 1).

Voiding Reflexes

When bladder volumes reach the micturition threshold, intense afferent activity originating in the bladder mechanoceptors triggers the micturition reflex, which consists of spino-bulbo-spinal reflex pathways passing through the pontine micturition center. Activation of the pontine micturition center induces both a firing in the sacral parasympathetic pathways leading to bladder contraction and secondarily to inhibition of the sympathetic and somatic pathways relaxing urethral and bladder outflow. Before reaching the pontine micturition center, afferent inputs from the spinal cord pass through an integrative relay center in the periacqueductal gray. This center functions as an "on-off" switch activated by afferent activity derived from bladder mechanoceptors, and it also receives inhibitory and excitatory inputs from the brain regions (Fig. 1).

Voiding is also facilitated by an urethrovesical reflex initiated by the stimulation of urethral afferents triggered by urine flow in the urethra, thereby enhancing bladder contractions.

The suppression of the striated urethral sphincter activity during micturition is mainly due to a direct pontine micturition center projection to sacral inhibitory interneurons in the dorsal gray commissure, also known as the intermediomedial cell column. These inhibitory dorsal gray commissure interneurons in turn inhibit sphincter motoneurons in Onuf's nucleus during micturition.

Historical Evolution of Functional Surgery in Lower Urinary Tract Dysfunction

Spinal Cord Stimulation

It was Budge who in 1858 opened up the concept of "micturition reflex" by stimulating the nervous system. Thanks to technical improvements made by Oersted in 1820 and especially Faraday in 1821, he was able to activate bladder contractions using an electrical stimulation in the sacral part of the spinal cord [130]. Over a century later in 1972, Friedman [64] performed selective bladder stimulation in animal models by implanting bipolar electrodes in the spinal cord. The preganglionic parasympathetic fibers that innervate the detrusor muscle emerge from the ventro-intermedial column of the sacral spinal cord, while the somatic efferent fibers that innervate the urethral sphincter come from the Onuf nucleus (anterior horn of 3rd and 4th sacral segments). The different localization of these two groups of motor neurons allowed selective bladder stimulation. Encouraged by these results, Grimes [69] operated five spinal cord injured patients by implanting two bipolar electrodes 2.5 mm deep at the level of S2. Four patients were then able to urinate by stimulation. Then Grimes and Nashold [68] analyzed a group of 10 patients: the clinical result was dramatically different depending on the position of the electrodes. Furthermore, low selectivity of this stimulation remains a major problem of this technique, which does not systematically avoid simultaneous contraction of the striated sphincter of the bladder. Sedan [134] made similar observations. Some are now re-assessing this abandoned technique, because it makes it possible to stimulate electively the motor neurons of the detrusor muscle, thus inducing efficient micturition without the need to perform a posterior rhizotomy [67, 167].

Intravesical Stimulation

In 1878, Saxtorph introduced the concept of direct stimulation of the bladder wall (and its nerve terminals) to induce a detrusor contraction and to activate urination in patients suffering from urinary retention [130]. In

1954, however, McGuire noted that the results differ depending on the position or the volume of the electrodes, as well as on the characteristics of stimulation [130]. From 1959, Boyce and Lathem [25] continued these efforts, as did Bradley who designed an implant system used initially in dog and then in human [27]. However, this technique was abandoned due to lack of encouraging results. Recently, Jiang and Linstrom [93] showed that intravesical stimulation could be used to activate a neurogenic bladder, especially in spina bifida patients.

Pelvic Nerve Stimulation

In 1957, Ingersoll [88] performed unilateral stimulation of a pelvic nerve. This technique, known as the Burgele-Ichim-Demetrescu technique [96], is theoretically possible and a few patients gained benefit from these implants (electro-stimulated micturition). However, the complexity in approaching the pelvic nerves and their fragility make this technique difficult. For some authors, it does not solve the problem of the simultaneous contraction of the detrusor and the sphincter produced by recurrent circuits, unless the pudendal nerves have been cut [12]. Mention should also be made of the less productive efforts of Hald in 1967 who tried to stimulate the detrusor muscle selectively through the pelvic nerve fibers [130].

Stimulation of Pelvic Floor Muscles

In 1963 Caldwell [130] performed the first stimulator implantation in a pelvic sphincter to treat urinary incontinence. However, it was observed shortly afterward that transrectal stimulation and transvaginal stimulation in women gave the same results. At present, the mechanism of these stimulations is known: the stimulated pudendal afferents activate the sympathetic inhibitor neurons, which in turn inhibit the central parasympathetic neurons, thereby reducing bladder hyperactivity [108].

Stimulation of Sacral Nerve Roots

Since 1971, it has been demonstrated in monkey and then in human beings that direct stimulation of the anterior sacral roots allows bladder emptying. The electrodes can be placed in the extra- or intradural space. Strong electrical stimulations cause simultaneous contraction of the detrusor and the striated sphincter of the urethra. However, due to its smooth muscle, detrusor contraction lasts longer than striated sphincter contraction, which relaxes intermittently, letting urine flow and thus protecting the upper urinary tract. GB Brindley pioneered the technique of adding a posterior rhizotomy, thereby improving bladder capacity and reducing bladder hyperreflexia. This is still the only technique for restoring bladder function, retention (continence) and emptying (micturition). It is especially useful for patients suffering from a complete spinal cord lesion who are not able to empty their bladder by conventional methods [30, 31, 32, 33, 158, 159].

Sacral Nerve Deafferentation

The goal of this process is to suppress the vesico-medullary reflex, which is responsible for bladder hyperreflexia (or overactivity), a condition closely related to incontinence. The principle of sacral deafferentation was introduced a century ago to reduce spasticity [46]. Some C fibers are responsible for maintaining this "short" reflex (vesico-medullary) [169]. Posterior rhizotomy can be performed at the level of the conus medullaris [126], intradurally in the lumbar region [158, 159], and in the radiculo-medullary junction (DREZotomy) [118, 145]. This type of destructive surgery must be utilized only in the case of complete sensitivo-motor function loss. Owing to the severe side effects of sacral alcoholization, this technique should not be recommended [115]. Sacral nerve thermocoagulation in the foramens, thus sectioning the thermosensitive C fibers but respecting the other sensitive and motor fibers, could be a less invasive alternative to treat bladder hyperreflexia, but it should be repetitive [102]. Recently, techniques involving denervation by intravesical instillation of some C fiber specific neurotoxins (capsaicin, resiniferatoxin) have been approved [52, 62].

Sacral Neuromodulation

In 1981, Tanagho and Schmidt in California attempted a procedure in paraplegic patients similar to Brindley's protocol, i.e. extradural stimulation of the sacral roots to induce a detrusor contraction, followed by posterior rhizotomy to eliminate bladder sphincter hyperactivity. Subsequently, they limited their work by using percutaneous puncture to stimulate the root of S3 without lesioning. In fact, they obtained an adverse effect, i.e. the inhibition of contraction [154]. This is how the term of sacral neuromodulation was coined: an electrical stimulation of the sacral roots was found to modify the pathologic behavior of a hyperactive bladder. The princeps articles reported an improvement in bladder hyperactivity in patients suffering from spinal cord lesions, but the method soon showed its efficacy in the treatment of idiopathic bladder hyperactivity, without any obvious neurological lesion. Thereafter, urologists widely used this technique instead of radical bladder surgery, thus allowing conservative treatment of some incontinent patients. Since 1997, the FDA has approved the utilization of this technique in urge incontinence, and since 1999 in cases of chronic retention. Recently the technique has proved efficient in some types of pelvic pain and fecal incontinence.

Methods and Techniques for Sacral Nerve Stimulation

Sacral Anatomy [107, 111]

The sacrum is normally composed of five modified vertebrae which are fused together. It is a triangular bone mass extending from the inferior vertebral column and containing the sacral and coccygeal nerves.

Posterior Sacrum

The skin in the sacral region is usually thick, and the subcutaneous tissue varies in thickness according to the habitus of the individual. It tends to be thinner than that found in the adjacent gluteal and lumbar regions. Situated deep below the superficial fascia are two layers of fibrous connective tissue (the thoracolumbar fascia and the tendon of the erector spinae muscle group of the deep back muscles). Deeper still are to be found a few muscle fibers of the erector spinae muscle group and many fibers of the inferior portion of the multifidus muscle layer. The left- and right-sided muscle groups are situated in a depression, whose medial and lateral walls are formed by the median sacral crest (spinous process) and the lateral sacral crest, respectively. The thickness of the tendon-muscle mass is approximately two centimeters in the region of the second sacral foramen, one centimeter near the third foramen, and 5 centimeters near the fourth foramen. Deep below the muscle mass, there is the periosteum covering the posterior surface of the bone. Components of the sacroiliac, sacrotuberous and sacrospinous ligaments are situated superiorly and laterally.

The dorsal surface of the sacrum is convex and irregular, with ridges and grooves. In the mid-line, there is the median sacral crest, consisting of three or four tubercles (rudimentary spinous process). At the inferior pole, the sacral hiatus is due to the failure of the fusion of the laminae of the fifth sacral vertebra. Laterally, the sacral crest just lateral to the sacral grooves comprises a row of four small tubercles representing the fusion of the articular processes. It forms the medial aspect of the posterior foramina. The lateral foramina correspond to the fusion of the transverse processes and are the site of insertion of the gluteus maximus muscle. The dorsal sacral foramina transmit the small dorsal rami of the sacral spinal nerves from the sacral canal to the deep back muscle compartment. The foramina are closed by a thin membrane. Small bony projections may be formed on the medial aspects of the foramina and are associated with muscle attachment points.

Sacral Foramen

The posterior foraminae become smaller from top to bottom. They appear to be equidistant in the vertical plane from the midline. Within the sacral foramina, there is abundant adipose tissue, particularly at the level of the third and fourth posterior foramina enclosing the nerve roots. Each nerve root follows an oblique course from top to bottom and from internally to externally. Each nerve root contains afferent and efferent parts of the somatic and parasympathic branches. A thin branch of each sacral root joins the skin surface and provides the buttock with topographic sensitivity. A foraminal arterial branch is always present lateral to the ventral nerve root, close to the inferolateral edge of each posterior sacral foramen. A veinous plexus is generally observed near the midline. The second foramen is nearly half filled by its nerve root along with its ganglion, which partially plugs the foramen. The third and fourth nerve roots occupy a relatively smaller proportion of their respective foramina.

The sacral foramina can be considered as a cylindrical space into which the sacral neuromodulation electrode is introduced. The upper sacrum tends to more curved than its lower part, especially in males. Needles can be inserted into the foramina as far as the anterior part of the foramen in order to reach the sacral root, at a wide range of angles in both the vertical and horizontal planes. For the third sacral cylinder, the angle is approximately 60 to 70 degrees to the posterior surface of the sacrum.

Anterior Sacrum

The ventral surface is concave in the vertical plan. There are four transverse lines on the surface which represent the original division of the bone into five separate vertebral bodies. Immediately anterior to the bone of the sacrum in the midline, there is the periosteum and continuation of the anterior longitudinal ligament. The piriformis muscle is attached to the sloping surfaces of the anterior foramina. Then, a layer contains portions of the pelvic nerve, components of the hypogastric plexuses, and blood vessels before the posterior pelvic viscera (rectum and lower sigmoid colon).

Localization of Sacral Foramen

Anatomical Landmarks

There are few methods to locate the posterior foramen. Usually, this depends on the positions of the posterior superior iliac spine, coccygeal tip,

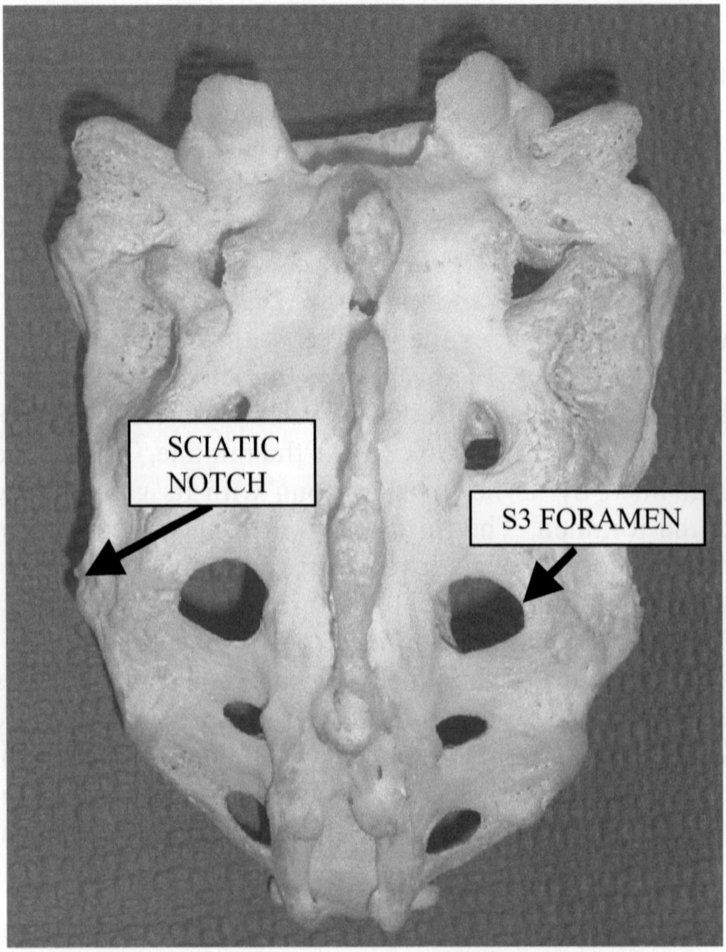

Fig. 2. Anatomical landmarks of S3 foramen

and midline. There are three sets of landmarks to confirm location of the S3 foramen, which is the elective foramen for neuromodulation.

– The S3 foramen is found by palpating the upper edge of the greater sciatic notch, 2 cm just lateral to the sacrum [thon, wju, 1991].

– Another technique [111] estimates the location of the S3 foramina approximately 2 cm from the midline, and 9 cm above the sacrococcygeal junction cephalad from the tip of the coccyx, identified by a knuckle-like protuberance at the apex of the sacrum (equidistant between apex of the sacrum and coccyx). However, this technique is sometimes difficult, especially in obese patients.

– The sacral crest, the region where the sacrum approaches the horizontal plane, corresponds to S4. From this point, the sacral spine curves downward to S3, which is located 2 cm above the S4 landmark.

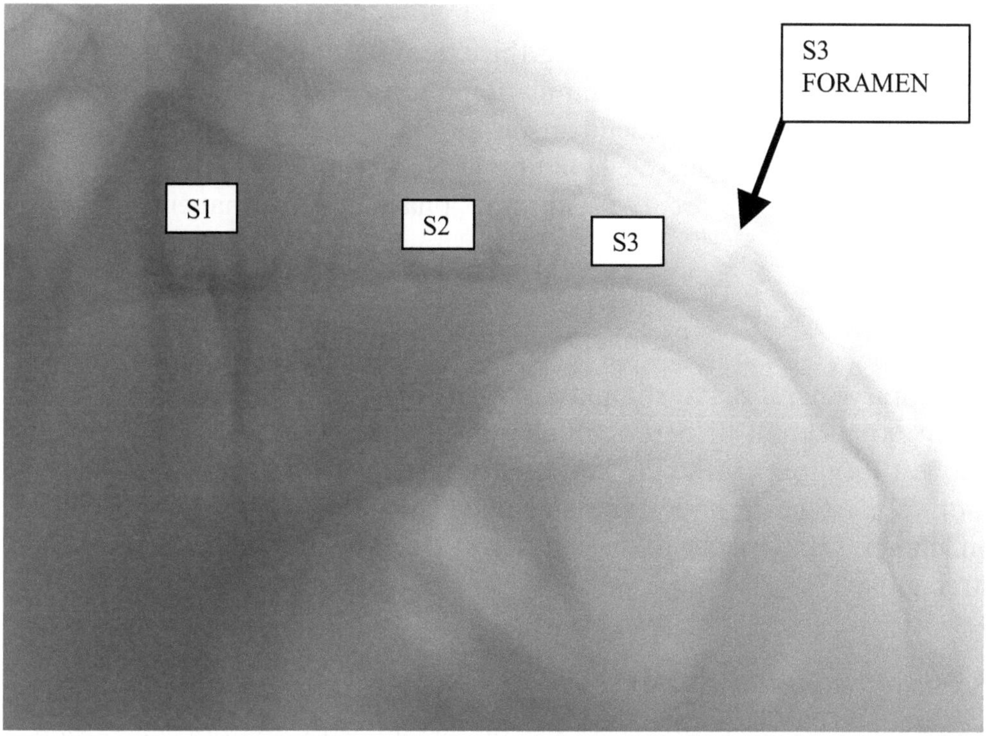

Fig. 3. Radiological landmarks of S3 foramen

Radiological Landmarks

The use of fluoroscopy is certainly the best approach for a quick and effective electrode placement, especially in overweight patients. Even though the sacral foramina may not be visualized fluoroscopically, interrelationships between fluoroscopically identifiable landmarks may be used to deduce their location. On anteroposterior radiographs of the pelvis, the interrupted line across the inferior aspect of the sacroiliac joint corresponds with the S3 foramen. On lateral views of the sacrum, the S3 foramen corresponds to the midpoint between the base of the sacrum and the tip of the coccyx.

Surgical Approach

Electrodes are generally placed in the third sacral foramen (S3). The electrode leads are subsequently attached to an implantable pulse generator. Patients undergoing sacral nerve stimulation must complete three phases of therapy.

Phase 1 or the "acute phase" involves a percutaneous test stimulation where a temporary electrode is placed in the S3 foramen and connected to an external pulse generator.

Phase 2 or the "sub-chronic phase" follows on from the acute phase. It involves monitoring and adjusting the external pulse generator to identify the optimal comfort level of stimulation and to evaluate therapy.

Phases 1 and 2 are dependent on an external generator and are considered as peripheral nerve evaluation (PNE).

During phase 3 or the "chronic phase", a permanent device is implanted.

Peripheral Nerve Evaluation

Temporary sacral nerve stimulation is the first step. It comprises the temporary application of neuromodulation as a diagnostic test to determine the best location for the implant and to control the integrity of the sacral root. This stage is important and may generate a number of technical difficulties [7]. Operators should always follow the manufacturer's instructions.

Preoperative Considerations:

The physician can verify electrode placement with anatomical or radiological landmarks (# chapter), and by analyzing motor or sensitive responses. There is considerable disagreement whether PNE can be performed outside of a hospital setting, with or without fluoroscopy, and with or without muscle responses [142]. However, neurologic patients may develop severe dysautonomia during electric sacral nerve stimulation [135, 136].

Material:

The testing hardware consists of a needle, test lead, test stimulator, interconnect cabling and a ground pad. A 20-gauge foramen needle with a beveled tip is used to gain access to the sacral nerve for placing the test stimulation lead. The stainless steel needle is depth-marked along its length (9 or 12 cm) and electrically isolated along its central part. The portion near the hub is exposed to allow connection to the test stimulator. By stimulating through the unisolated tip of the needle, the physician can determine the correct sacral nerve stimulation site for the test stimulation lead. For PNE, the test lead is a fluoro-polymer-coated, coiled, 11-stranded straight wire. An exposed metal tip at the distal end serves as an electrode. The lead contains its own stylet, which is removed once the correct position has been found. The external test stimulator is used both for patient screening and for intraoperative usage in determining lead placement thresholds. This provides output characteristics that are similar to those of the implantable neurostimulator and can be operated in either monopolar or bipolar modes. Amplitude control is accessible to the patient when it is being used as a screening device. The physician can set amplitude limits to ensure patient safety and the validity of the test. Finally, the ground pad (stuck to the patient's skin) provides the positive polarity in the electrical circuit during the test stimulation and the evaluation.

Surgery:

Anatomical orientation is much easier in a prone position. The patient must be comfortable, and a local anesthesia is administered to infiltrate the skin and subcutaneous tissues (particularly the periosteum).

Once the needle is in place, it is possible to determine nerve responses. Variations in neural anatomy may induce S3 motor responses whereas stimulation is given at S2 or S4. Consequently, the levels must be defined functionally as well as anatomically. In general, two levels of sacral nerve sites are tested to locate the optimum response. In most patients, stimulation of the S3 sacral nerve yields optimal results.

Typical S3 responses include the following: contraction of the levator ani muscles, causing a "bellows" contraction of the perineum (deepening and flattening of the buttock groove); plantar flexion of the big toe (and sometimes other toes) due to sciatic nerve stimulation and paresthesia in the rectum, perineum, scrotum or vagina.

Stimulation of S2 causes the following: rotation of the leg or hip, plantar flexion of the entire foot, contraction of the calf, contraction of the superficial pelvic floor, and a pulling sensation in the genital area and in the leg.

Stimulation of S4 causes activation of the posterior levator ani muscles, no motor response in the lower extremities, and pulling sensations in the rectum only.

The lead is then threaded through the needle cannula, and the foramen needle and lead stylet are removed. When the electrode is in place, the appropriate response is reconfirmed and the lead is coiled under the skin. Anterior/posterior and cross-table lateral X-rays of the sacral region provide documentation of the lead's position. This X-ray can serve as a reference for positioning during the implantation phase.

Duration of the "Sub-Chronic Phase":

As in the standard test stimulation procedure, the equipment is set up for 3 to 7 days of evaluation. At the end of the evaluation, the percutaneous lead extension is removed. If test stimulation is successful, a permanent device is implanted.

Implantation of Neurostimulator

The original technique for implanting a long-term sacral neuromodulator was described by Schmidt *et al.* [132]. The manufacturer's recommendations should be followed.

Implant Equipment:

Initially, neurostimulators were used for pain control (Itrel II®, MEDTRONIC). Then a specific neurostimulator (InterStim®, MEDTRONIC) was developed with the same technology. The chronic lead has four electrodes and a larger stimulation zone than the temporary test stimulation lead. Control equipment are used to adjust stimulation parameters (generally amplitude at 0.1 volts, rate at 10 to 14 pulses per second (Hz), and pulse width at 210 microseconds).

"Classic Surgery":

The patient is given a general anesthetic without long-acting muscle relaxants, which could block the motor responses needed to verify the effects of stimulation. During the implant procedure, the sacral region, buttocks and feet should also be visible to allow observation of motor responses. The patient is positioned facedown with slight hip flexion. Prior to incising, some surgeons provide a local anesthetic to prevent postoperative pain.

To implant and anchor the lead, the sacral foramen must first be located and fluoroscopy is recommended at the beginning of the procedure to help lead placement. Then a 5 cm midline or paramedial incision over the selected foramen is made. The skin and adipose tissue are dissected down to the glistening, white, fibrous lumbodorsal fascia. The fascia is then divided approximately 1.5 cm lateral from the midline, parallel to the spine over the appropriate foramen. The distal end of the lead is inserted into the foramen. Beginning at the distal tip of the lead, the four electrodes are numbered from zero to three; the numbered connector contacts correspond to these electrodes. Each of the four metal contacts should be tested and the lead repositioned to obtain the desired response. The optimal nerve responses are identified, the distal end of the lead is anchored to the lumbodorsal fascia and the lead is connected to the extension. The neurostimulator is placed in a subcutaneous pocket in the upper buttock. The neurostimulator may also be placed in a pocket in the abdomen (particularly in very thin patients). Postoperatively, the lead and neurostimulator placements are usually documented with X-rays.

Minimally Invasive Surgery:

Recently, Spinelli [148] reported a new technique of sacral nerve stimulation, characterized by a percutaneous approach to the sacral nerves resulting in minimal invasiveness of the procedure and the ability to have the patient awake during electrode placement. Under local anesthesia, it is possible to place a definitive quadripolar lead during the percutaneous test, which could reduce the risk of an inconclusive stimulation response. If test stimulation is successful, the pulse generator can be implanted under local anesthesia. Nevertheless, long-term evaluation of this technique is mandatory.

Unilateral or Bilateral Stimulation?

Since the original technique described by Tanagho and Schmidt, the unilateral sacral foramen electrode has been the gold standard for sacral neuromodulation [165]. Indeed, bilateral is not superior to unilateral sacral neuromodulation [127]. In rare cases, bilateral chronic sacral neuromodulation may prove necessary [84] particularly when unilateral percutaneous nerve evaluation fails [127].

It seems that bilateral stimulation does not increase the excitatory response but increases bladder inhibition at a lower stimulation intensity. Some authors have reported success with bilateral stimulation but the risk of complications is increased [84] and life of the device is significantly shorter.

Clinical Application of Sacral Neuromodulation

Indications

Neuromodulation of the sacral nerves is a therapeutic option for voiding dysfunction in patients who do not respond to the common non-invasive therapies and in whom disturbance in reflex coordination between the bladder, sphincter and pelvic floor is suspected. The rationale for using electrical stimulation techniques for the treatment of such voiding dysfunction is that this stimulation turns the neurological control mechanism back towards a more functional status. The main indications are urge incontinence, OAB syndrome, urinary retention and chronic pelvic pain.

OAB syndrome, which is also called urge syndrome or urgency-frequency syndrome, is characterized by urgency, with or without urge incontinence, usually with increased daytime frequency and nocturia, in the absence of local or metabolic factors explaining these symptoms [4]. In patients suffering from an OAB, sacral neuromodulation is an appealing therapeutic modality for symptoms refractory to conventional pharmacotherapy, and is relevant for both neurologic and non-neurologic causes.

In patients suffering from chronic urinary retention, sacral neuromodulation should be reserved for functional urinary retention without evidence of mechanical obstruction. Various indications such as Fowler's syndrome, spastic pelvic floor syndrome and bladder hypo/acontractility have been proposed.

Pelvic pain syndrome is the occurrence of persistent or recurrent episodic pelvic pain associated with symptoms suggestive of lower urinary tract, sexual, bowel or gynecological dysfunction, without any infection or other obvious pathology [4]. Chronic pelvic pain is defined as pain of a minimum of 6 months duration that is not related to any identifiable cause or etiology [125]. In patients suffering from chronic pelvic pain, sacral neuromodulation could be indicated when the symptoms are refractory to conventional pharmacotherapy after exclusion of obvious local pathological conditions.

Evaluation

Before the implantation of a neuromodulatory device, each patient should undergo a minimal investigation performed to confirm the pertinence of the indication, exclude any contraindications and to have baseline values.

The work-up for treatment by sacral neuromodulation must include careful assessment of past history with special emphasis on drugs influencing bladder function. A physical examination may be given to assess neurologic status, togther with a perineal examination with urodynamic investigation to assess bladder and sphincter function. To rule out any other lower urinary tract pathological conditions, urine culture can be performed to exclude urinary tract infection. Cytology and cystoscopy are helpful in ruling out carcinoma cystitis, and when indicated, imaging of the upper tract may be performed. It is recommended to perform MRI of the entire spinal cord to screen for neurologic diseases such as multiple sclerosis, a neoplasm, syringomyela, lipoma, etc.

For treatment of incontinence, the primary outcome measure should include a voiding diary recording the number of episodes of incontinence and micturition during a specified time. Recording the mean number of pads used per 24 hours may be helpful. For some authors, the quantification of the amount of urine lost during the pad test is also recommended [14]. Patient assessment of the severity of the symptoms can be recorded by a validated urinary incontinence outcome score, such as the Urogital Distress Inventory, the Bristol Female Lower Urinary Tract Symptoms or the Incontinence Impact Questionnaire [89, 90, 141, 168]. Many scores, such as the Short-form-36 (SF-36) and Beck Depression Inventory (BDI), may be used to evaluate the repercussion of the incontinence on quality of life [163]. Even if there is no systematic correlation between severity of clinical OAB symptoms and urodynamic parameters of detrusor overactivity, most authors recommended the use of cystometrograms to evaluate the responsiveness to sacral neuromodulation. The maximum cystometric capacity (volume at which the patient feels he/she can no longer delay micturition), the reflex volume (volume at which the first uninhibited contraction of the detrusor occurs), the sensation of bladder filling and the degree of bladder compliance (relationship between change in bladder volume and change in detrusor pressure) may reflect the extent of bladder activity.

In the treatment of urinary retention, the primary outcome measure should be the post-void residual urine [14]. Most authors recommended evaluating the mean voided volume and the mean number of intermittent catheterizations per 24 hours. Urodynamically, it seems reasonable to evaluate the urine flow with pressure flow studies (measuring the relationship between pressure in the bladder and urine during bladder emptying) or, at least by recording flow rate and voiding time, to assess bladder contractility during cystometry.

For painful bladder syndromes, the primary outcome measure should ideally be based on a validated pain assessment instrument. In addition, patients considered candidates for implantation may have benefited from psychological screening [14].

Pediatric Setting

There are some specific etiologies of urinary dysfunction in children, such as neurogenic bladder (myelomeningocele, occult spinal dysraphism, sacral agenesis, tethered cord syndrome, cord lipoma, cerebral palsy), non-neurogenic bladder (anatomic bladder exstrophy), functional bladder (enuresis, urinary infection), and non-neurogenic neurogenic bladder (Hinman syndrome). Various electric stimulation modalities are possible in children.

Intravesical electrical stimulation (IVES) is used to treat underactive detrusor, idiopathic or neurogenic in children. IVES is given by a catheter electrode in the bladder (cathode) with the anode attached to the suprapubic abdominal skin or the thoracic region. Continuous stimulation from 20 to 100 Hz is delivered (pulses = 0.2 to 10 ms, intensity from 0.1 to 64 mA). Sessions are 60 to 90 min daily over a period of 3 weeks to 3 months. For Gladh *et al.* [65], the frequency of urinary tract infections and incontinence decreased significantly and long-term normalization of voiding was obtained for 83% children with idiopathic problems and 40% with neurogenic problems. Another study [18] showed an absence of improvement in patients with myelomeningocele. This kind of stimulation is proposed as an alternative to clean intermittent catheterization.

Transcutaneous stimulation may be attempted to treat urinary urgency and incontinence in children. Current is delivered via skin electrodes for a short duration daily on a home treatment basis. Surface electrodes are placed at the level of the sacral root S3. Stimulation of 2 Hz is applied from 1 to 2 hours every day. For non-neurological bladder dysfunction, dryness improved in 73.3% of children, with a significant increase in mean voided volume [82]. 68% responded after 1 month of trial therapy with an increase in bladder capacity, decrease in urgency, and decrease in incontinence and/or better sensitivity. In the series of Hoebeke *et al.*, 56% of children were cured after 1 year. In a randomized controlled trial, no difference was found between the active spina bifida group and the placebo group [112].

Sacral neuromodulation was applied in children at the beginning of the nineties [153]. Contrary to adults, there has been no large randomized prospective study. Recently, a prospective randomized study [75] reported results of sacral neuromodulation in 42 patients (spina bifida in 38 cases) from 5 to 19 years old (mean age 11.9 years). Patients were compared (clinical examination, voiding diary, urodynamic evaluation) every three months for a minimal period of 12 months. Despite the improvement noted in implanted patients, the difference was not significant between the two groups. A multicenter study now seems to be necessary to increase the number of patients. The integrity of the nerves (even incomplete) is also a predictive factor of success.

The Overactive Bladder

Definition

Until the most recent definition of the International Continence Society (ICS), the term of bladder overactivity referred to urodynamic status. The bladder was considered as overactive when objectively shown to contract, spontaneously or on stimulation, during the filling phase of a cystometrogram while the patient is attempting to inhibit micturition [2]. Furthermore, in the first standardization report, the threshold of 15cmH20 was necessary to conclude that an uninhibited bladder was related to detrusor overactivity [11]. The definition currently endorsed by the ICS is that of a symptom syndrome suggestive of lower urinary tract dysfunction characterized by urgency, with or without urge incontinence, usually with increased daytime frequency and nocturia, in the absence of local or metabolic factors explaining these symptoms [2]. This overactive bladder syndrome can also be described as urge syndrome or urgency-frequency syndrome.

Diagnosis of Overactive Bladder

Clinical Parameters

The major role of cystometry in the diagnosis of overactive bladder (OAB) has recently been dissipated since overactive bladder is now taken to be a medical condition referring to the symptoms of frequency and urgency, with or without urge incontinence [2]. Thus the diagnosis of the OAB symptom complex is based upon the subjective perception of lower urinary tract dysfunction. However, as emphasized above, the OAB is a complex of symptoms that can be diagnosed as such only when there is no proven urinary tract infection or other obvious pathology.

Urgency is the complaint of a sudden compelling desire to pass urine that is difficult to defer [2]. Increased daytime frequency, or pollakiuria, is the complaint of voiding too often during the day [2]. Nocturia is having to get up one or more times at night to void [2]. Urge incontinence is characterized by a strong desire to void coupled with an involuntary loss of urine [2]. Since asking patients to record micturition and symptoms for a period of days provides invaluable information, measurement of lower urinary tract symptoms is based on a bladder diary. The bladder diary records the times of micturition and voided volumes, episodes of incontinence, pad usage and other information such as fluid intake, the degree of urgency and the degree of incontinence [2]. Validated questionnaires may also be useful for recording symptoms, their frequency, severity and bother, and the impact on quality of life [89, 168, 141].

Urodynamic Parameters [60]

Even if urodynamic testing is not yet required to define an OAB syndrome, it is often suggestive of urodynamically demonstrable detrusor overactivity. Detrusor overactivity is the urodynamic observation of involuntary detrusor contractions, whatever their amplitude, during the filling phase, which may be spontaneous or induced (Fig. 4).

The main interest of cystometry in patients suffering from OAB syndrome is to improve the diagnostic evaluation both by defining the underlying pathophysiology and by indicating treatment. However, there are controversial data concerning the correlation between OAB symptoms and urodynamic findings. Indeed, most authors do not recommend a filling and voiding study in the first-line treatment of OAB, but only in previously failed and complicated cases of OAB and prior invasive therapy.

Classification of Overactive Bladder

In some cases, detrusor overactivity may be further classified according to cause. Three main distinguishable types of detrusor overactivity are

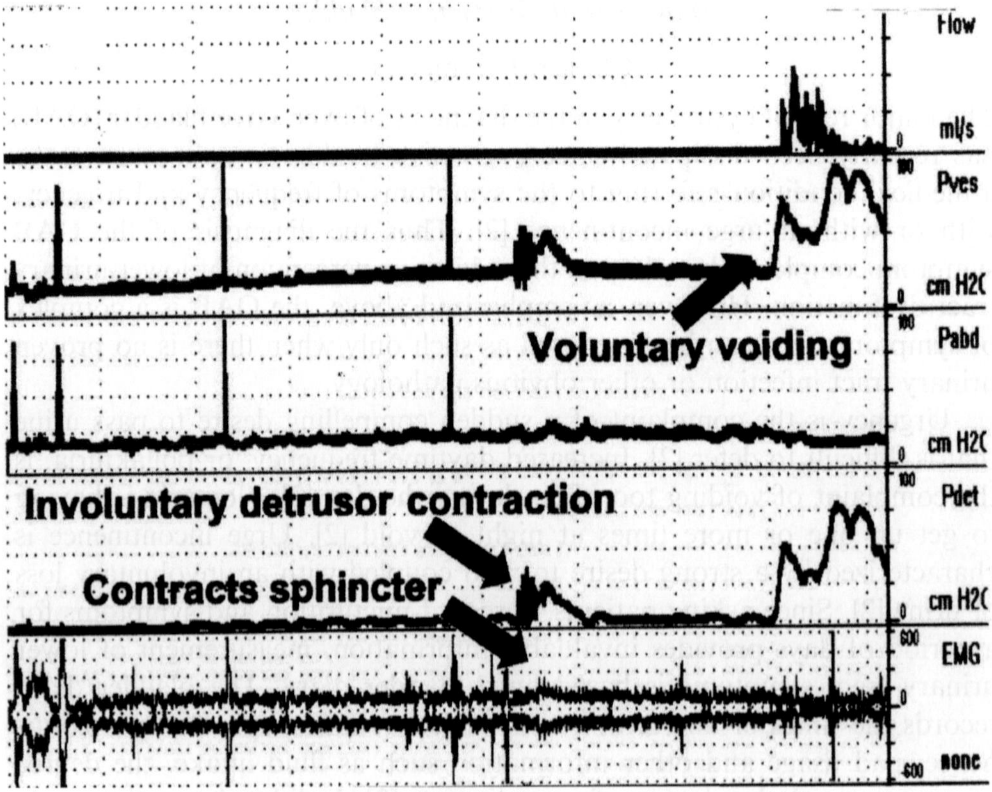

Fig. 4. Urodynamical characteristics of detrusor overactivity

usually considered [119]. The first is neurogenic detrusor overactivity (formerly detrusor hyperreflexia) in which there is a relevant neurological condition. Secondly, there is idiopathic detrusor overactivity (formerly detrusor instability) in which there is no defined cause. Finally, detrusor instability is related to bladder outlet obstruction or other conditions irrelevant for defining neurological causes.

Etiology

The different forms of overactivity may result from neurogenic or myogenic causes, or a combination of the two. These neurogenic and myogenic defects could be due to a wide variety of pathogenic conditions, which may be classified in 6 principal subtypes [119, 149].

1) Neurologic illness or injury, most commonly traumatic or medical spinal cord injury, demyelinating disease including multiple sclerosis, supraspinal disease such as stroke, Parkinson disease, tumor, degenerative disease or dementia. The neurogenic mechanism of OAB may be related to various changes in both peripheral and central neural pathways, such as

a decreased peripheral or central inhibitory control, an enhancement of excitatory transmission in the micturition reflex pathway, an increased primary afferent input from the lower urinary tract, the emergence of bladder reflexes that are resistant to central inhibition, or some combination of these factors [50].

2) Bladder outlet obstruction, which alters sensory and motor aspects of micturitional reflexes and points to an abnormal activity pattern of the detrusor cells characterized by a spontaneous mechanical activity, a hypersensitivity to acetylcholine with depressed responses to intrinsic nerve stimulation and increased sensitivity to direct electrical stimulation [Brading 1997].

3) A hypersensitivity-induced overactivity due to the emergence of a aberrant voiding reflex mediated by unmyelinated capsaicin-sensitive C-afferents, which may be caused by neurogenic disease or factors not yet fully understood [109, 50].

4) Urethral weakness due to smooth sphincter deficiency and pelvic relaxation in middle-aged and elderly women [56].

5) Detrusor hyperactivity and impaired contractility in elderly patients [123].

6) So-called idiopathic bladder overactivity, which has no defined cause but may be due to an unknown combination of some of the above-mentioned factors [57, 124].

Results of Sacral Neuromodulation

To date, effectiveness has been assessed by objective and subjective measures. In evaluating the effectiveness of sacral nerve stimulation, results are frequently discussed according to urge incontinence chronic urinary retention, and chronic pelvic pain.

However, subjective measures are difficult to implement because the definition tends to vary from what actual improvement occurs in patients.

Urge Incontinence

In a multicenter randomized controlled trial, Schmidt *et al.* [131] observed at six months after sacral nerve stimulator implantation in 34 patients followed up. Approximately three-quarters of the patients showed a clinically significant response with 50% reduction in the frequency of incontinent episodes. There was an improvement from the baseline average of 9.7 mean leaks per day to 2.6 per day after 6 months of treatment. Efficacy was defined as a clinical effect greater than 50% reduction in symptomes. At 18 months, 84% were clinically successful in eliminating heavy leaking episodes, 76% were successful in eliminating or reducing (by 50%) the number of pads and 47% were completely dry. In contrast, patients in the control group experienced either no improvement or worsening symptoms. Similar results were reported in a randomized controlled trial [164] with a significant decline in leakage episodes (improvement by 88%) and pad use (improvement by 90%) compared with baseline. Finally, 56% of patients were completely dry. Another prospective randomized multicentric study [79] showed a significant reduction in the number of daily voids from 16.9 to 9.3 at 6 months follow-up, and 56% of patients demonstrated a reduction of 50% or more in the number of voids.

Some case series studies report the change in the average number of incontinence episodes post-implantation compared to baseline and at 30 months the daily frequency of leaking episodes had significantly reduced from 10.9 to 4.2 [22]. Results from the Italian national register [147] show a decrease in mean incontinence episodes from 5.4 to 1.1 at 12 months follow up. Similar results were obtained with a reduction in leaking episodes from 6.4 to 2.0 per day at 24 months in a case series of 44 patients [1].

The reduction in pad consumption (mean reduction from 4.8 to 2.2 pads) was found to be statistically significant in some studies [19, 23, 165]. In a multicenter investigation [92], the number of pads used daily dropped significantly from 7.1 to 3.8 per day ($p < 0.0001$): 33% of patients were dry and 28% experienced a greater than 50% improvement in pad use. At least 61% of patients have excellent or good results.

Generally speaking, on urodynamic assessment, bladder capacity is found to increase statistically from baseline measures [84, 165]. Voided volume has also been found to increase [71, 84].

In the case of urge incontinence, the objective measures reported in the literature are not usually adapted to assessing the number of episodes of micturition, whereas pollakiuria is a major symptom impacting on quality of life.

Chronic Urinary Retention

In a multicenter randomized controlled trial, Grunewald *et al.* [72] observed that 69% of patients with chronic idiopathic urinary retention achieved complete micturition without catheterization (versus 9% without electric stimulation). The number of catheterizations decreased > 50% in 83% patients (versus 9% without electric stimulation). Another prospective, randomized multicenter study [94] investigated the efficacy of sacral neuromodulation in patients with chronic non obstructive urinary retention. Compared to the control group, implanted patients had statistically and clinically significant reductions in catheter volume per catheterization. 69% of patients eliminated catheterization at 6 months and 14% of patients had a 50% or greater reduction in catheter volume per catheterization. 83% of patients achieved sucessful results at 6 and 18 months compared to 9% of the control group.

In a case study, Elabbady *et al.* [56] reported a significant percentage increase in voided volume from 15% to 71% among seven patients. Grunewald *et al.* [71] and Hohenfellner *et al.* [84] reported significant increases in voided volume (respectively 490 and 334 ml). Decreases in mean residual volumes of 455 and 334 ml respectively were noted by the same authors. For Spinelli *et al.* [147] stimulation can decrease average residual volume from 227 to 108 ml; in their series, 50% of patients stopped catheterization and 13% needed it only once daily. In other studies, bladder capacity was not found to change significantly from baseline to post-implantation [56, 84, 139]. Reported success rates vary from 52% [156] to 82% [71] but there is no consensus regarding the definition of success.

Pain

Although frequent, chronic pelvic pain syndrome probably receives little attention from clinicians. It is a diagnostic and therapeutic challenge and is often related to psychological and psychosomatic disorders. Theoretically, neurogenic inflammation is responsible for neurogenic pain, as in a complex regional pain syndrome [13]. Trauma may also induce pain (fracture, nerve damage). Compared to dorsal column or peripheral nerve stimulations, some authors propose sacral nerve stimulation for the treatment of chronic pelvic pain syndrome. To date, few results have been reported for this technique but it is feasible. Aboseif *et al.* [1] analyzed a group of 41 patients with chronic pelvic pain associated with other voiding symptoms: stimulation decreased the severity of pain from 5.8 to 3.7 on their scale.

Long-Term Effectiveness

The results seem stable over time. Some authors [22] report a reduction in the benefit at 5 years. However, long-term studies are lacking and until now, there have not been any studies with control groups. In a multicentrer, prospective study, Siegel *el al.* [143] demonstrate that after three years, 56% of 41 urge incontinent patients showed greater than 50% reduction in leaking episodes per day with 32% of patients being completely dry. After two years, 34% of urgency-frequency patients showed greater than 50% reduction in voids per day, including 21% of patients who attained a normal range of voiding frequency. After 1.5 years, 70% of 42 patients with urinary retention showed greater than 50% reduction in catheter volume per catheterization, including 58% of patients who eliminated use of catheterization.

Impact on Quality of Life

When the SF-36 and BDI scales are administered, no significant improvement in quality of life is demonstrated but these scales are not powerful in the setting of urinary handicap. However, patients seemed satisfied by their device in one study, probably due to the decrease in the number of incontinence episodes [36]. Approximately two-thirds were satisfied by their implant. Elsewhere, quality of life results were superior for implanted patients on some components [131].

Results for Neurogenic Bladder

The first publications showed an improvement in bladder hyperactivity among spinal cord lesion patients [63, 154]. Bosch and Groen [20] showed that treatment of refractory urge incontinence by chronic S3 nerve stimulation was feasible in selected multiple sclerosis patients. The fact that no irreversible changes to the bladder or nerves occur is an advantage of this treatment option over destructive alternatives. However, the unpredictable evolution of the disease and particularly cognitive alterations are contreindicated in case of rapid evolution. In a case series, Chartier-Kastler *et al.* [41] reported 9 women with spinal diseases (including vascular myelitis, multiple sclerosis and traumatic spinal cord injury) undergoing neuromodulation. All patients reported an improvement of 75% in their visual analog scale at last follow-up (mean follow-up 43 months). In another case series, Hohenfellner *et al.* [83] evaluated patients with neurogenic bladder (complete or incomplete spinal cord lesions, inflammatory neuronal reaction, borreliosis, lumbar herniated disk). Patients not likely to benefit from the procedure were those with complete or almost complete spinal lesions, but incomplete spinal lesions seemed to be a potential indication [23, 83].

Predictive Factors for Sacral Neuromodulation

Percutaneous nerve evaluation gives an accurate identification of suitable candidates [8]. Generally, authors consider an improvement of more than 50% in voiding parameters for definitive implantation and PNE is positive in 40% of patients with neurogenic and idiopathic etiologies [128]. On the other hand, a negative PNE did not reliably predict the therapeutic efficacy of the implanted system in a recent study and 25% of patients needed more than one PNE [164]. In fact, the higher the patient's age, the greater the number of test failures; moreover, longer lasting complaints result in a higher risk of a negative test [128]. Patients with neurogenic bladder dysfunction had a four-fold higher probability of negative test results compared with patients having no obvious neurologic problems. Patients with urinary retention have a higher probability of having a negative test result compared with patients with urge incontinence. Urodynamic studies during test stimulation do not have any predictive influence.

Complications of Sacral Neuromodulation

Complications of Peripheral Nerve Evaluation (PNE)

Siegel *et al.* [143] noted 18.2% of adverse effects in 914 test stimulation procedures. The most common adverse events are lead migration from 11.8% [143] to 18.6% [143], technical problems (2.6%) and pain (2.1%). One surgical intervention (0.1%) was required to remove a test lead electrode that became dislodged during lead removal. Local infection and subcutaneous hematoma are rare [143].

Complications of Sacral Nerve Neuromodulation

For chronic sacral neuromodulation, complication rates range from 22 to 43% [19, 53, 138, 156] and re-operation rates from 6 to 50% [56, 99, 138, 156]. However, many studies do not discuss the complications arising from stimulation implants [21, 71, 156].

A prospective study was performed by the manufacturer Medtronic (Minneapolis, MN, USA) including 14 North American and 9 European centers [95]. Of the 633 patients enrolled in this study, 250 had been implanted with the sacral nerve stimulator system by the end of the reporting period, representing 6506 months of device experience. Of the 250 implanted patients, 157 (62.8%) experienced a total of 368 adverse events associated with the device for use of stimulation therapy. Of the reported 368 events, 56 (15.2%) required no intervention, 151 (41%) required non-surgical intervention and 161 (43.8%) required surgical intervention. Overall, 89.4% (329) events were fully resolved. In the 250 implanted patients, post-implant adverse events associated with the devices or use of stimulation were pain at the internal pulse generator site (14.2%), new pain (10.8%), suspected lead migration (9.1%), infection (7%), pain at lead site (5.5%), transient electric shock (5.6%), suspected device problem (2.2%), adverse change in bowel function (3%), technical problems (3.9%), persistent skin irritation (0.8%), change in menstrual cycle (0.9%), suspected nerve injury (0.4%), device rejection (0.4%) and others (14.1%).

Pain

Pain is a frequent adverse event occurring in 4% [22] to 29% [164] of patients. Little is known about the severity and treatment of pain related to device implantation. Frequently, no distinction is made between post-operative pain, pain associated with the device, referred pain, pain related to stimulation, neuropathic pain and psychological pain. In one study, placement in the upper buttock reduced the rate of revision surgery but not pain [95]. The symptoms of pain should always be thoroughly analyzed in order to treat it.

Infection

Any infection should always be detected and treated early. Removing the device either temporarily or definitively may prove necessary. Despite being a common complication of all implantable devices, few studies refer to this adverse event. No information is available in the literature concerning the etiology, severity or timing of infection. Mention has been made of skin irritation requiring device explantation [131]. Compared to the Brindley technique (sacral anterior root stimulation with posterior rhizotomy) which has a maximum 2.4 infection % rate [158], the mean of 6.1% related to sacral neuromodulation [143] appears to be too high. Progress in prevention and device modification is required.

Problem of Nerve Injury

To date, there have been no reports of permanent injury or nerve damage [131]. Sometimes nerve injury is suspected [143] and there is a potential risk.

The configuration of the electrode itself (incorrect fit to the nerve), surgical trauma, pressure caused by post-surgical edema, excessive scar formation and tension on the electrode cables are all potential contributors to neural damage [120]. The peripheral nerve may be affected adversely by chronic constriction and compression [103]. However, these risks are less important in the case of epineural electrodes than in intraneural ones [105]. In animal studies, excessive or prolonged stimulation may cause early axonal degeneration [114]. The risk of injury is also affected by the duration of continuous stimulation [5]. It is well known that needle insertion into the sacral foramen can result in damage to nerve root and vessels [107]. Because these structures are more likely to be found on the medial aspect of the foramen, injury can be minimized by using a more lateral foramen entry. Increasing the angle of needle entry in the vertical plan can increase the risk of injury to the vessels (venous plexus), and therefore that of hematoma and fibrosis [77]. The S2 foramen is nearly half filled by its nerve root and ganglion, which increases the likelihood of penetration during needle placement. On the other hand, the S3 and S4 foramina are filled mostly with fat and their nerve occupies a relatively smaller portion of the foramen [107]. It has been observed that the therapeutic efficacy of the implant sometimes becomes limited over time, and the potential formation of fibrosis between the electrode and target nerve has been suggested [83].

Technical Problems and Device-Related Complications

Bosch *et al.* [22] described difficulties in maintaining proper electrode positioning, breakage of the lead, fracture of the extension cable, electrode dislocation or malpositioning, early failure of pulse generator, contact lead point dysfunction and seroma around the generator site. However, device-related complications appear to be the most frequent. The following complications have been reported in patients undergoing sacral nerve stimulation for urinary urge incontinence:

– Device complications such as pain at the implant site [22], device rejection [143], early pulse generator failure [147], stimulation-dependent pain in leg or buttock [22] and current-related problems.

– Lead complications such as disturbed toe flexion, lead migration [22, 147], adverse changes in elimination function e.g. bowel (diarrhea) and urinary system [143], suspected nerve injury [143], lead site pain [143], transient electric shock [143] and fracture of the extension cable [22] or lead [147].

– Wound complications such as partial wound dehiscence of the sacral incision [22], hematoma [147], infection [147] or skin irritation [143].

Surgical Revision

More than one third of patients go to surgical revision [143], mostly for repositioning of the lead or the extension. Tempory removal with subsequent reimplantation is normally the result of infection or chronic pelvic pain. Repositioning of the internal pulse generator is performed to relieve pain at the site, or because the battery is dead. Permanent removal is to the result of infection, chronic intractable pain, or because the device has not proved satisfactory. Surgical revision does not appear to affect the overall degree of patient satisfaction [143], and it seems to decline with time [131].

Conclusion

Although relatively frequent, complications have until now received insufficient attention. Many patients require re-intervention to reposition or remove the device due to displacement, breakage or migration. However, the procedure is safe in experienced hands.

Therapeutic Alternatives and Developping Treatments in Refractory Urge Incontinence and Idiopathic Bladder Overactivity

Medical Therapeutics

Conservative therapies such as pelvic floor exercises, bladder retraining, electrical stimulation of the pelvic floor and pharmacotherapy involving anticholinergics, antispasmodics and tricyclic antidepressants are primary discussed. The use of pelvic floor muscle training with or without biofeedback for overactive bladder is suggested to inhibit detrusor muscle contraction by voluntary contraction of the pelvic floor at the same time, and to prevent sudden falls in urethral pressure by change in pelvic floor muscle morphology, position and neuromuscular function [17]. Some promising results have been reported, and these treatments are widely used, but there is still a need for high quality randomized trials on the effect of pelvic floor exercises on the inhibition of detrusor contraction. Detrusor overactivity current pharmacological treatment involve use of muscarinic receptor antagonists, but their therapeutic activity is limited by side effects resulting in the non continuance of treatment in a significant number of patients. More selective muscarinic antagonists (M3 receptor subtype) as dariferacin and vancamine or percutaneous treatment (oxybutinin patches) may reduce side effects of these treatments. The development of new drugs can proceed by targeting alternative pathways affecting detrusor overactivity [100]. Intravesical agents appear to be attractive alternatives to oral medication [62]. *Vanilloids drugs* as Capsaicin and resiniferatoxin have showed some promising results. Capsaicin-sensitive bladder afferents do contribute to hyperactivity of the bladder in neurogenic and non-neurogenic detrusor overactivity [43, 152]. Capsaicin is a specific neurotoxin that desensitizes C-fibres afferent neurons which may be responsible for the signals that trigger detrusor overactivity. Resiniferatoxin is a less pugent agent which desensitizes as capsaicin afferent C-fibres but fail to depolarize nerves and may show less local side effects such as pain associated with capsaicin. The role of alcoholic vehicle in acute pain and irritation associated with capsaicin has to be clarified [166], as the duration of effects which have been reported to be shorter after resiniferatoxin intravesical application [45, 104]. Efficacy of vanilloids has been shown not only in patients with detrusor hyperreflexia due to spinal cord disease [52], but also patients with bladder hypersensitivity and idiopathic detrusor instability [45, 51, 97, 144]. Botulinum toxin interacts with components of the presynaptic membrane of cholinergic nerves and inhibits the vesicular release of acetylcholine producing long lasting neuronal blockade. The effects of toxin lasts for up to 9 months and further injections may be required.

Surgical Alternatives

The other treatments available are more invasive and often irreversible surgical procedures. Surgical therapy should only be considered when all conservative methods have failed. Endoscopic approaches have been used in urgency incontinence [162]. *Overdistension of the bladder* is thought to reduce bladder distension by causing degeneration of unmyelinated C afferent small sensory fibers. This technique requires anaesthesia and have some complications including hematuria, urinary retention and bladder perforation in 5% to 10% [146]. Although effective in short term management, this procedure is usually temporary in symptomatic control. *Bladder myectomy* (autoaugmentation) has beeen proposed as an alternative to enterocystoplasty. Detrusor myectomy involves incising and removing the bladder muscle to allow bladder mucosa to form a pseudodiverticulum. Detrusor myectomy for treatment of refractory urge incontinence due to detrusor overactivity in both sexes has been reported to be successful in a small number of patients [106, 151]. These data need to be confirmed in larger group and on a long term experience. *Enterocystoplasty* results are similar, good results vary from 58% to 88%, with an average of 77% [80]. The goal of enterocystoplasty is to create a low-pressure, large capacity reservoir with low-filling pressure, which protects the upper tract from pressure related reflux and infection, as well as providing urinary continence. The most frequently used bowel segment is ileum followed by sigmoid colon. It is generally agreed that it is best to de-tubularize the intestine into a sphere to reduce disrupt peristaltic contractions and increase capacity of the bladder [29, 77, 106, 113]. A minimum of 10% of patients requires intermittent catheterization for bladder emptying. This procedure has a significant complication rate and should be evaluated carefully when applying. Urinary tract infection and mucus production which can cause either catheter or voiding obstruction are significant long term problems. Electrolyte abnormalities such as hyperchloremic acidosis are reported. Perforation has been reported in 2–6% of patients in the long term and surgical revision rates range from 15 to 36% [61, 80]. *Urinary diversion* is rarely needed but may be useful in patients with intractable detrusor instability with a very small reservoir. It may be use to treat pelvic pain that may be associated. However, it represents a last step intervention in the surgical management of incontinence resultant from detrusor overactivity.

The best surgical procedure for detrusor overactivity is still to be elucidated. Furthermore, the future of surgical treatment will depend on new developments in non-surgical therapy.

Perspectives in Electric Stimulation

Nonimplanted stimulators have been used for bladder inhibition. There is good evidence that the use of vaginal electrical stimulators can reduce the occurrence of symptoms of overactive bladder in about half of the patients treated [34]. Unfortunately, precise treatment protocols and patients selection remain unresolved as many different protocols and stimulators were used. *Pudendal nerve stimulation* positively impacts cystometric parameters in a significant number of patients with refractory detrusor instability. It may provide efficacious treatment for patients suffering from the symptoms of urgency-frequency and urge incontinence associated with overactive bladder syndrome [160]. Implantable microstimulator weighting less than 1 gram is now available. Electronics, rechargeable battery and stimulating electrode are integrated into a single implantable device. Implantation at Alcock's canal is performed using minimally-invasive procedure with local anaesthesia and an approach similar to the well-established transperineal pudendal block [35]. Patient sit on a custom cushion for approximately 20 minutes a day to recharge the microstimulator through radio waves. Treating clinician use bidirectional telemetry to program the neurostimulator and to retrieve information about the patient's use of the device. Patient use the remote control to stimulate or to turn the microstimulator on and off. Long term efficacy of this device need to be confirmed in larger group clinical trials.

Posterior tibial nerve electrical stimulation to inhibit detrusor activity have been used with success by Mc Guire *et al.* [116]. More recently, intermittent *posterior tibial nerve stimulation* was re-introduced as a treatment of patients presenting bladder overactivity [9]. A needle is inserted to allow 12 weekly outpatient treatment sessions, each lasting for 30 minutes. In this study a positive response was seen at 12 weeks follow-up in 60% of patients with a significant decrease in leakage episodes, number of pads used, voiding frequency and nocturia. Tolerance was excellent. Posterior tibial nerve stimulation may be technically less demanding and probably more cost-effective for management of lower urinary tract dysfunction. Percutaneous tibial nerve stimulation showed short term clinical efficacy in leakages, quality of life and significant cystometric data improvements in neurologic and non-neurologic patients suffering with detrusor overactivity [6, 157]. Further studies are needed to show if posterior tibial nerve stimulation may be use as a minimally invasive test to select candidates for definitive S3 neuromodulation or may constitute for some patients a long term therapeutic issue.

Cost of Sacral Neuromodulation

General Issues in Urge Incontinence Costs

Urinary incontinence and urinary retention are a costly illness that affect personal resources, medical treatment and quality of life. The overall prevalence of overactive bladder is similar between men (16.0%) and women (16.9%), but sex specific prevalence differed substantially by severity of symptomes [150]. Anatomic differences increase the frequency of urge incontinence linked to bladder overactivity among women compared with men. In women, prevalence of urge incontinence increase with age from 2.0% to 8.9%, and in men from 0.3% to 19%. Moreover, symptome occurrence is later in age in men. United States most recent estimates of the annual direct costs of incontinence in all ages are approximately $16 billion: $11 billion in community and $5 billion in nursing home (1994 dollars) [59]. These costs estimate increased by 250% over 10 years [86], greater than can be accounted by medical inflation. Data from the National Overactive Bladder Evaluation (NOBLE) survey in the United States had shown that the estimated total economic cost of overactive bladder was $12.02 billion in 2000, with $9.17 and $2.85 billion incurred in the community and institutions, respectively. NOBLE program in the US surveyed approximately 5,000 adults. The average cost per community-dwelling person with overactive bladder was $267 per year [87]. A prospective population study in US suggested that total urinary incontinence expense is as high as $16.3 billion, with 78% of cost issued from women and 22% from men. Overall surgical cost of incontinence (all techniques) represent 4 years of pads and care. Similar findings were shown in several non USA studies issued from European and Autralian studies [86]. Apart from the cost to the health system there is also the burden to the patient and his/her family [76]. Urinary incontinence may also affect individual's lost work time or interfere with job performance and productivity. Embarrassment, shame, need to change clothing altered social interactions, loss of self esteem, depression are frequent. Given the large number of patients (mainly younger women) and the relatively high prevalence of incontinence, future efforts to objectively quantify such impact is needed.

Expected Cost per Patient of Sacral Nerve Stimulator Implant Treatment

The initial expense of the therapy, especially measured over the 7–10 year life of any neurostimulator should be considered in relation to the potential savings to the healthcare system and to the effect on the patient's quality of life. Abrams *et al.* [3] examine the benefit-risk profile of neuromodulation in treating refractory urge incontinence and other voiding disorders. They feel that both efficacy and safety have improved beyond the earlier studies as the development of new percutaneous technology and the minimally invasive placement of leads have improved the technique.

No studies of the cost-effectiveness of sacral nerve stimulator implant treatment were found in the literature, but some economic data looking at direct costs up to 12th month post implantation are available. Cost of equipment (percutaneous nerve evaluation and sacral nerve stimulator) is approximately of 9000€ per patient. It may be higher if bilateral stimulation is chosen. Surgery and re-surgery costs have been evaluated nearly to 8000€ in Australian review [117]. Oral anticholinergic treatment cost is much lower (200€ per year) but neuromodulation is usually used in refractory cases to these therapies. Reduction of pads and laundry for urge incontinence and catheterization for urinary retention is the main factor of cost reduction [131]. Unfortunately, studies have showed significant consummation variability of these equipments and some may have overestimated this data [86]. Over six months, sacral nerve stimulator implant treatment has an estimated cost per patient initiated to treatment with percutaneous sacral nerve evaluation of approximately 18 700€. These costs include medical treatment and re-operations arising from complications, for both indications (incontinence, urinary retention) [117].

Capellano *et al.* [37] reported economical and social impact of sacral nerve modulation therapy in 62 patients with lower urinary tract dysfunction. These patients were enrolled in the economic session of the Italian Sacral Nerve Modulation Registry from February 2000 to September 2002. 41 were incontinent patients (61% female) and 21 were affected by chronic urinary retention (71% female). A quarterly based analysis comparing the baseline data to the last follow-up (12th month) was performed. Visits to the general practitioner decreased from 1.1 to 0.05 ($p < 0.01$), visits to the urologist did not change significantly from baseline (1.5 to 1.2). Diagnostic tests decreased from 2 to 0.8 ($p < 0.01$). In the use of pads a major change was observed from a daily use of 2.1 (three months expenses per patient of €120.96) to 0.5 (three months expenses per patient of €28.8) ($p = 0.08$); and for urinary retention the use of catheters decreased from 1.1 baseline (three months expenses per patient of €178.2) to 0.1 at 12 months (three months expenses per patient of €16.2) ($p = 0.09$). Costs of drug consumption decreased significantly ($p < 0.05$) from €47.24 to €10.53. This study suggests that sacral nerve modulation therapy is efficient to improve the economic management of patients with lower urinary tract dysfunction. Furthermore, the reduction in the use of pads and catheters has also a positive effect on quality of life and patients' social interaction [36].

It has been suggested that the implant may last for up to five years. Over the long term the total cost saving in incontinence products is likely to increase if the device continues to be effective, but further revision surgery may also be required. Optimal treatment can be define at the end of a fine tuning process which could take several months, during which specialist visits decrease progressively, allowing some cost reduction. A longer follow-up is therefore requested in order to evaluate long term costs after implant therapy is stabilized.

Conclusion

The technique of sacral neuromodulation is available in neuro-urologic situations in which there is an imbalance between the neurological systems which regulate retention and micturition. It is generally used to treat vesical overactivity with pollakiuria, which disturbs patients' quality of life. The need to urinate can even trigger urinary or fecal leakage. When the pharmacological arsenal has been tried to no avail, sacral neuromodulation remains an alternative to urologic interventions of the bladder. Neurogenic and idiopathic overactive bladder must be differentiated, and it is essential to considerer psychological factors affecting the patient. Urologic etiologies are a contra-indication for sacral neuromodulation. The major indication is bladder overactivity, followed by idiopathic chronic retention and chronic pelvic pains. Sacral neuromodulation is a mini-invasive technique but requires methodological rigor and a preliminary percutaneous test. When selection is performed, more than three-quarters of the patients showed a clinically significant response, but the results vary according to the author's mode of evaluation. From the economic point of view, the initial investment in the device is amortized in the mid-term by savings related to lower urinary tract dysfunction. Finally, this technique requires an attentive follow-up and adjustments to the electric parameters so as to optimize the equilibrium between the neurological systems.

Acknowledgements

The authors thank the Expert Committee for Sacral neuromodulation of the Medical School Hospital of Bordeaux, and particularly Doctor Elisabeth Marit-Ducamps, Uro-Neurologist, for her precious help in selecting patients, and Doctor Luc Soyeur from the Department of Urology.

References

1. Aboseif S, Tamaddon K, Chalfin S, Freedman S, Kaptein J (2002) Sacral neuromodulation as an effective treatment for refractory pelvic floor dysfunction. Urol 60: 52–56
2. Abrams P, Blaivas JG (1990) Fourth report on the standardization of terminology of lower tract function recommended by the International Continence Society. Int Urogynecol J 1: 45–58

3. Abrams P, Blaivas JG, Fowler CJ, Fourcroys JL, Macdiarmid SA, Siegel SW, Van Kerrebroeck P (2003) The role of neuromodulation in the management of urinary urge incontinence. BJU Int 91: 355–359
4. Abrams P, Cardozo L, Fall M, Griffiths D, Rosier P, Ulmsten U, van Kerrebroeck P, Victor A, Wein A (2002) The standardisation of terminology of lower urinary tract function: report from standardisation sub-committee of the international continence society. Neurourol Urodyn 21: 167–178
5. Agnew WF, McCreery DB (1990) Consideration for safety with chronically implanted nerve electrodes. Epilepsia 31: 27–32
6. Amarenco G, Sheik-Ismael S, Even-Schneider A, Raibaut P, Demaille-Wlodyka S, Parratte B, Kerdraon J (2003) Urodynamic effect of acute transcutaneous posterior tibial nerve stimulation in overactive bladder. J Urol 169: 2210–2215
7. Amundsen CL, Webster GD (2002) Sacral neuromodulation in an older, urge-incontinent population. Am J Obstet Gynecol 187: 1462–1465
8. Arlandis Guzman A, Alapont Alacreu JM, Bonillo Garcia MA, Ruiz Cerda JL, Martinez Agullo E, Jimenez Cruz F (2003) Peripheral nerve evaluation: indications, technique and results. Actas Urol Esp 27: 491–500
9. Balken MR, Vandoninck V, Gisolf KHW, Vergunst H, Kiemeney LALM, Debruyne FMJ, Bemelmans BLH (2001) Posterior tibial nerve stimulation as a neuromodulative treatment of lower urinary tract dysfunction. J Urol 166: 914–918
10. Barrington FJF (1925) The effects of lesions of the hind- and mid-brain on micturition in the cat. Quart J Exp Physiol Cogn Med 15: 81–102
11. Bates P, Bradley WE (1981) Fourth report on the standardization of terminology of lower tract function. Br J Urol 53: 333–335
12. Bauchet L, Segnarbieux F, Martinazzo G, Frerebeau P, Ohanna F (2001) Traitement neurochirurgical de la vessie hyperactive chez le blessé médullaire. Neurochirurgie 47: 13–24
13. Beard RW, Highman JH, Pearce S, Reginald PW (1984) Diagnosis of pelvic varicosities in women with chronic pelvic pain. Lancet 27: 946–949
14. Blaivas JG (2001) Chronic sacral neuromodulation. J Urol 166: 546
15. Blok BF, Holstege G (1994) Direct projections from the periacqueductal gray to the pontine micturition center (M-region). An anterograde and retrograde tracing study in the cat. Neurosci Lett 166: 93–96
16. Blok BFM (2002) Central pathways controlling micturition and urinary incontinence. Urol 59: 13–17
17. BØ K, Berghmans LCM (2000) Nonpharmacologic treatments for overactive bladder-pelvic floor exercises. Urol 55 S5A: 7–11
18. Boone TB, Roehrborn CG, Hurt G (1992) Transurethral intravesical electrotherapy for neurogenic bladder dysfunction in children with myelodysplasia: a prospective, randomised clinical trial. J Urol 148: 550–554
19. Bosch JLH, Groen J (1995) Sacral (S3) segmental nerve stimulation as a treatment for urge incontinence in patients with detrusor instability: Results of chronic electrical stimulation using an implantable neural prosthesis. J Urol 154: 504–507

20. Bosch JLH, Groen J (1996) Treatment of refractory urinary urge incontinence with sacral spinal nerve stimulation in multiple sclerosis patients. Lancet 348: 717–719
21. Bosch JLH, Groen J (1997) Seven years of experience with sacral (S3) segmental nerve stimulation in patients with urge incontinence due to detrusor instability of hyperreflexia. Neurourol Urodyn 16: 426–427
22. Bosch JLH, Groen J (2000) Sacral Nerve Neuromodulation in the Treatment of Patients with Refractory Motor Urge Incontinence: Long-Term Results of a Prospective Longitudinal Study. J Urol 163: 1219–1222
23. Bosch JLHR, Groen J (1998) Neuromodulation: Urodynamic effects of sacral (S3) spinal nerve stimulation in patients with detrusor instability or detrusor hyperreflexia. Behav Brain Res 92: 141–150
24. Bower WF, Moore KH, Adams RD (2001) A pilot study of the home application of transcutaneous neuromodulation in children with urgency or urge incontinence. J Urol 166: 2420–2422
25. Boyce WH, Lathem JE, Hund LD (1964) Research related to the developpement of an artificial electric stimulator for the paralysed human bladder. J Urol 91: 45–51
26. Brading AF (1997) A myogenic basis for the overactive bladder. Urology 36: 57–67
27. Bradley WE, Chou SN, French LA (1963) Futher experience with the radio transmitter receiver unit for the neurogenic bladder. J Neurosurg 20: 953–960
28. Bradley WE, Conway CJ (1966) Bladder representation in the pontine mesencephalic reticular formation. Exp Neurol 16: 237–249
29. Bramble FJ (1982) The treatment of adult enuresis and urge incontinence by enterocystoplasty. Br J Urol 54: 693–696
30. Brindley GS (1994) The first 50 patients with sacral root stimulator implants: general desciption. Paraplegia 32: 795–805
31. Brindley GS (1995) The first 50 sacral anterior root stimulators: implant failures and their repair. Paraplegia 33: 5–9
32. Brindley GS, Polkey CE, Ruston DN (1982) Sacral anterior root stimulators of bladder control in paraplegia. Paraplegia 28: 365–381
33. Brindley GS, Polkey CE, Ruston DN, Cardozo L (1986) Sacral anterior root stimulators for bladder control in paraplegia, the first 50 cases. J Neurol Neurosurg and Psych 49: 1104–1114
34. Brubaker L (2000) Electrical stimulation in overactive bladder. Urol 55: 17–23
35. Buller JL, Cundiff GW, Noel KA, Van Rooyen JA, Leffler KS, Ellerkman RM, Bent AE (2002) An implantable microstimulator for the treatment of overactive disorders in females. XVIIth Congress of the European Association of Urology, Birmingham UK
36. Capellano F, Bertapelle P, Spinelli M, Cartanzaro F, Carone R, Zanollo A, De Seta F, Giardello G for the Italian Group of Sacral Neuromodulation (GINS) (2001) Quality of life assessment in patients who undergo sacral neuromodulation implant for urge incontinence: an additional tool for evaluating the outcome. J Urol 166: 2277–2290

37. Capellano F, Bertapelle P, Spreafico L, Del Popolo G, Kocjancic E, Donelli A, Ponzi P, Giardello G, Caprari F, Catanzaro F (2003) Economical and social impact of sacral nerve stimulation therapy in 62 patients with lower urinary tract dysfunction. International Continence Society 33rd Annual Meeting 2003 Proceedings Abstract Book, pp 58–59
38. Caraballo R, Bologna RA, Lukban J, Whitmore KE (2001) Sacral nerve stimulation as a treatment for urge incontinence and associated pelvic floor disorders at a pelvic floor center: a follow-up study. Urol 57: 121
39. Chai TC, Steers WD (1996) Neurophysiology of micturition and continence. Urol Clin North Am 23: 221–236
40. Chapple CR (2000) Muscarinic receptor antagonists in the treatment of overactive bladder. Urol 55: 33–46
41. Chartier-Kastler EJ, Bosch JL, Perrigot M, Chancellor MB, Richard F, Denys P (2000) Long-term results of sacral nerve stimulation (S3) for the treatment of neurogenic refractory urge incontinece related to detrusor hyperreflexia. J Urol 164: 1476–1480
42. Chartier-Kastler EJ, Denys P, Chancellor MB, Haertig A, Bussel B, Richard F (2001) Urodynamic monitoring during percutaneous sacral nerve neurostimulation in patients with neurogenic detrusor hyperreflexia. Neurol Urodyn 20: 61–71
43. Cheng CL, Ma CP, de Groat WC (1995) Effects of capsaicin on micturition and associated reflexes in chronic spinal rats. Brain res 678: 40–48
44. Cruz F (1998) Desensitisation of bladder sensory fibers by intravesical capsaicin or capsaicin analogs: a new strategy for treatment of urge incontinence in patients with spinal detrusor hyperreflexia or bladder hypersensitivity disorders. Int Urogynecol J 9: 214–229
45. Cruz F, Silva C, Ribeiro M, Aveilino A (2002) The effect of intravesical resiniferatoxin in neurogenic forms of bladder overactivity: preliminary results of a randomized controlled trial. Neurourol Urodyn 21: 692–697
46. Dahms SE, Tanagho EA (1998) The impact of sacral root anatomy on selective electrical stimulation for bladder evacuation. World J Urol 16: 322–328
47. de Groat WC (1997) A neurologic basis for the overactive bladder. Urol 50: 36–52
48. De Groat WC, Araki I, Vizzard MA, Yoshiyama M, Yoshimura N, Sugaya K, Tai C, Roppolo JR (1998) Developmental and injury induced plasticity in the micturition reflex pathway. Behav Brain Res 92: 127–140
49. De Groat WC, Nadelhaft I, Milne RJ, Booth AM, Morgan C, Thor K (1981) Organization of the sacral parasympathetic reflex pathways to the urinary bladder and large intestine. J auton nerv Syst 3: 135–160
50. De Groat WC, Kawatatni T, Hisamitsu T (1997) Mechanisms underlying the recovery of urinary bladder function following spinal cord injury. J Auton Nerv Syst 1990 30: 71–78
51. De Ridder D, Baert L (2000) Vanilloids and the overactive bladder. BJU Int 86: 172–180
52. De Seze M, Wiart L, Ferrier J, De Seze MP, Joseph P, Barat M (1999)

Intravesical instillation of capsaicin in urology: a review of the literature. Eur Urol 36: 267–277

53. Dijkema HE, Weil EH, Mijs PT, Janknegt RA (1993) Neuromodulation of sacral nerves for incontinence and voiding dysfunctions. Clinical results and complications. Eur Urol 24: 72–76
54. Ebraheim NA, Lu J, Galluch D, Yang H, Yeasting RA (2000) Location of the first and second sacral nerve roots in relation to pedicle screw placement. Am J Orthop 29: 873–877
55. Ebraheim NA, Lu J, Yang H, Heck BE, Yeasting RA (1997) Anatomic considerations of the second sacral vertebra and dorsal screw placement. Surg Radiol Anat 19: 353–357
56. Elabbady A, Hassouna M (1994) Neural stimulation for chronic voiding dysfunctions. J Urol 152: 2076–2080
57. Elbadawi A, Yalla SV, Resnick NM (1993) Structural basis of geriatric voiding dysfunction. III. Detrusor overactivity. J Urol 150: 1668–1680
58. Fall M, Lindstöm S, Mazières L (1990) A bladder-to-bladder cooling reflex in the cat. J Physiol (Lond) 427: 281–300
59. Fantl J, Newman D, Colling J, Delancey J, Keeys C, Mcdowell B (1996) Urinary incontinence in adults: acute and chronic management. Clinical practice guideline, N°2. Rockville, Maryland: AHCPR
60. Flisser AJ, Blaivas JG (2002) Role of cystometry in evaluating patients with overactive bladder. Urol 60: 33–42
61. Flood HD, Malhotra SJ, O'Connell HE, Ritchey MJ, Bloom DA, Mc Guire EJ (1995) Long term results and complications using augmentation cystoplasty in reconstructive urology. Neurourol Urodynam 14: 297–309
62. Fowler CJ (2002) Intravesical treatment of overactive bladder. Urol 55: 60–64
63. Fowler CJ, Van Kerrebroeck PE, Nordenbo A, Van Poppel H (1993) Treatment of lower urinary tract dysfunction in patients with multiple sclerosis. Committee of the European Study Group of SUDIMS (Sexual and Urological Disorders in Multiple Sclerosis). J Neurol Neurosurg Psych 55: 986–989
64. Friedman H, Nashold BS, Senechal P (1972) Spinal cord stimulation and bladder function in normal and paraplegic animal. J Neurosurg 36: 430–437
65. Gladh G, Mattsson S, Lindstrom S (2003) Intravesical electrical stimulation in the treatment of micturition dysfunction in children. Neurourol Urodyn 22: 233–242
66. Grill WM, Bhadra N, Wang B (1999) Bladder and urethral pressures evoked by microstimulation of the sacral spinal cord in cats. Brain Res 31: 19–30
67. Grill WM, Craggs MD, Foreman RD, Ludlow CL, Buller JL (2001) Emerging clinical applications of electrical stimulation: opportunities for restoration of function. J Rehabil Res Dev 38: 641–653
68. Grimes JH, Nashold BS (1974) Clinical application of electronic bladder stimulation in paraplegics. Br J Urol 46: 653–657
69. Grimes JH, Nashold BS, Currie DP (1973) Chronic electrical stimulation of the paraplegic bladder. J Urol 109: 242–245

70. Groen LJHR, Bosch J, Schroder FH (1993) Neuromodulation (sacral segmental nerve stimulation) as a treatment for urge incontinence in patients with bladder instability. J Urol 367A
71. Grünewald V, Hofner K, Thon W (1999) Sacral electrical neuromodulation as an alternative treatment option for lower urinary tract dysfunction. Res Neurol Neurosc 14: 189–193
72. Grünewald V, Jonas U, and the MDT-103 Multicenter Study Group (1999) Sacral electrical nerve stimulation for treatment of severe voiding dysfunction. J Urol 275: abstract 1064
73. Guérin J, Bioulac B (1979) The anatomical and physiological organization of motor activity in the spinal cord. Anat Clin 1: 267–289
74. Habler HJ, Janig W, Koltzenburg M (1990) Activation of unmyelinated afferent fibres by mechanical stimuli and inflammation of the urinary bladder in the cat. J Physiol (Lond) 425: 545–562
75. Haddad M, Guys JM, Planche D, Louis-Borrione C (2003) Sacral nerve modulation in children's neurogenic bladder: results of a prospective study. XVIth International symposium of paediatric surgical research. Marseille, France
76. Hannestad YS, Rortveit G, Sandvik H, Hunskaar S (2000) A community-based epidemiological survey of female urinary incontinence: the Norwegian EPICONT study. Epidemiology of incontinence in the county of Nord-Trondelag. J Clin Epidemiol 53: 11150–1157
77. Hasan ST, Marshall C, Robson WA, Neal DE (1995) Clinical outcome and quality of life following enterocystoplasty for idiopathic detrusor instability and neurogenic instability. Br J Urol 76: 551–557
78. Hassouna M, Elhilali MM (1991) Role of the sacral root stimulator in voiding dysfunction. World J Urol 9: 145–148
79. Hassouna M, Siegel S, Lycklama A, Nyeholt A *et al* (2000) Sacral neuromodulation in the treatment of urge-incontinence symptome: a multicenter study on efficacy and safety. J Urol 163: 1849–1854
80. Herschorn S, Bosh R, Brushini H, Hanus T, Low A, Shick E (2002) Surgical treatment of urinary incontinence in men. In: Incontinence. In: Abrams P, Cardozo L, Khoury S, Wein A (eds) Health Publications Ltd, Plymouth
81. Herschorn S, Hewitt RJ (1998) Patient perspective of long term outcome of augmentation cystoplasty for neurogenic bladder. Urol 52: 672–678
82. Hoebeke P, Van Laecke E, Everaert K, Renson C, De Paepe H, Raes A, Vande Walle J (2001) Transcutaneous neuromodulation for the urge syndrome in children: a pilot study. J Urol 166: 2416–2419
83. Hohenfellner M, Humke J, Hampel C, Dhams S, Matzel K, Roth S, Thuroff JW, Schultz-Lampel D (2001) Chronic sacral neuromodulation for treatment of neurogenic bladder dysfunction: Long-term results with unilateral implants. Urol 58: 887–892
84. Hohenfellner M, Schultz-Lampel D, Dahms S, Matzel KE, Thuroff JW (1998) Bilateral chronic sacral neuromodulation for treatment of lower urinary tract dysfunction. J Urol 160: 821–824
85. Holstege G, Griffiths D, de Wall H, Dalm E (1986) Anatomical and

physiological observations on supraspinal control of bladder and urethral sphincter muscles in the cat. J Compl Neurol 250: 449–461
86. Hu TW, Moore K, Subak L, Versi E, Wagner T, Zinner N, Ouslander J (2002) Economics of incontinence. In: Incontinence. In: Abrams P, Cardozo L, Khoury S, Wein A (eds) Health publications Ltd, Plymouth, pp 967–983
87. Hu TW, Wagner TH, Bentkover JD, Leblanc K, Piancentini A, Stewart WF, Corey R, Zhou SZ, Hunt TL (2003) Estimated economic costs of overactive bladder in the United States. Urol 61: 1123–1128
88. Ingersoll EH, Jones LL, Hegre ES (1957) Effect on urinary bladder of unilateral stimulation of pelvic nerves in the dog. Am J Pphysiol 189: 167
89. Jackson S (1997) The patient with an overactive bladder-symptoms and quality-of-life issues. Urol 50: 18–22
90. Jackson S, Donovan J, Brookes S, Ecford S, Swithinbank L, Abrams P (1996) The Bristol Female Lower Urinary Tract Symptoms Questionnaire: development and psychometric testing. Br J Urol 77: 805–812
91. Jänig W, Morrion JFB (1986) Functional properties of spinal visceral afferents supplying abdominal and pelvic organs, with special emphasis on visceral nociception. Prog Brain res 67: 87–114
92. Janknegt RA, Hassouna MM, Siegel SW, Schmidt RA, Gajewski JB, Rivas DA, Elhilali Mammilla DC, Van Kerrebroeck PE, Dijkema HE, Lycklama A, Nyeholt AA, Fall M, Jonas U, Catanzaro F, Fowler CJ, Oleson KA (2001) Long-term effectiveness of sacral nerve stimulation for refractory urge incontinence. Eur Urol 39: 101–106
93. Jiang CH, Lindstrom S (1999) Prolonged enhancement of the micturition reflex in the cat by repetitive stimulation of bladder afferents. J Physiol 517: 599–605
94. Jonas U, Fowler C, Chancellor C *et al* (2001) Efficacy of sacral nerve stimulation for urinary retention: results 18 months after implantation. J Urol 165: 15–19
95. Jonas U, Van Den Hombergh U (2001) Complications of sacral nerve stimulation. In: Jonas U, Grunewald V (eds) New perspectives in sacral nerve stimulation for control of lower urinary tract dysfunction. Martin Dunitz Ltd, London
96. Kaeckenbeek B (1979) Electro-stimulation de la vessie des paraplégiques. Technique de Burghele-Ichim-Demetrscu. Arch Urol Bel 47: 139–140
97. Kim DY, Chancellor MB (2000) Intravesical neuromodulatory drugs: capsaicin and resiniferatoxin to treat the overactive bladder. J Endourol 14: 172–180
98. Kingsley RE (2000) The autonomic nervous system. In: Conciste of neuroscience, 2nd edn. Lippincott Williams and Wilkins, Baltimore, pp 471–487
99. Koldewijn E, Meuleman E, Bemelmans B, van Kerrebroeck P, Debruyne F (1999) Neuromodulation effective in voiding dysfunction despite high reoperation rate. J Urol 161: 984A
100. Kumar V, Templemen L, Chapple CR, Chess-Williams R (2003) Recent developements in the management of detrusor overactivity. Curr Opin Urol 13: 285–291

101. Kuo HC (2003) Effectiveness of intravesical resiniferatoxin for anticholinergi treatment refractory detrusor overactivity due to nonspinal cord lesions. J Urol 170: 835–839
102. Lagarrigue J, Lazorthes Y, Verdie JC, Alwan A, Sarramon JP, Rossignol G (1979) Thermocoagulation percutanée des racines sacrées dans le traitement des neuro-vessies spastiques. Neurochirurgie 25: 91–95
103. Larsen JO, Thomsen M, Haughland M, Sinklaer T (1998) Degeneration and regeneration in rabbit peripheral nerve with long-term nerve cuff electrode implant: a stereological study of myelinated and unmyelinated axons. Acta Neuropathol 96: 365–378
104. Lazzeri M, Beneforte P, Turini D (1997) Urodynamic effects of intravesical resiniferatoxin in humans: preliminary results in stable and unstable detrusor. J Urol 158: 2093–2096
105. Lefurge T, Goodall E, Horch K *et al* (1991) Chronically implanted intrafascicular recording electrodes. Ann Biomed Eng 19: 197–207
106. Leng WW, Blalock HJ, Frederiksson WH, English SE, Mc Guire EJ (1999) Enterocystoplasty or detrusor myomectomy? Comparison of indications and outcomes for bladder augmentation. J Urol 161: 758–763
107. Liguoro D, Viejo-fuertes D, Midy D, Guerin J (1999) Th posterior sacral foramina: an anatomical study. J Anat 195: 301–304
108. Lindstrom S, Fall M, Carlsson CA, Erlandson BE (1983) The neurophysiological basis of bladder inhibition in response to intravaginal electrical stimulation. J Urol 129: 405–410
109. Maggi CA, Barbanti G, Santicioli P *et al* (1989) Cystometric evidence that capsaicin-sensitiv nerve modulate the afferent branchs of micturion reflex in human. J Urol 142: 150–154
110. Mallory B, Steers WD, de Groat WC (1989) Electrophysiological study of micturition reflexes in rats. Am J Physiol 257: 410–421
111. Mamo GA (2002) Anatomy of the sacral region. In: Jonas U, Grunewald V (eds) New perspectives in sacral nerve stimulation for control of lower urinary tract dysfunction. Martin Dunitz Ltd, London
112. Marshall DF, Boston VE (1997) Altered bladder and bowel function following cutaneous electrical field stimulation in children with spina bifida—interim results of a randomized double-blind placebo-controlled trial. Eur J Pediatr Surg 1: 41–43
113. Mc Inerney PD, De Souza N, Thomas PJ, Mundy AR (1995) The role of urodynamic studies in the evaluation of patients with augmentation cystoplasties. Br J Urol 76: 475–478
114. McCreery DB, Agnew WF, Yuen TGH, Bullara LA (1995) Damage in peripheral nerve from continuous electrical stimulation: comparison of two stimulus waveforms. Med Biol Eng Comput 30: 109–114
115. McGuire EJ, Savastano JA (1984) Urodynamic findings and clinical status following vesical denervation procedures for control of incontinence. J Urol 132: 87–88
116. McGuire EJ, Zhang SC, Horwisnki ER, Lytton B (1983) Treatment of motor and sensory detrusor instability by electric stimulation. J Urol 129: 78–79

117. Medicare Service Advisory Committee (MSAC): Application 1009 Assessment report. Sacral nerve stimulation for refractory urinary urge incontinence or urinary retention. Camberra June 2000: 1–45
118. Mertens P, Sindou M (2003) Traitement de la vessie hyperactive par drezotomie microchirurgicale sacrée. Neurochirurgie 49: 399–403
119. Mostwin JL (2002) Pathophysiology: the varieties of bladder overactivity. Urology 60: 22–27
120. Naples GG, Mortimer JT, Yuen TGH (1990) Overview of peripheral nerve electrode design and implantation. In: Agnew WF, McCreery DB (eds) Neural prostheses: fundamental studies. Prentice Hall, Englewood Cliffs, pp 107–145
121. Nieuwenhuys R, Voogd J, Van Huijzen C (1988) The human central nervous system, 3rd edn. Springer-Berlin Heidelberg New York Tokyo
122. Nour S, Svarer C, Kristensen JK, Paulson OB, Law I (2000) Cerebral activation during micturition in normal men. Brain 123: 781–789
123. Ouslander JG (2002) Geriatric considerations in the diagnosis and management of overactive bladder. Urol 60: 50–55
124. Payne CK (1998) Epidemiology, pathophysiology, and evaluation of urinary incontinence and overactive bladder. Urology 5: 3–10
125. Robinson JC (1993) Chronic pelvic pain. Curr Opin Obstet Gynecol 5: 740–743
126. Sarrias M, Sarrias F, Borau A (1993) The «barcelona» technique. Neurourol Urodyn 12: 495–496
127. Scheepens WA, de Bie RA, Weil EH, van Kerrebroeck PE (2002) Unilateral versus bilateral sacral neuromodulation in patients with chronic voiding dysfunction. J Urol 168: 2046–2050
128. Scheepens WA, Jongen MMGJ, Nieman FHM, De Bie RA, Weil EHJ, Van Kerrebroeck PEV (2002) Predictive factors for sacral neuromodulation in chronic lower urinary tract dysfunction. Urol 60: 598–602
129. Scheepens WA, Van Koeveringe GA, De Bie RA, Weil EH, Van Kerrebroeck PE (2002) Long-term efficacy and safety results of the two-stage implantation technique in sacral neuromodulation. BJU Int 90: 840–845
130. Schlote N, Tanagho EA (2002) Electrical stimulation of the lower urinary tract: historical overview. In: Jonas U, Grunewald V (eds) New perspectives in sacral nerve stimulation for control of lower urinary tract dysfunction. Martin Dunitz Ltd, London
131. Schmidt R, Jonas U, Oleson K, Janknegt RA, Hassouna MM, Siegel SW, Van Kerrebroeck PEV (1999) For the sacral nerve stimulation study group. Sacral nerve stimulation for treatment of refractory urinary urge incontinence. J Urol 162: 352–357
132. Schmidt RA, Senn E, Tanagho EA (1990) Functional evaluation of sacral nerve root integrity: Report of a technique. Urology 35: 388–392
133. Schurch B, Stohrer M, Kramer G, Schmid DM, Gaul G, Hauri D (2000) Botulinum-A toxin for treating detrusor hyperreflexia in spinal cord injured patients: a new alternative to anticholinergic drugs? Preliminary results. J Urol 164: 692–697

134. Sedan R, Bourdhis A, Sarrazin C, Barre E, Lazorthes Y, Sarramon JP, Lagarrigue J, Leandri P (1977) Résultats à long terme de la stimulation électrique du cône médullaire dans les problèmes de vessie neurogénique 23: 47–54
135. Sesay M, Vignes JR, Liguoro D, Guérin J (2002) L'hyperréflexie autonome induite par la stimulation des racines sacrées est détectée par l'analyse spectrale de l'ECG. Can J Anaesth 49: 936–941
136. Sesay M, Vignes JR, Stockle M, Mehsen M, Boulard G, Maurette P (2003) L'analyse spectrale de l'intervalle R-R de l'ECG permet une détection précoce des réponses vagales aux stimuli neurochirurgicaux. Ann Fr Anesth Reanim 22: 421–424
137. Sethia KK, Webb RJ, Neal DE (1991) Urodynamic study of ileocystoplasty in the treatment of idiopathic detrusor instability. Br J Urol 67: 286–290
138. Shaker HS, Hassouna M (1998) Sacral nerve root neuromodulation: An effective treatment for refractory urge incontinence. J Urol 159: 1516–1519
139. Shaker HS, Hassouna M (1998) Sacral root neuromodulation in idiopathic nonobstructive chronic urinary retention. J Urol 159: 1476–1478
140. Shefchyk SJ (2001) Sacral spinal interneurones and the control of urinary bladder and urethral striated sphincter muscle function. J Physiol 533: 57–63
141. Shumaker SA, Wyman JF, Uebersax JS, McClish D, Fantl JA (1994) Health related QOL measures for women with urinary incontinence: the Incontinence Impact Questionnaire and the Urogenital Distress Inventory. Qual Life Res 3: 291–306
142. Siegel SW (2002) Sacral nerve stimulation: PNE. In: Jonas U, Grunewald V (eds) New perspectives in sacral nerve stimulation for control of lower urinary tract dysfunction. Martin Dunitz Ltd, London
143. Siegel SW, Catanzaro F, Dijkema HE, Elhilali MM, Fowler CJ, Gajewski JB, Hassouna MM, Janknegt RA, Jonas U, van Kerrebroeck PE, Lycklama a Nijeholt AA, Oleson KA, Schmidt RA (2000) Long-term results of a multicenter study on sacral nerve stimulation for treatment of urinary urge incontinence, urgency-frequency, and retention. Urol 56: 87–91
144. Silva C, Ribeiro MJ, Cruz F (2002) the effects of intravesical resiniferatoxin in patienst with idiopathic detrusor instability suggests that involuntary detrusor contractions are triggered by C-Fiber input. J Urol 168: 575–579
145. Sindou M (1995) Microsurgical DREZtomy (MDT) for pain, spasticity, and hyperactive bladder, a 20-year experience (1995) Acta Neurochir (Wien) 137: 1–5
146. Smith ARB, Daneshgari F, Dmochowski R, Ghoniem G, Jarvis G, Nitti V, Paraiso M (2002) Surgical treatment of urinary incontinence in women. In: Incontinence. In: Abrams P, Cardozo L, Khoury S, Wein A (eds) Health Publications Ltd, Plymouth, pp 823–863
147. Spinelli M, Bertapelle P, Capellano F, Zanollo A, Carone R, Catanzaro F, Giardiello G, De Seta F, Gins Group (2001) Chronic sacral neuromodulation in patients with lower urinary tract symptoms: results from a national register. J Urol 166: 541–545
148. Spinelli M, Giardiello G, Arduini A (2003) New percutaneous technique of

sacral nerve stimulation has high initial success rate: Preliminary results. Eur Urol 43: 70–74

149. Staskin DR, Wein AJ, Andersson KE, Bauer SB, Blaivas JG, Burgio KL, Cardozo L, Chapple CR, Dmochowski RR, Gupta S, Mostwin JL, Ouslander JG, Weiss JP, King L (2002) Consensus statement. Urol 60: 1–6
150. Stewart WF, Van Trooyen JB, Cundiff GW, Abrams P, Herzog AR, Corey R, Hunt TL, Wein AJ (2003) Prevalence and burden of overactive bladder in the United States. World J Urol 20: 327–336
151. Swami KS, Feneley RCL, Hammonds JC, Abrams P (1998) Detrusor myectomy for detrusor overactivity: a minimum 1-year follow-up. Br J Urol 81: 68–72
152. Szallasi A, Fowler CJ (2002) After a decade of intravesical vanilloid therapy: still more question than answers. The Lancet Neurol 1: 167–172
153. Tanagho (1992) Neuromodulation in the management of voiding dysfunction in children. J Urol 655–657
154. Tanagho EA, Schmidt RA (1982) Bladder pacemaker: scientific basis and clinical future. Urol 20: 614–619
155. Tanagho EA, Schmidt RA (1988) Electrical stimulation in the clinical management of the neurogenic bladder. J Urol 140: 1331–1339
156. Thon WF, Baskin LS, Jonas U *et al* (1991) Neuromodulation of voiding dysfunction and pelvic pain. World J Urol 9: 138–141
157. Vandoninck V, van Balken MR, Finazi Agro E, Petta F, Micali F, Hesakkers JPFA, Debruyne FMJ, Kiemeney LALM, Bemelmans BLH (2003) Percutaneous tibial nerve stimulation in the treatment of overactive bladder: urodynamic data. Neurourol urodyn 22: 227–232
158. Vignes JR, De Seze M, Sesay M, Barat M, Guérin J (2003) Neurostimulation des racines sacrées antérieures avec rhizotomies postérieures (Technique de Brindley). Neurochirurgie 49: 383–394
159. Vignes JR, Liguoro D, Sesay M, Barat M, Guérin J (2001) Dorsal rhizotomy with anterior sacral root stimulation for neurogenic bladder. Stereotact Funct Neurosurg 76: 243–245
160. Vodusek DB (1988) Detrusor inhibition on selective pudendal nerve stimulation in the perineum. Neurourol Urodyn 6: 389–393
161. Wagner TH, Wu TW (1998) Economic costs of urinary incontinence in 1995. Urol 5: 355–361
162. Wang SC, Mc Guire EJ, Bloom DA (1988) A bladder pressure management system for myelodysplasia-clinical outcome. J Urol 140: 1499–1502
163. Ware JE, Sherboune CD (1992) The MOS 36-item Short-Form Health Survey (SF-36). Med Care 30: 473–483
164. Weil EH, Ruiz-Cerda JL, Eerdmans PH, Janknegt RA, Bemelmans BL, van Kerrebroeck PE (2000) Sacral root neuromodulation in the treatment of refractory urinary urge incontinence: a prospective randomized clinical trial. Eur Urol 37: 161–171
165. Weil EH, Ruiz-Cerda JL, Eerdmans PH, Janknegt RA, van Kerrebroeck PE (1998) Clinical results of sacral neuromodulation for chronic voiding dysfunction using unilateral sacral foramen electrodes. World J Urol 16: 313–321

166. Wiart L, Joseph PA, Petit H, Dosque JP, de Seze M, Brochet B, Deminiere C, Ferriere JM, Mazaux JM, N'Guyen P, Barat M (1998) The effects of capsaicin on the neurogenic hyperreflexic detrusor. A double-blind placebo controlled study in patients with spinal cord disease. Preliminary results. Spinal Cord 36: 95–99
167. Woodford BJ, Carter RR, McCreery D, Bullara LA, Agnew WF (1996) Histopathologic and physiologic effects of chronic implantation of microelectrodes in sacral spinal cord of the cat. J Neuropathol Exp Neurol 55: 982–991
168. Wyman JF, Harkins SW, Fantl JA (1990) Psychological impact of urinary incontinence in the community dwelling population. J Am Geriatr Soc 38: 282–288
169. Yoshimura N (1999) Bladder afferent pathway and spinal cord injury: possible mechanisms inducing hyperreflexia of the urinary bladder. Prog Neurobiol 57: 583–606

Offprint from
Advances and Technical Standards in Neurosurgery, Vol. 30
Edited by J.D. Pickard

Prevention and Treatment of Postoperative Pain with Particular Reference to Children

A. Chiaretti and A. Langer

Paediatric Intensive Care Unit, Catholic University Medical School, Rome, Italy

With 7 Figures

Contents

Abstract

Pain therapy is an important aspect of medical practice for patients of all ages, to optimize care, to obtain an adequate quality of life and to improve their general conditions. Pain is among the most prevalent symptoms experienced by patients undergoing surgery. The success of post-operative pain therapy depends on the ability of the clinician to assess the presenting problems, identify and evaluate pain syndromes and formulate a plan for comprehensive continuing care. The prevalence of acute pain has led to the need to develop techniques for the assessment and management of this symptom in order to focus the attention on an interdisciplinary therapeutic approach (including pharmacologic, cognitive-behavioral, psychologic and physical treatment) and on the timing of different interventions (pre and postoperative). In this chapter we describe the principal therapeutic approaches to control pain in post-operative patients, such as non-opioid, opioid and adjuvant analgesics with particular attention in paediatric age. Moreover we report the principal scales to assess the pain intensity in the post-operative period. The need of a multidisciplinatory team and of a pre and postoperative pain management program represents an important goal in order to obtain effective pain relief and optimize pediatric care and rapid recovery. The introduction of a perioperative team service will improve the approach to pain management programs and it is considered the most important challenge for future.

Keywords: Pain; pain assessment; analgesics drugs; patient controlled analgesia; childhood.

Introduction

Pain is among the most prevalent symptoms experienced by patients undergoing surgery. The success of postoperative pain therapy depends on the ability of the clinician to assess the presenting problems, identify and evaluate pain syndromes and formulate a plan for comprehensive continuing care. This requires familiarity with a range of therapeutic options and an approach to optimise medical care that is responsive to the changing needs of the patient. The formulation of an effective therapeutic strategy for the management of post-operative pain is predicated on a comprehensive assessment of the patient. The assessment should clarify the characteristics of the pain and its impact on function and psychological well-being, identify the pain syndrome and infer the putative mechanisms that may underlie the pain. In addition, the assessment should evaluate both the nature and extent of the underlying disease and identify concurrent problems that are contributing, or may soon contribute, to patient or family distress. The particular therapeutic strategy that evolves from this information depends on the goals of care. These goals are diverse, but can generally be grouped into two broad categories:

1. optimizing comfort
2. maximising function

The relative priority of these goals provides an essential context for therapeutic decision making. The therapeutic strategy should address a prioritized problem list that best serves both the current goals of the patient and the anticipated problems that would benefit from advanced planning. Most postoperative patients can attain satisfactory relief of pain through an approach that incorporates primary treatments, systemic analgesic therapy and, at times, other non-invasive techniques (such as psychological or rehabilitative interventions). Some patients whose pain is refractory to this approach benefit from invasive anaesthetic or other treatments. Such patients should have access to specialists in pain management, who can provide additional expertise in addressing these complex problems. Differently from the adult patients in paediatric age it is more difficult to assess and treat efficaciously the pain and postoperative pain in childhood have been undertreated or not treated. In some areas this practice still exists and is a likely reflection of persistence of myths related to the infant's ability to perceive pain. Such myths include the lack of ability to perceive pain, remember painful experiences and other reasons (Box 1). Recent evidences have documented the deleterious physiologic effects of pain and the beneficial results of efficacious postoperative analgesia both in adult patients and in children. Due to the increasing prevalence of both acute and chronic pain in the paediatric age new techniques for pain management have been developed. In 2001, the American Academy of Paediatrics and the American Pain Society issued a statement to ensure human and competent treatment of pain and suffering in all children and adolescents in order to focus the attention on an interdisciplinary therapeutic approach, including pharmacologic, cognitive-behavioural, psychologic and physical treatments (Box 2) [1]. There is a growing awareness of the effects of unrelieved pain

Box 1. *Reasons for the inadequate management of acute pain in children*

- Idea that pain is merely asymptom and not harmful in itself
- Mistaken impression that analgesia makes accurate diagnosis difficult or impossible
- Fear of the potential for addiction to opioids
- Concerns about respiratory depression and other opioid-related side effects
- Lack of understanding of the pharmacokinetic of analgesic drugs
- Prescriptions for opioids which include the use of inappropriate doses
- Thinking that opioids must not be given more often than 4 hourly
- Patient's difficulties in communicating their need for analgesia

Box 2. *Clinical practice and acute pain: guidelines and major goals*

Guidelines

- A collaborative, interdisciplinary approach to pain control, including all members of the healthcare team
- Assessment and frequent reassessment of the patient's pain
- Use both drug and non-drug therapies to control and/or prevent pain

Major goals

- Reduce the incidence and severity of patient's postoperative pain
- Educate patients about the need to communicate regarding unrelieved pain, so they can receive prompt evaluation and effective treatment
- Enhance patient comfort and satisfaction
- Contribute to fewer postoperative complications and shorter stays after surgical procedures

in children and the need to provide effective pain relief, especially with regard to acute pain. Some principles can be extended to all forms of acute pain, but some of them are particularly decisive in postoperative pain management both in adults and in children (Box 3) [2]. Several studies documented that in children undertreatment after surgery is more common than in adults leading to unnecessary distress and suffering for children and their families [3]. The need of a multidisciplinary team and of a pre and post-operative pain management program represent an important goal in order to obtain effective pain relief and optimize medical care and rapid recovery after post-operative procedures. The introduction of a perioperative team service and the utilization of pain management programs will represent an important challenge for the future.

Box 3. *Priciples of safe and effective acute post-operative pain management*

- Adverse physiological and psychological effects result from unrelieved severe pain
- Proper assessment and pain control require patient involvement
- Pain is best treated early, because established, severe pain is more difficult to treat
- While it is not possible to completely alleviate all pain in the postoperative period, it should be possible to reduce pain to a tolerable or comfortable level
- Postoperative analgesia should be planned preoperatively, with consideration given to the type of surgery, perioperative use of analgesics and regional anaesthetic techniques
- Frequent assessment of pain intensity and charting of analgesia
- Adequate education of all involved in pain management, including the patient
- Formal programmes, protocols and guidelines covering acute pain management

Acute Pain Assessment in Paediatric Age

The pain experience includes physiological, sensory, affective, behavioural, cognitive and sociocultural components. While in adults is more easy to assess the pain simptoms, in children pain assessment should consider age, cognitive level and the presence of eventual disability, type of pain and the situation in which pain is occurring. McGrath on the subject of assessment of pain in children states: "Measurement of pain should be distinguished from the assessment of pain. Measurement refers to the application of a specific metric to a specific element of pain, usually the intensity of pain. Assessment is a much broader endeavor that includes the measurement of various elements that impact on the pain experience" [4]. Despite this consideration, there are some commonly used methods of measurement of pain that have been proved to be reliable. Observational and behavioural measures consider child's reaction to pain. Self-report measures rely on the child's description of his experience of pain. Biological measures consider some physiologic parameters that may be modified by the presence of pain, such as heart and respiratory rates, blood pressure, etc. [5]. In infants and non-verbal children, self-report measures are unavailable, but behavioural indices (motor responses, vocalization, facial expressions, crying and complex behavioural responses such as the sleep-wake patterns) can be easily evaluated to assess pain. Different behavioural scales have been validated by several studies that enrolled infants and neonates [6, 7]. Behavioural

ITEM	BEHAVIOR	SCORE	DEFINITION
Cry	*No cry*	1	Child is not crying
	Moaning	2	Child is moaning or quietly vocalizing;silent cry
	Crying	2	Child is crying, but the cry is gentle or whimpering
	Scream	3	Child is in a full-lunged cry;sobbing:may be scored with/whitout complaint
Facial	*Composed*	1	Neutral facial expression
	Grimace	2	Score only if negative facial expression
	Smiling	0	Score only if definite positive facial expression
Child verbal	*None*	1	Child not talking
	Other complaints	1	Child complaints, but not about pain
	Pain complaints	2	Child complaints about pain
	Both complaints	2	Child complaints about pain and about other things
	Positive	0	Child makes any positive statement or talks about other things without complaint
Body	*Neutral*	1	Body (not limbs) is at rest; torso is inactive
	Shifting	2	Body is is in motion in a shifting or serpentine fashion
	Tense	2	Body is arched or rigid
	Shivering	2	Body is shuddering or shaking involuntarily
	Upright	2	Child is in a vertical or upright position
	Restrained	2	Body is restrained
Touch	*Not touching*	1	Child is not touching or grabbing at wound
	Reach	2	Child is reaching for but not touching wound
	Touch	2	Child is gently touching wound or wound area
	Grab	2	Child is grabbing vigorously at wound
	Restrained	2	Child's arms are restrained
Legs	*Neutral*	1	Legs may be in any position but are relaxed
	Squirming/ kicking	2	Definitive uneasy or restless movements in the legs or striking out with feets
	Drawn up/tensed	2	Legs tensed and/or pulled up tightly to body and kept there
	Standing	2	Standing, crouching, or kneeling
	Restrained	2	Child's legs are being held down

Fig. 1. CHEOPS Score. CHEOPS pain score: SUM (points for all 6 parameters), Minimum score: 4 (min pain); Maximum score: 13 (max pain)

parameters, even if non-specific, may be usefully associated to physiologic parameters such as heart rate, cardiac rate, arterial blood pressure, transcutaneous oxygenation and palmar sweating [8–10]. The Children's Hospital of Estern Ontario Pain Scale (CHEOPS) is one of the commonest scales used for postoperative pain management (Fig. 1) [11]. Parents who are able to assess behavioural changes related to discomfort or pain may help differentiate pain from anxiety or stress due to other causes [12, 13]. Children aged 3 to 7 years are increasingly able to describe pain characteristics. Observational scales as well as self-report scales represent useful tools to assess pain in this period of life. Composite measures of pain have been developed combining behavioural and biological items, such as the Objective Pain Scale and the Comfort Scale (Figs. 2, 3). The Objective Pain Scale is used to assess both physiologic parameters and behavioural changes in children that may be modified by the presence of pain or discomfort after procedures and/or postoperative interventions [14, 15]. The Comfort Scale is used to assess the level of sedation and distress in the paediatric intensive care unit, but recent studies have validated this measurement method also in procedural and postoperative pain [16, 17]. Self-

Parameter	Finding	Points
Systolic blood pressure	increase < 20% of preoperative blood pressure	0
	increase 20-30% of preoperative blood pressure	1
	increase > 30% of preoperative blood pressure	2
Crying	not crying	0
	responds to age appropriate nurturing (tender loving care)	1
	does not respond to nurturing	2
Movements	no movements relaxed	0
	restless moving about in bed constantly	1
	thrashing (moving wildly)	2
	rigid (stiff)	2
Agitation	asleep or calm	0
	can be comforted to lessent the agitation (mild)	1
	Cannot be comforted (hysterical)	2
Complains of pain	Asleep	0
	states no pain	0
	Cannot localize	1
	localizes pain	2

Fig. 2. Objective Pain Scale (*OPS*). Minimum score: 0; Maximum score: 10, Maximum score if too young to complain of pain: 8, The higher the score the greater the degree of pain

report measures of pain represent the gold standard in older children who can describe the subjective pain experience [18, 19]. These methods include different strategies such as routine and direct questioning, verbal and non verbal methods (i.e. pictorial scales) and self rating scales. Visual Analogue

ALLERTNESS	Time					
Deeply asleep	1					
Lightly asleep	2					
Drowsy	3					
Fully Awake and alert	4					
Hyper-alert	5					
CALMNESS/AGITATION						
Calm	1					
Slightly anxious	2					
Anxious	3					
Very anxious	4					
Panicky	5					
RESPIRATORY RESPONSE						
No coughing and no spontaneous respiration	1					
Spontaneous respiration with little or no response to ventilation	2					
Occasional cough or resistance to ventilator	3					
Actively breathes against ventilator or coughs regularly	4					
Fights ventilator; coughing or choking	5					
PHYSICAL MOVEMENT						
No movement	1					
Occasional, Slight movement	2					
Frequent, Slight movement	3					
Vigorous movement limited to extremities	4					
Vigorous movement including torso and head	5					
BLOOD PRESSURE (MAP) BASELINE						
Blood pressure below baseline	1					
Blood pressure consistently at baseline	2					
Infrequent elevations of 15% or more (1-3)	3					
Frequent elevations of 15% or more (more than 3)	4					
Sustained elevation ≥ 15%	5					
HEART RATE BASELINE						
Heart rate below baseline	1					
Heart rate consistently at baseline	2					
Infrequent elevations of 15% or more above baseline (1-3) during observation period	3					
Frequent elevations of 15% or more above baseline (more than 3)	4					
Sustained elevation ≥ 15%	5					
MUSCLE TONE						
Muscles totally relaxed; no muscle tone	1					
Reduced muscle tone	2					
Normal muscle tone	3					
Increased muscle tone and flexion of fingers and toes	4					
Extreme muscle rigidity and flexion of fingers and toes	5					
FACIAL TENSION						
Facial muscles totally relaxed	1					
Facial muscle tone normal; no facial muscle tension evident	2					
Tension evident in some Facial muscles	3					
Tension evident throughout Facial muscles	4					
Facial muscles contorted and grimacing	5					

Adapted from: McCaffery M. Beebe A. *Pain: Clinical Manual for Nursing Practice*. St. Louis, MO: CV Mosby Co. 1989.
Used with permission. May be duplicated and used in clinical practice.

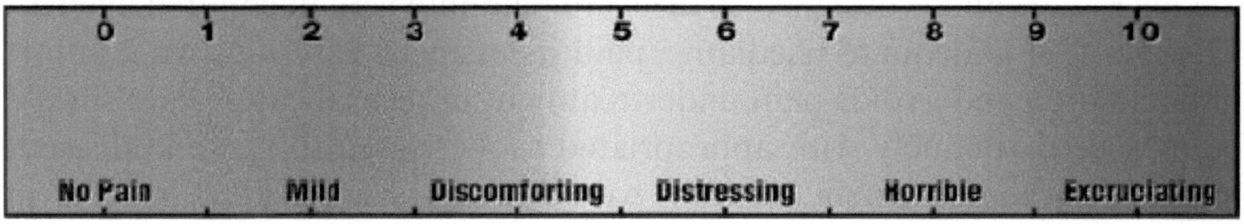

This scale incorporates a visual analogue scale, a descriptive word scale and a colour scale all in one tool

Fig. 4. Visual Analogue Scale (VAS) and Facial Pain Scale

Scale (VAS) and Facial Pain Scale are two of the commonest self rating scales to assess pain intensity in children (Fig. 4) [20, 21]. In the VAS children rate the intensity of pain on a 10 cm line anchored at one end by a label such as "no pain" and at the other end "severe pain". The scores are obtained by measuring the distance between the "no pain" and the patient's mark, usually in millimetres. The VAS has many advantages: it is simple and quick to score, avoids imprecise descriptive terms and provides many measuring points. Disadvantages are represented by the need of concentration and coordination, which can be difficult post-operatively or in children with neurological disorders. Self reported measures require a cognitive and linguistic development related to the capacity to answer to different questions. They are reliable to monitor pain relief in every single patient, while are less specific and effective if utilized to compare different patients. Self reported measures include categorical scales that use words (from four to five) to describe the magnitude of pain. However, in specific categories of patients, they are not useful. Faces scales represent another form of self reported measures: faces express different amounts of distress. The Facial Pain Scale is the commonest used in young children who may have difficulty with more cognitively demanding instruments. The original scale was composed by seven faces without an absolute meaning, but related to children's experience [20]. Different versions exist, based anyway on the same measurement principle [21, 22]. In figure 4 we report one of them more used in the clinical practice. The Oucher Scale is a variant of the faces scale and is designed to measure pain intensity in children aged 3–12 years [23]. Adequate paediatric pain assessment can improve comfort in ill children and avoids pain undertreatment in several cases. Pain should be measured routinely with appropriated tools related to age and disease. Simple pain measurement methods would improve not only pain relief in children, but would also decrease nurses and health professional workload and create a common language and an adequate communication among the medical and nurse staffs [24].

Specific Aspects of Post-Operative Pain

The perception of acute and post-operative pain is a complex interaction that involves sensory, emotional, and behavioural factors. The role of psychological factors must always be considered to be an important component in the perception and expression of post-operative pain. The biological processes involved in our perception of acute pain are no longer viewed as a simple "hard-wired" system with a pure stimulus-response relationship. Trauma to any part of the body, and nerve damage in particular, can lead to changes within other regions of the nervous system, which influence subsequent responses to sensory input. There is increasing recognition that long-term changes occur within the peripheral and central nervous system following noxious input. This plasticity of the nervous system then alters the body's response to further peripheral sensory input. Based on these mechanisms pain relief in postoperative patients represents an important therapeutical aspect, leading to significant physiological benefit. Surgical trauma are associated with an injury response or "inflammatory response". Part of the inflammatory response is the release of intracellular contents from damaged cells and inflammatory cells such as macrophages, lymphocytes and mast cells. Nociceptive stimulation also results in a neurogenic inflammatory response with the release of substance P, neurokinin A and calcitonin gene-related peptide (CGRP) from the peripheral terminals of nociceptive afferent fibres. Release of these peptides results in a changed excitability of sensory and sympathetic nerve fibers, vasodilatation, and extravasation of plasma proteins. These interactions result in the release of several inflammatory mediators such as potassium, serotonin, bradykinin, substance P, histamine, cytokines, nitric oxide and products from the cyclooxygenase and lipooxygenase pathways of arachidonic acid. These chemicals then act to sensitize high-threshold nociceptors which results in the phenomenon of peripheral sensitisation [5]. Following sensitization, low-intensity mechanical and thermal stimuli which would not normally cause pain are now perceived as painful. This zone of "primary hyperalgesia" surrounding the site of injury is caused by peripheral changes and is a feature that is commonly observed following surgery and other forms of trauma. Following injury, there is an increased responsiveness to normally innocuous mechanichal stimuli (allodynia) in a zone of "secondary hyperalgesia" in uninjured tissue surronding the site of injury. These changes are believed to be a result of processes that occur in the dorsal horn of the spinal cord following injury. This is the phenomenon of central sensitisation [5]. Several changes have been noted to occur in the dorsal horn with central sensitisation. Firstly, there is an expansion in receptive field size so that a spinal neuron will respond to stimuli that would normally be outside the region that respond to nociceptive stimuli. Secondly, there is an increase in the magnitude and duration of the response to stimuli that are above threshold in strength. Lastly, there is a reduction in threshold so that stimuli that are not normally noxious activate neurons that normally transmit nociceptive informations. These changes may be important both in acute pain states such as post-operative pain and in the development of chronic pain syndromes. Transmission of nociceptive information is subject to modulation at several levels of the neuraxis including the dorsal horn. Afferent impulses arriving in the dorsal horn initiate inhibitory mechanisms which limit the effect of subsequent impulses. Inhibition occurs through the effect of local inhibitory interneurons and descending pathways from the brain. In the dorsal horn incoming nociceptive messages are modulated by endogenous and exogenous agents that act on opiod, alpha-adreno-, GABA, and glycine receptors located at pre- and post-synaptic sites. Opioids are widely used and generally efficacious in the management of post-operative pain. Opioid receptors are found both pre- and postsynaptically in the dorsal horn, although the majority are located presynaptically. Activation of presynaptic opioid receptors results in a reduction in the release of neurotransmitters from the nociceptive primary afferent. Activation of alpha-adrenoceptors in the spinal cord has an analgesic effect either by endogenous release of noradrenaline by descending pathways from the brain stem or by exogenous spinal administration of agents such as clonidine. There are a number of alpha-adrenoceptor subtypes and the development of selective alpha-adrenoceptor subtype agonists has the potential to provide effective new analgesic agents with reduced side effects. Both GABA and glycine are involved in inhibition of nociceptive input, and loss of their inhibitory action can result in features of neuropathic pain. Descending inhibition involves the action of endogenous opiod peptides as well as other neurotransmitters, including serotonin, noradrenaline and GABA. Many of the traditional strategies available in acute and post-operative pain management such as the use of opioids and non-opiod drug administration, such as NSAIDs, act via these inhibitory mechanisms. Opioids have traditionally been viewed as centrally acting drugs. However, there is now evidence for the action of endogenous opioids on peripheral sites following tissue damage. Opioid receptors are transported toward the central terminal in the dorsal horn and toward the periphery. These peripheral receptors then become active following local tissue damage. This occurs with unmasking of opioid receptors and the arrival of immunocompetent cells that possess opioid receptors and have the ability to synthesize opioid peptides. This has led to an interest in the peripheral administration of opioids following surgery or topical administration of morphine. NSAIDs are commonly used for peripheral analgesia and one of their actions is a reduction in the inflammatory response. Agents such as aspirin and other NSAIDs provide their anti-inflammatory action by blocking the cyclooxigenase pathway. Cyclooxigenase exists in two forms, COX1 and COX2. While COX1 is always present in tissues, including the gastric mucosa, COX2 is induced by inflammation. This presents an opportunity for the development of agents that have a selective anti-inflammatory effect without gastric side effects. Selective COX2 inhibitor drugs (e.g. rofecoxib, celecoxib) that may offer analgesia with less gastrointestinal toxicity than NSAIDs have been developed. Besides the peripheral action of NSAIDs, there is increasing evidence that they exert their analgesic effect through central mechanisms [5]. The discovery of the changes associated with the phenomenon of peripheral and central sensitization has led to attempts to prevent these changes occurring. It was hoped that steps which would reduce or abolish noxious imput to the spinal cord during a painful event such as surgery would reduce or minimize spinal cord changes and thereby lead to reduced pain postoperatively. This concept has led to an increasing interest in the use of pre-emptive analgesia. Preemptive analgesia is based on the administration of an analgesic such as opioids and NSAIDs before a painful stimulus generates, so as to prevent the subsequent rebound mechanism [26]. Opioids and NSAIDs have been used alone or in combination and have been administered locally, epidurally, intrathecally or sistemically. Several studies have purported to show that pre-emptive analgesia results in reduced pain, decreased analgesic requirements, improved morbidity and decreased hospital stay [27–30]. However, pre-emptive analgesia may also be important in reducing the incidence and prevention of chronic pain states but further studies are necessary to address this important question. Improvement of post-operative pain control can be achieved by better education for all staff concerned postoperative pain relief and by making the assessment and recording of pain levels part of the routine management of each patient. The best strategy is to reduce or eliminate pain and discomfort with a minimum of side effects. A multidisciplinary acute pain service can ensure an adequate pain assessment and relief using different tools in order to reduce post-operative course with earlier discharge from hospital (Box 4) [31, 32].

Box 4. *Organizational aspects of an anaesthesiology-based postoperative pain programme*

1. Education
- Anaesthetists
- Surgeons
- Nurses
- Patients and families

2. Areas of regular administrative activity
- Mainteinance of clear lines of communication
- Evaluation of equipment (e.g. pumps)
- Economic issues
- Continuous quality improvement
- Pain management-related research

3. Collaboration with nursing services
- Nursing policies and procedures
- Nurses in-service and continuing education
- Definition of roles in patient care
- Continuous quality improvement
- Research activities

4. Elements of documentation
- Preprinted orders
- Procedures
- Protocols
- Bedside pain management flow sheets
- Daily consultation notes
- Educational packages

Post-Operative Pain Management

There is evidence that patients benefit from the use of multimodal, or balanced, analgesia after surgery. NSAIDs, paracetamol, local anaesthetics, adjuvant drugs, and opioids are employed in combination to improve pain relief (Table 1). Multimodal analgesia employs a variety of drugs, given by different routes, to achieve analgesia, with a reduction in the incidence and severity of side effects. NSAIDs contribute significantly to multimodal analgesia and postoperative recovery of the patient by minimizing opioid side effects including the inevitable opioid-induced gastrointestinal stasis that delays the resumption of normal enteral nutrition after surgery. However, the effect on morbidity and mortality has been disappointing in some studies, demonstrating that very good pain control is not automatically associated with an improvement in outcome. Recent studies have suggested

Table 1. *Scientific Evidence for Pharmacological Interventions to Manage Postoperative Pain in Adult Patients*

Intervention	Level of evidence	Comments
NSAIDs		
Oral (alone)	I	Effective for mild to moderate pain. Relatively contraindicated in patients with renal disease and risk or actual coagulopathy. Risk of coagulopathy, gastrointestinal bleeding and other risk factors should be carefully sought
Oral (adjunct to opioid)	I	Potentiating effect resulting in opioid sparing. Caution as above
Parenteral (Ketorolac)	I	Effective for moderate to severe pain. Useful where opioids contraindicated or to produce "opioid sparing", especially to minimize respiratory depression, sedation, and gastrointestinal stasis. Best used as part of a multimodal analgesia regimen
Opioids		
Oral	IV	As effective as parenteral in appropriate doses. Use as soon as oral medication tolerated.
Route of choice		
Intramuscular	I	Has been the standard parenteral route, but injections painful and absorption unreliable. Hence, avoid this route when possible.
Subcutaneous	I	Preferable to intramuscular because of patient comfort and a reduced risk of needlestick injury
Intravenous	I	Parenteral route of choice after major surgery. Suitable for titrated bolus or continuous administration. Significant risk of respiratory depression with inappropriate dosing
PCA (systemic)	I	Intravenous or subcutaneous routes recommended. Good steady level of analgesia. Popular with patients but requires special infusion pumps and staff education.
Epidural and intrathecal		When suitable, provides good analgesia. Risk of respiratory depression (as with opioids by other routes), but (as with opioids by other routes), but sometimes delayed in onset. Requires careful monitoring. Use of infusion pumps requires additional equipment and staff education. Expensive if infusion pumps are employed

Table 1. (*Continued*)

Intervention	Level of evidence	Comments
Local anaesthetics		
Epidural and intrathecal	I	Indications in particular settings. Effective regional analgesia. May blunt "stress response" and aid recovery. Opioid sparing. Addition of opioid to local anaesthetic may improve analgesia. Risks of hypotension, weakness, numbness. Requires careful monitoring. Use of infusion pumps requires additional equipment and staff education. Expensive if infusion pumps are employed
Peripheral nerve block	I	Plexus block, peripheral nerve block and infiltration. Effective regional analgesia. Opioid sparing

that the use of multimodal analgesia after major surgery may improve recovery and thus reduce costs of hospital stay. Several authors have proposed that the "pain-free state" should be employed as a fundamental component of an aggressive regimen of postoperative mobilization and early oral feeding in a process of acute rehabilitation after surgery. Multimodal analgesia employing NSAIDs to minimize opioid requirements has the particular advantage over unimodal systemic opioid administration. In addition, by using non-opioid drugs as part of a balanced analgesic plan, the patient can return to normal enteral nutrition much more quickly by avoiding the undesiderable opioid problems of gastrointestinal stasis, nausea and vomiting. The best approach to post-operative pain therapy is based on pharmacologic protocols, using all drugs involved in post-operative pain relief (Table 1). In fact, a correct use of drugs for pain should control symptoms and achieve a good outcome. As the World Health Organization guidelines support there are two main goals to consider [25]: Pain therapy must be assessed "By the Patient" and "By the Ladder".

By the Patient

Different factors may alter the amount of pain suffered by the individual patient. The general conditions, the patient himself, his disease and psychological factors are important factors to consider in order to start an adequate pain management (Box 5a, 5b). Severe pain can cause a number of changes in an individual behaviour, including increased self absorption

Box 5.

a. Psychological factors affecting pain response
- Cultural differences
- Cognitive appraisal
- Fear and anxiety
- Neuroticism and extroversion
- Perceived control of events
- Coping style
- Attention/distraction

b. Psychological methods for reducing pain
- Placebo and expectation
- Psychological support
- Sensory information
- Relaxation training
- Cognitive coping strategies

and withdrawal from interpersonal contact. Fear and anxiety are the major emotional concomitants of acute pain and are especially pronounced when associated with fear of death. Severe acute pain that remains unrelieved for days may lead to depression and helplessness as a result of patients experiencing a loss of control over their environment. It is now generally agreed that unrelieved severe acute pain exacerbates premorbid tendencies for anxiety, hostility, depression, or preoccupation with health. In a few cases, the inability to cope with pain may create an acute psychotic reaction. However, acute pain is one of the important factors contributing to the development of delirium in intensive care units. For all these reasons psychological approaches are an integral part of the medical care of the patient with pain (Box 5b). All patients can benefit from psychological assessment and support and some are good candidates for specific psychological therapy. Cognitive-behavioural interventions can help some patients decrease the perception of distress engendered by the pain through the development of new coping skills and the modification of thoughts, feelings and behaviours. Relaxation methods may be able to reduce muscular tension and emotional arousal or enhance pain tolerance. Other approaches reduce anticipatory anxiety that may lead to avoidant behaviours or lessen the distress associated with the pain. Approaches that give patients more control are likely to be successful in reducing anxiety and decreasing the requirement for pain and medication. Patient-controlled analgesia (PCA) is a highly successful example (see below). Successful implementation of these approaches in the postoperative patients requires a cognitively intact patient and a dedicated, well-trained clinician.

By the Ladder

Analgesic pharmacotherapy is the mainstay of postoperative pain management. Although concurrent use of other interventions is valuable in many patients and essential in some, analgesic drugs are needed in almost every case. The guiding principle of analgesic management is the individualization of therapy. Through a process of repeated evaluations, drug selection and administration is individualized so that a favourable balance between pain relief and adverse pharmacological effects is achieved and maintained (Table 1). An expert committee convened by the World Health Organization (WHO) has proposed a useful approach to drug selection for acute and chronic pain states, which has become known as the 'analgesic ladder' (World Health Organization 1986) (Fig. 5). The World Federation of Societies of Anaesthesiologist (WFSA) has been developed to treat acute and post-operative pain. Initially, pain can be expected to be severe and may need strong analgesics in combination with local anaesthetic blocks and peripherally acting drugs to be controlled (Fig. 6). When combined with appropriate dosing guidelines, this approach is capable of providing adequate pain relief to patients. Emphasizing that pain intensity

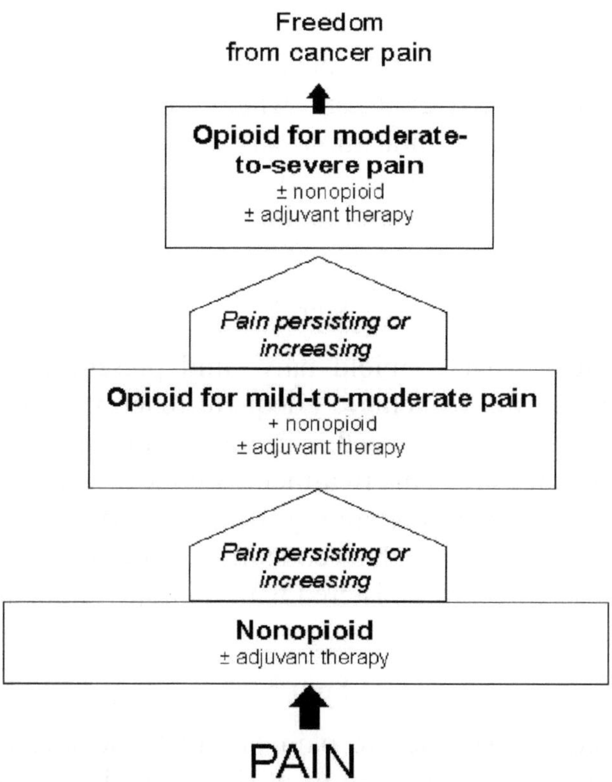

Fig. 5. WHO guidelines for pain therapy

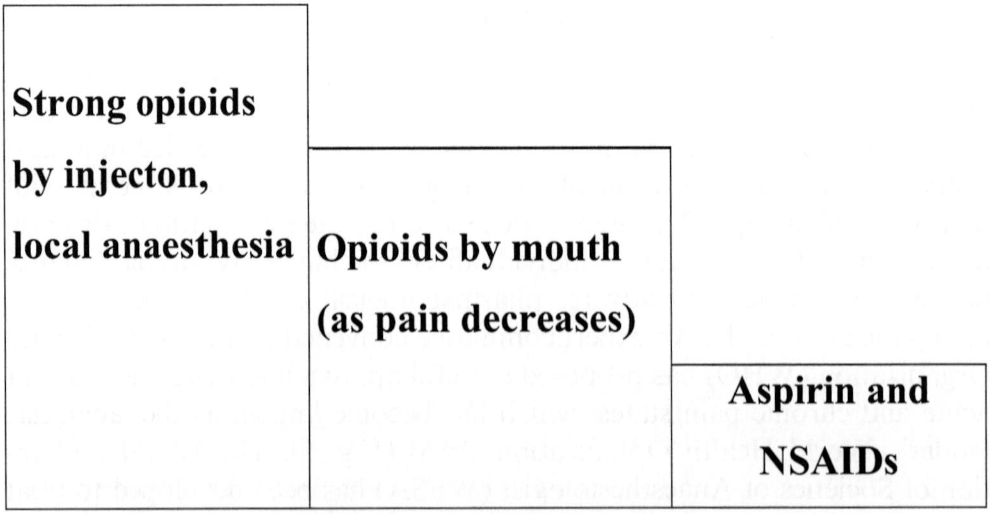

Fig. 6. WFSA analgesic ladder

should be the prime consideration in analgesic selection, the approach advocates three basic steps:

Step 1

Patients with mild to moderate post-operative-related pain should be treated with a non-opioid analgesic, which should becombined with an adjuvant drug if a specific indication exists. For example, a patient with mild to moderate arm pain caused by fracture may benefit when a tricyclic antidepressant is added to acetaminophen.

Step 2

Patients who are relatively opioid naive and present with moderate to severe pain, or who fail to achieve adequate relief after a trial of a non-opioid analgesic, should be treated with an opioid conventionally used to treat pain of this intensity. This treatment is typically accomplished by using a combination product containing a non-opioid (e.g. aspirin or acetaminophen) and an opioid (such as codeine, oxycodone or propoxyphene). This drug can also be co-administered with an adjuvant analgesic.

Step 3

Patients who present with severe pain or fail to achieve adequate relief following appropriate administration of drugs on the second rung of the 'analgesic ladder' should receive an opioid agonist conventionally used for pain of this intensity. This drug may also be combined with a non-opioid analgesic or an adjuvant drug. Recently, the evidence of the long-term efficacy of this approach has been the subject of criticism. Nonetheless, the approach remains widely used and has been strongly endorsed.

Based on clinical convention, analgesic drugs can be divided into three groups:

1. the non-opioid analgesics
2. the opioid analgesics
3. the adjuvant analgesics (which are drugs with other primary indications that can be effective analgesics in specific circumstances).

Non-Opioid Analgesics

The non-opioid analgesics acetylsalicylic acid, acetaminophen and the nonsteroidal anti-inflammatory drugs (NSAIDs) constitute a heterogeneous group of compounds that differ in chemical structure but share many pharmacological actions (Table 2). These drugs are useful alone for mild to moderate pain (step 1 of the analgesic ladder) and provide additive analgesia when combined with opioid drugs in the treatment of more severe pain [33–42]. Acetylsalicylic acid is a potent inhibitor of cyclooxygenases which is used frequently in medical care. Acetylsalicylic acid may be used as solution or as salt for very fast absorption, distribution, and pain relief. The inevitable irritation of the gastric mucosa may be acceptable in otherwise healthy patients. Acetylsalicylic acid should not be used in pregnant women (bleeding, closure of ductus arteriosus) or children before puberty (Reye's syndrome). Acetaminophen (or paracetamol) is a specific drug with characteristics similar to NSAIDs. Paracetamol has analgesic and antipyretic properties and is devoid of the side effects typical of the NSAIDs [33, 34]. The administration of paracetamol in children and infants for postoperative pain after minor surgery is a well established and safe treatment option, if appropriately used. However, if paracetamol is dosed according to traditional recommendations (about 2 mg/kg body weight) frequently a sufficient analgesic effect cannot be achieved immediately after painful interventions [38]. Recently, a higher initial dose (40 mg/kg body weight) was suggested for effective postoperative pain control [44]. Current recommendations also involve appropriate timing and route of administration of paracetamol to be most effective under different clinical circumstances. The rectal route of administration is unreliable for eliciting an analgesic effect and the oral route is to be prefer. The risk for liver toxicity appears to be very low if the daily paracetamol dose does not exceed 90 mg/kg body weight in healthy children and if specific risk factors of the individual patient are always considered [44]. The NSAIDs can be categorized into four different groups:

Table 2. *Non-Opioid Analgesics*

Chemical class	Generic name
Non-acidic	Acetaminophen
	Nabumetone
	Nemuselide
	Meloxicam
Acidic	
Salicylates	Aspirin
	Diflunisal
	Choline magnesium trisalycilate
	Salsalate
Proprionic acids	Ibuprofen
	Naproxen
	Fenoprofen
	Ketoprofen
	Flurbiprofen
	Oxaprosin
Acetic acids	Indomethacin
	Tolmetin
	Sulindac
	Diclofenac
	Ketorolac
Oxicams	Piroxicam
Fenamates	Mefenamic acid
	Mecolofenamic acid

a. NSAIDs with low potency and short elimination halflife
b. NSAIDs with high potency and short elimination halflife
c. NSAIDs with intermediate potency and elimination halflife
d. NSAIDs with high potency and long elimination halflife

a. NSAIDs With Low Potency and Short Elimination Halflife

The prototype of this group is ibuprofen. The bioavailability of ibuprofen is complete; the elimination is always fast even in patients with severe impairment of the liver or kidney function. Ibuprofen is used in single doses between 200 mg and 0.8 g. Ibuprofen (at low doses) appears particularly useful for treatment of acute and post-operative pain. It may also be used in chronic rheumatic diseases. Ibuprofen is also used as a pure S-enantiomer and this enantiomer is a direct COX-inhibitor. It has not been proven whether the use of the pure S-enantiomer offers any benefit.

b. NSAIDs With High Potency and Short Elimination Halflife

The drugs of this group are standard in the therapy of rheumatic pain. The most widely used compound is diclofenac, which is less active on COX1 than on COX2. This is taken as a reason for the low incidence of gastrointestinal side effects. The limitations of diclofenac result from the usual galenic formulation consisting of a monolythic acid-resistant encapsulation. This may cause retarded absorption of the active ingredient. Moreover, diclofenac encounters first-pass metabolism, which limits oral bioavailability (about 50%). The higher incidents of liver toxicity with diclofenac may result from first-pass metabolization. This group contains important drugs such as indometacin and ketoprofen. All of them show high oral bioavailability and good effectiveness in post-operative pain relief.

c. NSAIDs With Intermediate Potency and Elimination Halflife

This group of drugs is intermediate in potency and speed of elimination. Some forms of migraine and post-operative pain appear as adequate indications for diflunisal and naproxen.

d. NSAIDs With High Potency and Long Elimination Halflife

The fourth group consists of the oxicam drugs (meloxicam, piroxicam, and tenoxicam). These compounds owe their slow elimination to slow metabolization together with a high degree of enterohepatic circulation. The long half-life (days) does not make these oxicam drugs of first choice for acute and post-operative pain. Their main indications is inflammatory pain likely to persist for days (i.e. bone metastases). The high potency and long persistence in the body may be the reason for the higher incidence of serious adverse effects in the gastrointestinal tract and the kidney.

Unlike opioid analgesics, the non-opioid analgesics have a ‘ceiling’ effect for analgesia and produce neither tolerance nor physical dependence. Some of these agents, like acetylsalicylic acid and the NSAIDs, inhibit the enzyme cyclo-oxygenase and consequently block the biosynthesis of prostaglandins, inflammatory mediators known to sensitize peripheral nociceptors [43–45]. A central mechanism is also likely and appears to predominate in acetaminophen analgesia, because its action on PGE2 synthesis. The safe administration of the non-opioid analgesics requires familiarity with their potential adverse effects. Acetylsalicylic acid and the other NSAIDs have a broad spectrum of potential toxicity. Bleeding diathesis due to inhibition of platelet aggregation, gastroduodenopathy (including peptic ulcer disease) and renal impairment are the most common [46–50]. Less common adverse effects include confusion, precipitation of cardiac failure and exacerbation of hypertension. Particular caution is required in the administration of these agents to patients at increased risk of adverse effects, including the elderly and those with blood clotting disorders, predilection to peptic ulceration, impaired renal function and concurrent corticosteroid therapy [51–55]. Of the NSAIDs, the drugs that are relatively selective cyclo-oxygenase-2 inhibitors (e.g. nabumetone, nemuselide and meloxicam) and those that are non-acetylated salicylates (choline magnesium trisalicylate and salsalate) are preferred in patients who have a predilection to peptic ulceration or bleeding; these drugs have less effect on platelet aggregation and no effect on bleeding time at the usual clinical doses. The development of NSAIDs that are fully selective cyclo-oxygenase-2 inhibitors may provide additional agents with favourable safety profiles that may be preferred in the treatment of the medically frail. Acetaminophen rarely produces gastrointestinal toxicity and there are no adverse effects on platelet function; hepatic toxicity is possible, however, and patients with chronic alcoholism and liver disease can develop severe hepatotoxicity at the usual therapeutic doses. The optimal administration of non-opioid analgesics requires an understanding of their clinical pharmacology. There is no certain knowledge of the minimal effective analgesic dose, ceiling dose or toxic dose for any individual patient with post-operative pain. These doses may be higher or lower than the usual dose ranges recommended for the drug involved. These observations support an approach to the administration of NSAIDs that incorporates both low initial doses and dose titration. Through a process of gradual dose escalation, it may be possible to identify the ceiling dose and reduce the risk of significant toxicity [56]. Several days are needed to evaluate the efficacy of a dose when NSAIDs are used in the treatment of grossly inflammatory lesions, such as arthritis. Since failure with one NSAID can be followed by success with another, sequential trials of several NSAIDs may be useful to identify a drug with a favourable balance between analgesia and side effects [57–60]. Table 3 shows the most commonly NSAIDs used in adults and in children for postoperative pain relief.

Opioid Analgesics

Postoperative pain of moderate or greater intensity should generally be treated with a systemically administered opioid analgesic [1, 5]. The need for analgesia largely depends on the magnitude of the surgical trauma. Generally, the greater the magnitude of surgery, the greater the postoperative discomfort. Major surgery typically requires more aggressive and complex pain management techniques to achieve optimal analgesia. Major surgery usually requires postoperative pain therapy with opioids associated with other drugs, such as oral or parenteral NSAIDs and local anaes-

Table 3. *NSAIDs Commonly Used for Postoperative Pain Relief in Adult and Pediatric Patients*

Drug	Pediatric dosage	Adult dosage	Notes
Acetaminophen or Paracetamol	10–15 mg/kg every 4–5 hr os 20–40 mg/kg every 6 hr rectally or Bolus 20 mg/kg + 15 mg/kg every 4 hr os Bolus 40 mg/kg + 20 mg/kg every 6 hr	325–650 mg every 4–6 hr (max 4 gr/day) os	No gastroenteric or hematologic side effects, No antinflogistic effect
Ibuprofen	5–10 mg/kg every 6–8 hr	200 mg every 3–4 hr os	gastroenteric or hematologic side effects, antinflogistic effect
Naproxen	5 mg/kg every 8–12 hr	0,5–1 gr/day	gastroenteric or hematologic side effects, antinflogistic effect
Ketorolac	Bolus: 1–3 mg/kg every 8 hr Drip: 0.20 mg/kg/hr	10 mg every 4–6 hr os (max 40 mg/day) 10–30 mg every 4–6 hr im or iv (max 90 mg/day)	Renal and hepatic toxicity
Acetylsalicylic acid	10–15 mg/kg every 6–8 hr	0,5–1 gr every 4–6 hr os	Reye's syndrome (children), gastroenteric or hematologic side effects

thetics, administered by different ways (wound infiltration, peripheral nerve block, epidural or iv). Opioids should be used in a multimodal balanced analgesia approach that minimizes opioid requirement and the degree of their side effects [70, 71]. Optimal use of opioid analgesics requires a sound understanding of the general principles of opioid pharmacology, the pharmacological characteristics of each of the commonly used drugs and principles of administration. Fear of potential side effects has limited their use in many countries; this cultural phenomenon seems now to be overcame by the effective opioid titration with the use of incremental doses and a careful monitoring of side effects: this has largely increased their use both in adult patients and especially in children [68, 72]. The mechanism of action of opioid analgesics depends on the interaction of these molecules with specific receptors to which they bind and their intrinsic activity at that receptor [5]. The receptors have a pharmacologic nomenclature: μ (1 and 2), δ, κ. All opioids exert their effects by activating one or more of these receptors. Analgesia involves activation of mu_1 receptors in the brain and *kappa* receptors in the spinal cord. Mu_2 receptors are involved in respiratory depression and intestinal constipation. The contribution of *delta* receptors to analgesia in unclear, and may be more closely related to euphoria. The actions of opioids on receptors can vary depending on the location within the body. For example, a particular opioid may act as an antagonist at the *kappa* receptors in the brain, but as an agonist at the same type of receptors in the large intestines. Activation of Mu_1, Mu_2, and *delta* receptors close potassium channels, while *kappa* receptors are linked to calcium channels. Humans that have become tolerant to activation of one receptor type are not necessarily tolerant to the others.

Opioid Classification

Based on their interactions with the various receptor subtypes, opioid compounds can be divided into agonist, partial agonist, and mixed agonist-antagonist drugs (Table 4). The pure agonist drugs (Table 5) are most commonly used in clinical pain management, both in adult patients and in children (Table 6). The mixed agonist-antagonist opioids (pentazocine, nalbuphine, butorphanol and dezocine) and the partial agonist opioids (buprenorphine) play a minor role in the management of post-operative pain because of the existence of a ceiling effect for analgesia, the potential for precipitation of withdrawal in patients physically dependent to opioid agonists and, in some cases, the problem of dose-dependent psychotomimetic side effects that exceed those of pure agonist drugs. The pure agonist opioid drugs appear to have no ceiling effect for analgesia. As the dose is raised, analgesic effects increase until either analgesia is achieved or the patient loses consciousness. This increase in effect occurs as a log-linear

Table 4. *Opioid Classification*

Agonists	Partial agonists	Mixed agonist/antagonists
Morphine	Buprenorphine	Pentazocine
Codeine		Butorphanol
Oxycodone		Nalbuphine
Hydrocodone		Dezocine
Dihydrocodeine		
Heroin		
Oxymorphone		
Meperidine		
Levorphanol		
Hydromorphone		
Methadone		
Fentanyl		
Sufentanil		
Alfentanil		
Propoxyphene		

function: dose increments on a logarithmic scale yield linear increases in analgesia. In practice, it is the appearance of adverse effects, including confusion, sedation, nausea, vomiting or respiratory depression, that imposes a limit on the useful dose. The overall efficacy of any drug in a specific patient will be determined by the balance between analgesia and side effects that occurs during dose escalation.

'Weak' Versus 'Strong' Opioids

The division of opioid agonists into 'weak' versus 'strong' opioids was incorporated into the original 'analgesic ladder' proposed by the WHO. This distinction was not based on a fundamental difference in the pharmacology of the pure agonist opioids, but rather reflected the customary manner in which these drugs were used. This explains the observation that some opioids that were customarily used for moderate pain (step 2 of the analgesic ladder), such as oxycodone, are also used for severe pain in selected patients. Indeed, the controlled-release formulation of oxycodone is now widely used in the management of severe pain. Conversely, low-dose formulations of controlled-release morphine are suitable for the management of pain of moderate severity. Weak opioids are indicated in mild to moderate pain, usually associated to other drugs such as paracetamol. A weak opioid should be added to, not substituted for, a non opioid and it's important not to "kangaroo" from weak opioid to weak opioid. If a weak

Table 5. *Opioid Agonist Drugs*

Drug	Dose (mg) equianalgesic to 10 mg morphine IM	P.O.	Half-life (hrs)	Duration of action (hrs)	Comments
Codeine	130	200	2–3	2–4	Usually combined with a non-opioids
Oxycodone	15	30	2–3	2–4	Usually combined with a non-opioids
Propoxyphene	100	50	2–3	2–4	Usually combined with a non-opioids. Norpropoxyphene toxicity may cause seizures
Morphine	10	30	2–3	3–4	Multiple routes of administration available. Controlled release available. M6G accumulation in renal failure
Hydromorphone	2–3	7.5	2–3	2–4	No known active metabolites. Multiple routes available
Methadone	10	3–5	15–190	4–8	Plasma accumulation may lead to delayed toxicity. Dosing should be initiated on a p.r.n. basis. When switching to Methadone from another opioid, potency may be much greater than expected; the dose of Methadone should be lowered by 75–90% to account for this
Meperidine	75	300	2–3	2–4	Low oral bioavailability. Normeperidine toxicity limits utility. Containdicated in patients with renal failure and those receiving MAO inhibitors
Oxymorphone	1	10 (p.r)	2–3	3–4	No oral formulation available. Less histamine release
Heroin	5	60	0.5	3–4	High-solubility morphine prodrug
Levorphanol	2	4	12–15	4–8	Plasma accumulation may lead to delayed toxicity
Fentanyl transdermal	Empirically, transdermal fentanyl 100 μg/h = 2–4 mg/h intravenous morphine			48–72	Patches available to deliver 25, 50, 75 and 100 μg/h

Table 6. *Opioids Commonly Used for Postoperative Pain Relief in Children*

Drug	Iv/sc starting dose	Oral starting dose	Notes
Codeine	—	0.5–1 mg/kg every 3–4 hr	Nausea, vomiting
Idromorphone	Bolus: 0.015 mg/kg every 2–4 hr Drip: 0.006 mg/kg/hr	0.06 mg/kg every 3–4 hr	Nausea, vomiting, urinary retention
Morphine	Bolus: 0.05–0.1 mg/kg every 2–4 hr Drip: 0.03 mg/kg/hr	0.15–0.3 mg/kg every 4 hr	Nausea, vomiting, urinary retention, pruritus
Fentanyl	Bolus: 0.5–1 γ/kg every 1–2 hr Drip: 0.5–3.0 γ/kg/hr	—	Nausea, vomiting, urinary retention, pruritus, respiratory depression
Remifentanyl	Bolus: 0.1–0.5 γ/kg every 1 h Drip: 0.1–0.25 γ/kg/min	—	Nausea, vomiting, urinary retention, pruritus, respiratory depression
Sufentanyl	Bolus: 0.2 γ/kg every 1 h Drip: 0.1–0.5 γ/kg/min	—	Respiratory depression, haemodynamic alterations

opioid is inadequate when given regularly, the right step is to change to strong opioids.

Factors in Opioid Selection

The factors that influence opioid selection in post-operative pain states include pain intensity and the presence of co-existing disease.

Pain Intensity

Patients with moderate pain are conventionally treated with a combination product containing acetaminophen or aspirin plus codeine, dihydrocodeine, hydrocodone, oxycodone and propoxyphene. The doses of these combination products can be increased until the customary maximum dose of the non-opioid co-analgesic is attained (e.g. 4000 mg acetaminophen). Beyond this dose, the opioid contained in the combination product could be increased as a single agent or the patient could be switched to an opioid conventionally used for severe pain. New opioid formulations may improve the convenience of drug administration for patients with moderate pain.

These include controlled-release formulations of codeine, dihydrocodeine, oxycodone and tramadol. In some countries controlled-release morphine is available as a 10 mg tablet, which may also be used to treat moderate pain in the opioid-naive patient. The opioid drugs available to treat severe pain vary from country to country. Many countries provide clinicians with numerous options. In the United States, for example, patients who present with severe pain can be treated with morphine, hydromorphone, oxycodone, oxymorphone, fentanyl, methadone or levorphanol. As discussed previously, the agonist-antagonist opioids (e.g. pentazocine) are not preferred in the management of post-operative pain. Similarly the pharmacological characteristics of meperidine limit its role in the postoperative patients because its important side effects as tremulousness, multifocal myoclonus and, occasionally, seizures. Selective toxicity of meperidine can also occur following administration to patients receiving monoamine oxidase inhibitors. This combination may produce a syndrome characterized by hyperpyrexia, muscle rigidity and seizures, which may occasionally be fatal. The pathophysiology of this syndrome is related to excess availability of serotonin at the 5HT^ receptor in the central nervous system. Some patients will require sequential trials of several different opioids before a drug which is effective and well tolerated is identified. The frequency with which this strategy is needed is unknown, but it is estimated to be in the range of 15–30% of patients. The existence of different degrees of incomplete crosstolerance to various opioid effects (analgesia and side effects) may explain the utility of these sequential trials. To date, there are no data to suggest a specific order for opioid rotation. It is strongly recommended that clinicians be familiar with at least three opioid drugs used in the management of severe pain and have the ability to calculate appropriate starting doses using equianalgesic dosing data (Table 5).

Co-Existing Disease

Pharmacokinetic studies of meperidine, pentazocine and propoxyphene have revealed that liver disease may decrease the clearance and increase the bioavailability and half-lives of these drugs. These changes may eventuate in plasma concentrations higher than normal. Although mild or moderate hepatic impairment has only minor impact on morphine clearance, advanced disease may be associated with reduced elimination. Patients with renal impairment may accumulate the active metabolites of propoxyphene (norpropoxyphene), meperidine (normeperidine) and morphine (morphine-6-glucuronide). In the setting of renal failure or unstable renal function, titration of these drugs requires caution and close monitoring. If adverse effects appear, a switch to an alternative opioid is often recommended.

Selecting the Appropriate Route of Systemic Opioid Administration

Opioids should be administered by the least invasive and safest route capable of providing adequate analgesia.

Non-Invasive Routes

The oral route of opioid administration is the preferred approach in routine practice. Alternative routes are necessary for patients who have impaired swallowing or gastrointestinal dysfunction, those who require a very rapid onset of analgesia and those who are unable to manage either the logistics or side effects associated with the oral route. For highly tolerant patients, the inability to prescribe a manageable oral opioid programme due to an excessive number of tablets or volume of oral solution may be an indication for the use of a non-oral route. For patients who do not require very high opioid doses, non-invasive alternatives to the oral route of opioid administration include the rectal, transdermal and sublingual routes. Rectal suppositories containing oxycodone, hydromorphone, oxymorphone and morphine have been formulated and controlled-release morphine tablets can also be administered per rectum. The potency of opioids administered rectally is believed to approximate oral administration. Fentanyl is the only opioid available as a transdermal preparation. The fentanyl transdermal system consists of a drug reservoir that is separated from the skin by a copolymer membrane that controls the rate of drug delivery to the skin surface such that the drug is released into the skin at a nearly constant amount per unit time. There is some interindividual variability in fentanyl bioavailability by this route and this phenomenon, combined with large differences in elimination pharmacokinetics, necessitates dose titration in most cases. Transdermal patches capable of delivering 25, 50, 75 and 100 μg/h are available. Multiple patches may be used simultaneously for patients who require higher doses. At the present time, the limitations of the transdermal delivery system include its cost and the requirement for an alternative short-acting opioid for breakthrough pain. Sublingual absorption of any opioid could potentially yield clinical benefit, but bioavailability is very poor with drugs that are not highly lipophilic and the likelihood of an adequate response is consequently low. Sublingual buprenorphine, a relatively lipophilic partial agonist, can provide adequate relief of mild to moderate postoperative pain. Both fentanyl and methadone are relatively well absorbed through the buccal mucosa and sublingual administration of an injectable formulation is occasionally performed in the relatively opioid-naive patient who transiently loses the option of oral dosing. Overall, however, the sublingual route has limited value due to the lack of formulations, poor absorption of most drugs and the inability to deliver high doses or prevent swallowing of the dose. An oral transmucosal formulation of fentanyl, which incorporates the drug into a candy base, is under evaluation. Studies in cancer patients suggested that it is useful and that it can provide rapid and very effective relief of breakthrough pain.

Invasive Routes

For patients undergoing a trial of systemic drug administration, a parenteral route must be considered when the oral route is precluded or there is need for rapid onset of analgesia, or a more convenient regimen. Repeated parenteral bolus injections, which may be administered by the intravenous (IV), intramuscular (IM) or subcutaneous (SC) routes, may be useful in some patients but are often compromised by the occurrence of prominent 'bolus' effects (toxicity at peak concentration and/or pain breakthrough at the trough). Repetitive IM injections are a common practice, but they are painful and offer no pharmacokinetic advantage; their use is not recommended. Repeated bolus doses without repeated skin punctures can be accomplished through the use of an indwelling IV or SC infusion device. To deliver repeated SC injections, a 27-gauge infusion device (a 'butterfly') can be left under the skin for up to a week. Intravenous bolus administration provides the most rapid onset and shortest duration of action. Time to peak effect correlates with the lipid solubility of the opioid and ranges from 2–5 minutes for methadone to 15–30 minutes for morphine and hydromorphone. This approach is commonly applied in two settings:

1. to provide parenteral opioids to patients who already have venous access and are unable to tolerate oral opioids;
2. to treat very severe pain, for which IV doses can be repeated at an interval as brief as that determined by the time to peak effect, if necessary, until adequate relief is achieved.

Continuous parenteral infusions are useful for many patients who cannot be maintained on oral opioids. Long-term infusions may be administered IV or SC. In practice, the major indication for continuous infusion occurs among patients who are unable to swallow or absorb opioids. Continuous infusion is also used in some patients whose high opioid requirement renders oral treatment impractical. Continuous SC infusion is often used for ambulatory postoperative patients. A range of pumps is available, which vary in complexity, cost and ability to provide patient-controlled 'rescue doses' as an adjunct to a continuous basal infusion. Opioids suitable for continuous SC infusion must be soluble, well absorbed and non-irritant. Experience has been reported with heroin, hydromorphone, oxymorphone, morphine and fentanyl. Methadone appears to be relatively irritating and is not recommended. To maintain the comfort of an infusion site, the SC infusion rate should not exceed 5 cc/hr. Patients who require high doses may benefit from the use of concentrated solutions. In selected cases, concentrated opioid solutions can be compounded specifically for continuous SC infusion. Subcutaneous infusion, like repeated SC bolus injections, can usually be administered using a 27-gauge 'butter-fly' needle. The infraclavicular and anterior chest sites provide the greatest freedom of movement for patients, but other sites may be used. A single infusion site can usually be maintained for 5–7 days. Occasional patients develop focal erythematous swelling at the site of injection; this appears to be a common complication with methadone and has also been described with morphine and hydromorphone. Continuous SC delivery of drug combinations may be indicated when nausea, anxiety or agitation accompanies pain. An antiemetic, neuroleptic or anxiolytic may be combined with an opioid, provided that it is non-irritant, miscible and stable in combined solution. Experience has been reported with metoclopromide, haloperidol, scopolamine, cyclizine, methotrimeprazine, chlorpromazine and midazolam. In some circumstances, continuous IV infusion may be the most appropriate way of delivering an opioid. The need for very large doses, or treatment with methadone, may suggest the utility of this approach. If continuous IV infusion is to be continued on a long-term basis, a permanent central venous port is recommended.

Scheduling of Opioid Administration

The schedule of opioid administration should be individualized to optimize the balance between patient comfort and convenience. 'Around the clock' dosing and 'as needed' s dosing both have a place in clinical practice.

'Around the Clock' Dosing

Patients with severe post-operative pain generally benefit from scheduled 'around the clock' dosing, which can provide the patient with continuous relief by preventing the pain from recurring. Clinical vigilance is required, however, when this approach is used in patients with no previous opioid exposure and when administering drugs that have long half-lives (methadone or levorphanol) or produce metabolites with long half-lives (e.g. morphine-6-glucuronide and norpropoxyphene). In the latter situations, delayed toxicity may develop as plasma drug (or metabolite) concentrations rise toward steady state levels. Most patients who receive an 'around the clock' opioid regimen should also be provided a so-called 'rescue dose', which is a supplemental dose offered on an 'as needed' basis to treat pain that breaks through the regular schedule. The frequency with which the rescue dose can be offered depends on the route of administration and the time to peak effect for the particular drug. Oral rescue doses are usually offered up to every 1–2 hours and parenteral doses can be offered as frequently as every 15–30 minutes. The integration of 'around the clock' dosing with 'rescue doses' provides a method for safe and rational stepwise dose escalation, which is applicable to all routes of opioid administration. Patients who require more than 4–6 rescue doses per day should generally undergo escalation of the baseline dose. The quantity of the rescue medication consumed can be used to guide the dose increment. Controlled-release preparations of opioids can lessen the inconvenience associated with the use of 'around the clock' administration of drugs with a short duration of action. Currently, controlled-release formulations are available for administration by the oral, transdermal and rectal routes. The largest experience has been reported with oral controlled-release morphine preparations with 8–12 hours' duration of effect. Other controlled-release formulations include once-daily morphine preparations, controlled-release morphine suppositories and liquid suspension, transdermal fentanyl, and controlled-release tablets of oxycodone, hydromorphone, codeine and dihydrocodeine. Clinical experience suggests that controlled-release formulations should not be used to rapidly titrate the dose in patients with severe pain. The time required to approach steady-state plasma concentration after dosing is initiated or changed (at least 24 hours) may complicate efforts to rapidly identify the appropriate dose. Repeat-dose adjustments for patients with severe pain are performed more efficiently with short-acting preparations, which may be changed to a controlled-release preparation when the effective 'around the clock' dose is identified.

'As Needed' Dosing

In some situations, opioid administration on an 'as needed' basis, without an 'around the clock' dosing regimen, may be beneficial. In the opioid-naive patient, 'as needed' dosing may provide additional safety during the initiation of opioid therapy, particularly when rapid dose escalation is needed or therapy with a long half-life opioid such as methadone or levorphanol is begun. 'As needed' dosing may also be appropriate for patients who have rapidly decreasing analgesic requirement or intermittent pain separated by pain-free intervals.

Patient-Controlled Analgesia

Patient-controlled analgesia (PCA) generally refers to a technique of parenteral drug administration in which the patient controls an infusion device that delivers a bolus of analgesic drug 'on demand' according to parameters set by the physician. Use of a PCA device allows the patient to overcome variations in both pharmacokinetic and pharmacodynamic factors by carefully titrating the rate of opioid administration to meet individual analgesic needs. Although is should be recognized that the use of oral 'rescue doses' is, in fact, a form of PCA, the term is not commonly applied to this situation. Long-term PCA in postoperative patients is most commonly accomplished via the intravenous route using an ambulatory infusion device. In most cases, PCA is added to a basal infusion rate and acts essentially as a rescue dose. Rare patients have benefited from PCA alone to manage episodic pain characterized by an onset so rapid that an oral dose could not provide sufficiently prompt relief. Long-term intravenous PCA can be used for patients who require doses that cannot be comfortably tolerated via the subcutaneous route or in those who develop local reactions to subcutaneous infusion. PCA has also been applied to spinally administered opioids and non-opioid approaches such as nitrous oxide. In pediatric age PCA is recommended for children of 8 years or more, without disabilities, in whom moderate to severe pain is anticipated for 24 hours or more. Most children over the age of 7 years understand the PCA concept, and sometimes even younger children can learn to use PCA, but some may not have the cognitive or emotional resources to use it. In children as young as 5 or 6 years PCA has also been used, however pain relief is not always satisfactory because of poor patient understanding. In these patients Nurse or Parent Controlled Analgesia (NCA/PCA) represent a more suitable modality of drug administration. As continuous infusion, PCA allows a steady analgesic serum concentrations with safety and efficacy in pain control (Fig. 7) [90]. The use of a background infusion of opioids in PCA therapy is controversial. It might provide better analgesic during sleep but this is not strongly supported by literature. However it may increase the occurrence of adverse effects such as nausea and respiratory depression [87, 88]. Morphine is the most common drug used in PCA, followed by Fentanyl and Hydromorphone [88–91]. The selection of opioid used in PCA is perhaps critical than the appropriate selection of parameters such as bolus dose, lockout and background infusion rate (Table 7) [91]. PCA dosage regimens must be individualized on the basis of pain intensity and monitoring pain parameters must be age appropriate. Monitoring involves measurements of respiratory rate, level of sedation and oxygen saturation. Efficacy of PCA therapy is assessed by self-reporting, visual analogue scales, faces pain scales and usage pattern. The effectiveness of analgesic techniques may be limited by the incidence and severity of adverse effects; potential adverse effects of PCA therapy, including respiratory depression, nausea, vomiting, and pruritus, can be prevented or controlled by the use of adjuvant drugs and by careful titration. The patient should be instructed in the use of PCA prior to coming to operating room or even in the anaesthetic room before induction. Clinicians must become aware on age-related and developmental differences in

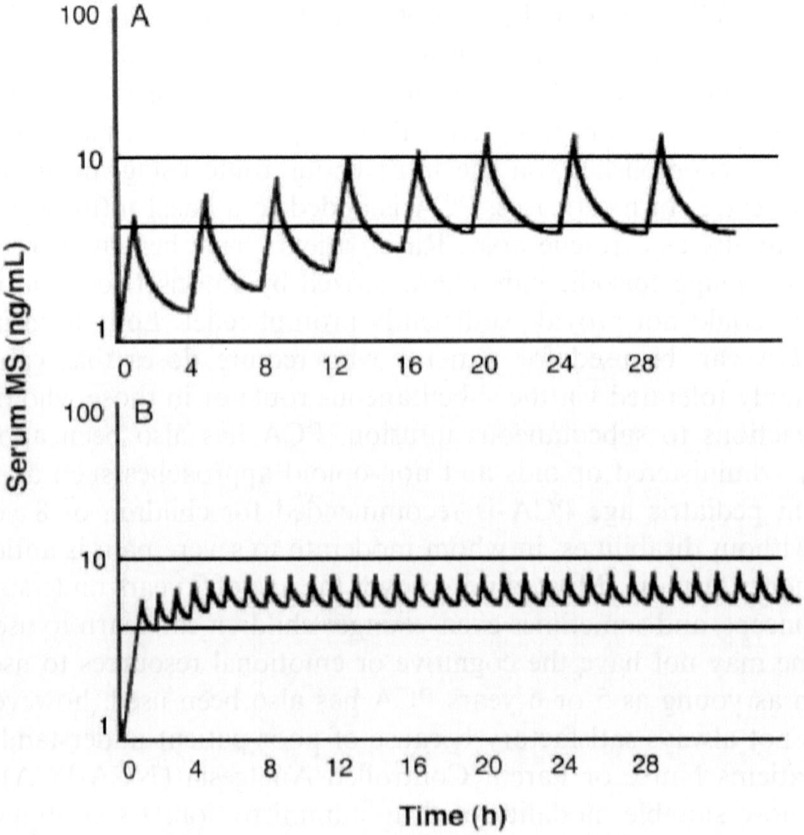

Fig. 7. Opioids plasma concentration following bolus or PCA administration. (A) bolus infusion; (B) PCA administration

Table 7. *PCA protocol with morphine*

PCA protocol	Purpose	Initial dose recomandations (Morphine)
Loading dose	Obtain immediate pain control	0.05 to 0.1 mg/kg max 10 mg
Background infusion (basal rate)	To mantain pain control	0.01 to 0.02 mg/kg/hr
Interval dose (PCA dose)	A bolus interval dose to tritate pain control by the patient himself	0.01 to 0.02 mg/kg
Lockout	To prevent overdose	6–15 minutes
4 hours maximum	To prevent overdose	0.25 to 0.35 mg/kg

the pharmacokinetic, pharmacodynamic and monitoring parameters for the patients with PCA therapy. To date, safety and efficacy of PCA also in paediatric patients has been established and a role of this procedure has been proposed in postoperative pain management as well as burns, oncology and palliative care.

Management of Opioid Adverse Effects

Successful opioid therapy requires that the benefits of analgesia clearly outweigh treatment-related adverse effects. This implies that a detailed understanding of adverse opioid effects and the strategies used to prevent and manage them are essential skills for all involved in postoperative pain management. The pathophysiological mechanisms contributing to adverse opioid effects are incompletely understood. The appearance of these effects depends on a number of factors, including patient age, extent of disease, concurrent organ dysfunction, prior opioid exposure, the route of drug administration, and the adverse drug interactions. The potential for additive side effects and serious toxicity from drug combinations must be recognized. The sedative effect of an opioid may add to that produced by numerous other centrally acting drugs, such as anxiolytics, neuroleptics and antidepressants. Likewise, drugs with anticholinergic effects probably worsen the constipatory effects of opioids. As noted previously, a severe adverse reaction, including excitation, hyperpyrexia, convulsions and death, has been reported after the administration of meperidine to patients treated with a monoamine oxidase inhibitor. The most frequent side effects of opioid drugs are represented by respiratory depression, nausea and vomiting, urinary retention, and physical dependence.

Respiratory Depression

Respiratory depression is potentially the most serious adverse effect of opioid therapy. Although these drugs may impair all phases of respiratory activity (rate, minute volume and tidal exchange), a compensatory increase in respiratory rate may obscure the degree of respiratory effect. This phenomenon explains the observation that patients who appear to have normal respiration during opioid therapy may be predisposed to respiratory compromise if any pulmonary insult occurs. Clinically significant respiratory depression is always accompanied by other signs of central nervous system depression, including somnolence and mental clouding. Respiratory compromise accompanied by tachypnoea and anxiety is never a primary opioid event. With repeated opioid administration, tolerance appears to develop rapidly to the respiratory depressant effects of the opioid drugs. As a result, opioid analgesics can be used in the management of postoperative pain without significant risk of respiratory depression. Indeed, clinically important respiratory depression is a very rare event in the post-operative patient whose opioid dose has been titrated against pain. When respiratory depression occurs in such patients, alternative explanations (e.g. pneumonia or pulmonary embolism) should be sought. Opioid-induced respiratory depression can occur, however, if pain is suddenly eliminated (such as may occur following neurolytic procedures) and the opioid dose is not reduced. This latter observation suggests that patients whose respiratory function is well compensated following repeated opioid administration do not entirely lack opioid effect on respiration, but rather have respiratory function that reflects a balance between ongoing opioid effects and factors that increase the respiratory drive, including pain, anxiety and alertness. When respiratory depression occurs in patients on opioid therapy, administration of the specific opioid antagonist naloxone usually improves ventilation. Naloxone is a potent pure semisynthetic opioid antagonist and it is used to reduce the effects of opioids and treat opioid overdoses. It has a high affinity for morphine receptors sites and reverses the effect of opioid analgesics by displacement. The degree of displacement is dose related [5]. When respiratory depression is observed, an initial dose of naloxone 2–4 μg/kg should be given and repeated to a total of 10 μg/kg. Duration of action of naloxone is shorter than the most opioids and a continuous infusion may be required to mantein reversal. Naloxone can precipitate a severe abstinence syndrome and should be administered only if strongly indicated. If the patient is bradypnoeic but readily arousable and the peak plasma level of the last opioid dose has already been reached, the opioid should be withheld and the patient monitored until improved. If severe hypoventilation occurs (regardless of the associated factors that may be contributing to respiratory compromise) or the patient is bradypnoeic and unarousable, naloxone should be administered. In the comatose patient, it may be prudent to place an endotracheal tube to prevent aspiration following administration of naloxone.

Nausea and Vomiting

Opioids may produce nausea and vomiting through both central and peripheral mechanisms. These drugs stimulate the medullary chemoreceptor trigger zone, increase vestibular sensitivity and have effects on the gastrointestinal tract (including increased gastric antral tone, diminished motility and delayed gastric emptying). In ambulatory patients, the incidence of nausea and vomiting has been estimated to be 10–40% and 15–40%, respectively. The likelihood of these effects is greatest at the start of opioid therapy. With the initiation of opioid therapy, patients should be informed that nausea can occur and that it is usually transitory and controllable.

Routine prophylactic administration of an antiemetic is not necessary, except in patients with a history of severe opioid-induced nausea and vomiting, but patients should have access to an antiemetic at the start of therapy if the need for one arises. Anecdotally, the use of prochlorperazine and metoclopramide has usually been sufficient. In patients with more severe or persistent symptoms, the most appropriate antiemetic treatment may be suggested by the clinical features. For nausea associated with early satiety, bloating or postprandial vomiting, all of which are features of delayed gastric emptying, metoclopramide is the most reasonable initial treatment. Patients with vertigo or prominent movement-induced nausea may benefit from the use of an antivertiginous drug such as scopolamine or meclizine. If signs of neither gastroparesis nor vestibular dysfunction are prominent, treatment is usually began with a neuroleptic, such as prochlorperazine or metoclopramide. Drug combinations are sometimes used and, in all cases, doses are escalated if initial treatment is unsuccessful. If these drugs are ineffective at relatively high doses, other options include trials of alternative opioids or treatment with antihistamines (e.g. hydroxyzine), other neuroleptics (e.g. haloperidol, chlorpromazine or droperidol), benzodiazepines (e.g. lorazepam) or steroids (e.g. dexamethasone) or the new serotonin antagonists (e.g. ondansetron).

Urinary Retention

Opioid analgesics increase smooth muscle tone and can occasionally cause bladder spasm or urinary retention (due to an increase in sphincter tone). This is an infrequent problem that is usually observed in elderly male patients. Tolerance can develop rapidly but catheterization may be necessary to manage transient problems. Rare patients appear to benefit from co-administration of either a cholinomimetic drug (e.g. bethanecol) or an a-adrenergic antagonist (e.g. terazocin).

Physical Dependence

Physical dependence is a pharmacological property of opioid drugs defined by the development of an abstinence (withdrawal) syndrome following either abrupt dose reduction or administration of an antagonist. Despite the observation that physical dependence is most commonly observed in patients taking large doses for a prolonged period of time, withdrawal has also been observed in patients after low doses or short duration of treatment. Occasionally, patients who are switched from a pure agonist opioid to transdermal fentanyl will develop an abstinence syndrome within the first 24 hours, presumably as a result of a delay in establishing blood levels after the transdermal system is placed. Physical dependence rarely becomes a clinical problem if patients are warned to avoid abrupt discontinuation of the drug; a tapering schedule is used if treatment cessation is indicated and opioid antagonist drugs (including agonist-antagonist analgesics) are avoided.

Adjuvant Analgesics

The term 'adjuvant analgesic' describes a drug that has a primary indication other than pain but is analgesic in some conditions. A large group of such drugs, which are derived from diverse pharmacological classes, is now used to manage non-malignant pain. In the post-operative patients, these drugs may be combined with primary analgesics in any of the three steps of the 'analgesic ladder' to improve the outcome for patients who cannot otherwise attain an acceptable balance between relief and side effects. The potential utility of an adjuvant analgesic is usually suggested by the characteristics of the pain or by the existence of another symptom that may be amenable to a non-analgesic effect of the drug. Whenever an adjuvant analgesic is selected, differences between the use of the drug for its primary indication and its use as an analgesic must be appreciated. Because the nature of dose-dependent analgesic effects has not been characterized for most of these drugs, dose titration is reasonable with virtually all. Low initial doses are appropriate given the desire to avoid early side effects. The use of low initial doses and dose titration may delay the onset of analgesia, however, and patients must be forewarned of this possibility to improve compliance with the therapy. There is great interindividual variability in the response to all adjuvant analgesics. Although patient characteristics, such as advanced age or coexistent major organ failure, may increase the likelihood of some (usually adverse) responses, neither favourable effects nor specific side effects can be reliably predicted in the individual patient. Furthermore, there is remarkable intraindividual variability in the response to different drugs, including those within the same class. These observations suggest the potential utility of sequential trials of adjuvant analgesics. The process of sequential drug trials, like the use of low initial doses and dose titration, should be explained to the patient at the start of therapy to enhance compliance and reduce the distress that may occur if treatments fail. In the management of postoperative pain, adjuvant analgesics can be broadly classified based on conventional use. The adjuvant drugs more frequently used in post-operative pain are corticosteroids, topical and local anaesthetics, neuroleptics and benzodiazepines.

Corticosteroids

Corticosteroids are among the most widely used adjuvant analgesics. They have been demonstrated to have analgesic effects in different conditions to significantly improve quality of life and to have beneficial effects on appetite, nausea, mood and malaise. The mechanism of analgesia produced by these drugs may involve anti-oedema effects, anti-inflammatory effects and a direct influence on the electrical activity in damaged nerves. The relative risks and benefits of the various corticosteroids are unknown and dosing is largely empirical. In the United States, the most commonly used drug is dexamethasone, a choice that gains theoretical support from the relatively low mineralocorticoid effect of this agent. Dexamethasone has also been conventionally used for raised intracranial pressure and spinal cord compression. Prednisone, methylprednisolone and prednisolone have also been widely used for other indications. Patients who experience pain and other symptoms may respond favourably to a relatively small dose of corticosteroid (e.g. dexamethasone 1–2 mg twice daily). In some settings, however, a high-dose regimen may be appropriate. Although high steroid doses are more likely to lead to adverse effects, clinical experience with this approach has been favourable. Although the effects produced by corticosteroids in patients with postoperative pain are often very gratifying, side effects are potentially serious and increase with prolonged usage. The varying constellations of adverse effects associated with brief or prolonged administration or with the withdrawal of these drugs following long-term use are widely appreciated. The risk of peptic ulcer is approximately doubled in patients chronically treated with corticosteroids. Several risk factors for peptic ulceration have been identified: relatively high dose, previous history of peptic ulceration, and concurrent administration of an NSAID. In general, the combined administration of a corticosteroid and an NSAID should be avoided. Patients who are predisposed to peptic ulcer disease can be considered for ulcer prophylaxis. Active peptic ulcer disease and systemic infection are relative contraindications to the use of corticosteroids as adjuvant analgesics.

Topical and Local Anaesthetics

Local anaesthetics are amazing drugs now commonly used in prevention and management of post-operative pain. Injected into tissue, around a nerve or for a regional block, they produce reversible block. For some operations, as inguinal hernia repair, there is proven advantage of regional over general anaesthesia. The use of local anaesthetics can produce reduced blood loss, faster surgery, reduced morbidity and faster rehabilitation. Local infiltration, blockade of peripheral nerves and plexuses, epidural blockade and regional analgesia represent the most frequent techniques adopted. Lidocaine and Bupivacaine are the most common local anaesthetics used in clinical practice. Particular attention to maximum drug dosing is required; excessive doses can cause seizures, cardiac depression and rhythm anomalies [5, 92]. Often local anaesthetic are combinated with epidural opioids to provide reliable analgesia in several pain contexts and extradural infusions of these drugs are used widely now for post-operative analgesia. Epidural local anaesthetics and opioids have been used for many years in the management of acute post-operative pain, and trauma. Several studies have confirmed synergism between local anaesthetics and opioids and support what has been observed clinically; that low doses of local anaesthetic and opioid can produce good analgesia. The mechanism of the synergy is not know. It may be that the local anaesthetic, by reducing the afferent input, is moving the opioid dose-respone to the right. Clinical observations suggest that chronic infusion of these two drugs can produce selective blockade, blocking pain fibers while leaving other sensory input intact. The adverse effects of these two drug classes are different. Epidural local anaesthetics can produce hypotension because of sympathetic blockade. Epidural opioids can produce delayed respiratory depression, urinary retention, priritus, nausea and vomiting. The epidural combination of these two drugs can produce pain relief, and the synergism between the drug classes offers the potential of effective analgesia at low doses of the components, minimizing the adverse effects of both. A clear demonstration of the advantage of the combination of local anaesthetic and opioid was seen in a comparison of 0.125% bupivacaine in saline, diamorphine 0.5 mg in 15 ml and diamorphine mixed with 0.125% bupivacaine infused for pain after major gynaecological surgery. The combination produced significantly superior analgesia to either of its component alone, without major side effects. Giving the diamorphine intravenously with epidural bupivacaine was significantly less effective than giving the same dose epidurally in combination with epidural bupivacaine. Three strategies in dosage of combination of these drugs are discernible: the low, the intermediate, and the high. High doses (bupivacaine 0.5% 25 mg/h and morphine 0.5 mg/h) were used to produce analgesia immediately after upper abdominal surgery but a some risk. Lower doses (bupivacaine 0.1% 4 mg/h and morphine 0.4 mg/h) did not provide total pian relief after major surgery, as thoracotomy. The issue of the minimum effective dose is of great importance, and unfortunately may have to be defined for particular circumstances. Topical formulations are useful for needle procedures, including EMLA, a cream containing an eutecthic mixture of 2 local anaesthetics (lidocaine 2.5% and prilocaine 2.5%). It is very effective in numbing the skin and the tissues just underneath the skin. Topical local anaesthetics can be used in the management of painful cutaneous and mucosal lesions and as a premedication prior to skin puncture. However, the depth of the skin which becomes numb is dependent upon how long the cream is left on. The maximum depth is about six to seven millimeters, after the cream has been left on the skin for two hours. This medication has been successfully used for a number of painful procedures, as bone marrow aspiration and lumbar puncture; the cream should be applied from 30 min to 1 hour before the shot or needle procedure [93]. Satisfactory numbing of the skin occurs 1 hour after application, reaches a maximum at 2 to 3 hours (1 hour for children less than 3 months), and lasts 1 hours after removal. EMLA has beeen proven to be safe, with low plasma local anaesthetic concentration. Mild side effects generally disappear spontaneously within 1 or 2 hours (skin paleness, redness, a changed ability to feel hot or cold, swelling, itching, and rash). It should not be used in children affected by a rare condition of congenital or idiopathic methaemoglobinemia, or in infants under the age of 12 months who are receiving treatment with methaemoglobin-inducing agents [93].

Neuroleptics

The role of neuroleptic drugs in the management of postoperative pain is limited. Methotrimeprazine is a proven analgesic and has been useful in bedridden patients with postoperative pain who experience pain associated with anxiety, restlessness or nausea. In this setting, the sedative, anxiolytic and antiemetic effects of this drug can be highly favourable and side effects, such as orthostatic hypotension, are less of an issue. Methotrimeprazine may be given by continuous SC administration, SC bolus injection or brief IV infusion (administration over 20–30 minutes). A prudent dosing schedule begins with 5–10 mg every 6 hours or a comparable dose delivered by infusion, which is gradually increased as needed. Most patients will not require more than 20–50 mg every 6 hours to gain the desired effects. Given their potential for serious toxicity and the limited evidence in support of analgesic efficacy, other neuroleptics should be used only for the treatment of delirium and nausea.

Benzodiazepines

There is little evidence that benzodiazepines have meaningful analgesic properties in most clinical circumstances and, indeed, there is some evidence that they may, in some circumstances, antagonize opioid analgesia. These drugs may play a role in the management of anxiety and muscle spasm.

Conclusions

Acute and post-operative pain has emerged as an important issue because ethics aspects and associated morbidity and mortality. Substantial progress in understanding peripheral, spinal cord and brain mechanisms involved in acute post-operative pain continues to be made with important consequences for treatment. The diagnosis and treatment of the cause of acute pain must always have high priority and post-operative pain management is an important goal in order to optimise medical care. Improved understanding of the pharmacology of the analgesics and the development of new techniques for analgesic administration have greatly enhanced the ability of medical doctors to success manage patients in pain. For some post-operative conditions the success of pharmacological strategies is remarkable, especially in adult patients. Even for children and adolescent with the most severe pain early evidence shows that it may be possible to reduce the impact of pain on the lives of the patients and their families. More action is necessary. Firstly, more paediatric centres are needed, to develop specific post-operative pain programmes. Secondly, collaboration between centres will be necessary to provide large enough samples of patients with the various pain conditions, considering the lack of data on this field. Finally, we must considerer that the incidence of post-operative pain in children is similar to that of adults but that our knowledge of how to help children cope with acute pain is underdeveloped. The psychological and physiologic uniqueness of children must not be forgotten. Cooperation and communication between the anaesthesiologist, surgeon, and paediatrician are essential for successful anaesthesia and pain management. The introduction of acute pain services has been shown to improve post-operative pain relief, but it is foreseeable that their role should expand and integrate into general perioperative care (Box 4). For these reasons the alleviation of pain and anxiety in post-operative patients is actually a high priority of all post-operative services and all persons involved in perioperative management of these patients are very much a part of "continuity of care" concept to obtain effective pain relief.

References

1. American Academy of Pediatrics, American Pain Society (2001) The assessment and management of acute pain in infants, children, and adolescents. Pediatrics 108: 793–797
2. Schechter NL, Blankon V, Pachter LM *et al* (1999) The ouchless place: no pain, children's gain. Pediatrics 99: 890–894
3. Zacharias M, Watts D (1998) Pain relief in children. BMJ 316: 1552
4. McGrath PA (1991) Assessment of pain in children. In: Bond MR, Charlton JE, Woolf CJ (eds) Proceedings of the VIth World Congress on Pain. Elsevier, Amsterdam
5. Wall PD, Melzack R (1999) Textbook of pain, 4th edn
6. Craig KD, Whitfield MF, Grunau RV, Linton J, Hadjistavropoulos HD (1993) Pain in the preterm neonate: behavioural and physiological indices. Pain 52: 287–299

7. Cote JJ, Morse JM, James SG (1991) The pain response of the postoperative newborn. J Adv Nurs 16: 378–387
8. McGrath PA (1995) Pain in the pediatric patient: practical aspects of assessment. Pediatr Ann 24:126–133, 137–138
9. Beyer JE, McGrath PJ, Berde CB (1990) Discordance between self-report and behavioral pain measures in children aged 3–7 years after surgery. J Pain Symptom Management 5: 350–356
10. Fuller BF, Conner DA (1995) The effects of pain on infant behaviours. Clin Nurs Res 4: 253–273
11. McGrath PJ (1998) Behavioural measures of pain. In: McGrath PJ (ed) IASP Press Measurement of Pain in Infants and Children. Seattle
12. Romsing J, Hertel S, Moller-Sonnergaard J, Rasmussen M (1996) Postoperative pain in children: comparison between ratings of children and nurses. J Pain Symptom Management 11: 42–46
13. Romsing J, Walther-Larsen S (1996) Postoperative pain in children: a survey of parents' expectations and perceptions of their children's experiences. Paediatr Anaesth 6: 215–218
14. Ambuel B, Hamlett KW, Marx CM, Blumer JL (1992) Assessing distress in pediatric intensive care environments: the COMFORT Scale. J Pediatr Psychol 17: 95–109
15. Van Dijk M, de Boer JB, Koot HM, Tibboel D *et al* (2000) The reliability and validity of the Comfort scale as a postoperative pain instrument in 0 to 3 years-old infants. Pain 84: 367–377
16. Norden J, Hannallah RS, Getson P, O'Donnell R, Kelliher G, Walker N (1991) Concurrent validation of an objective pain scale for infants and children. Anesthesiology 75: 312–316
17. Norden J, Hannallah RS, Getson P, O'Donnell R, Kelliher G, Walker N (1991) Reliability of an objective pain scale in children. Anesth Analgesia 72: S199
18. Myron AD, Yaster FA (2000) Acute pain in children. Pediatr Clin North Am 47: 22–26
19. Tyler DC, Tu A (1993) Toward validation of pain measurement tools for children: a pilot study. Pain 52: 301–309
20. Bieri D, Reeve RA, Champion GD, Addicoat L, Ziegler JB (1990) The Faces Pain Scale for the self-assessment of the severity of pain experienced by children: development, initial validation, and preliminary investigation for ratio scale properties. Pain 41:139–150
21. Hicks CL, von Baeyer CL, Spafford PA, van Korlaar I, Goodenough B (2001) The Faces Pain Scale revised: toward a common metric in pediatric pain measurement. Pain 93:173–183
22. Chambers CT, Giesbrecht K, Craaig KD, Bennett SM, Huntsman E (1999) A comparison of facial scales for measurement of pediatric pain: children's and parents' ratings. Pain 83: 25–35
23. Hain RDW (1997) Pain scales in children: a review. Palliative Med 11: 341–350
24. Tanube P (1995) Recognising pain as a component of the primary assessment; adding D for discomfort to the ABCs. J Emerg Nurs 21: 299–304

25. WHO (1998) Cancer pain relief and palliative care in children
26. Breivik H (2001) Opioids in cancer and chronic non-cancer pain therapy-indications and controversies. Acta Anaest Scand 45:1059–1066
27. Chiaretti A, Viola L, Pietrini D, Piastra M *et al* (2000) Preemptive analgesia with tramadol and fentanyl in pediatric neurosurgery. Child's Nerv Syst 16: 93–100
28. Møiniche S (2002) A qualitative and quantitative systematic review of preemptive analgesia for postoperative pain relief. The role of timing of analgesia. Anesthesiology 96: 725–741
29. McQuay HJ (1995) Preemptive analgesia: a systematic review of clinical studies. Ann Med 27: 249–258
30. Berry FA (1998) Preemptive analgesia for postoperative pain. Paediatr Anaesth 8:187–188
31. Feldman D, Reich N, Foster JM (1998) Pediatric anesthesia and postoperative analgesia. Pediatr Clin North Am 45: 1525–1537
32. Maxwell LG, Yaster M (2000) Perioperative management issues in pediatric patients. Anesth Clin North Am 18: 601–632
33. Collins SL, Edwards JE, Moore RA, McQuay HJ (2003) Single dose dextropropoxyphene, alone and with paracetamol (acetaminophen), for postoperative pain (Cochrane review). In: The Cochrane Library, Issue 2. Oxford
34. Moore A, Collins S, Carroll D, McQuay H, Edwards J (2003) Single dose paracetamol (acetaminophen), with and without codeine, for postoperative pain (Cochrane review). In: The Cochrane Library, Issue 2. Oxford
35. Collins SL, Moore RA, McQuay HJ, Wiffen PJ, Edwards JE (2003) Single dose oral ibuprofen and diclofenac for postoperative pain (Cochrane review). In: The Cochrane Library, Issue 2. Oxford
36. Collins SL (2003) Nonsteroidal anti-inflammatory drugs for postoperative pain: a focus on children. Paediatr Drugs 5:103–123
37. Chambliss CR, Anand KJ (1997) Pain management in the pediatric intensive care unit. Curr Opin Pediatr 9: 246–253
38. Hyllestead RA (2002) Comparative effect of paracetamol, NSAIDs or their combination in postoperative pain management: a qualitative review. Br J Anaesth 88: 199–214
39. Kost-Byerly S (2001) Risks and benefits of nonsteroidal anti-inflammatory drugs in children: a comparison with paracetamol. Paediatr Drugs 3: 817–858
40. Litalien C, Jacqz-Aigrain E (2002) New concepts in acute and extended postoperative pain management in children. Anesth Clin North Am 20: 115–135
41. Camu F, Van de Velde A, Vanlersberghe C (2001) Nonsteroidal anti-inflammatory drugs and paracetamol in children. Acta Anaesth Belg 52: 13–20
42. Sutters KA, Shaw BA, Gerardi JA, Hebert D (1999) Comparison of morphine patient-controlled analgesia with and without ketorolac for postoperative analgesia in pediatric orthopedic surgery. Am J Orthop 28: 351–358
43. Vetter TR, Heiner EJ (1994) Intravenous ketorolac as an adjuvant to pediatric patient-controlled analgesia with morphine. J Clin Anesth 6: 110–113

44. Mantzke US, Brambrink AM (2002) Paracetamol in childhood. Current state of knowledge and indications for a rational approach to postoperative analgesia Anaesthesist 51:735–746
45. Lieh-Lai MW, Kauffman RE, Uy HG *et al* (1999) A randomized comparison of ketorolac tromethamine and morphine for postoperative analgesia in critically ill children. Crit Care Med 27: 2786–2791
46. Anderson B (1998) What we don't know about ketorolac in children. Paediatr Anaesth 8: 451–460
47. Forrest JB (1997) Ketorolac for postoperative pain management in children. Drug Safety 16: 309–329
48. Harley EH, Dattolo RA (1998) Ibuprofen for tonsillectomy pain in children: efficacy and complications. Otolaryngol Head Neck Surg 119: 492–496
49. Autret-Leca E (2003) General overview of the use of ibuprofen in paediatrics. Int J Clin Pract [Suppl] 135: 9–12
50. Chiaretti A, Simeone E, Langer A, Butera G *et al* (1997) Analgesic efficacy of ketorolac and fentanyl in pediatric intensive care. Pediatr Med Chir 19: 419–424
51. Moiniche S, Romsing J, Dahl JB, Tramer MR (2003) Nonsteroidal antiinflammatory drugs and the risk of operative site bleeding after tonsillectomy: a quantitative systematic review. Anesth Analg 96: 68–67
52. Rusy LM, Houck CS, Sullivan LJ *et al* (1995) A double-blind evaluation of ketorolac tromethamine versus acethaminophen in pediatric tonsillectomy: analgesia and bleeding. Anesth Analg 80: 226–229
53. Lesko SM, Mitchell AA (1999) The safety of acetaminophen and ibuprofen among children younger than two years old. Pediatrics 104: 952
54. Wolfe MM (1999) Gastrointestinal toxicity of nonsteroidal antiinflammatory drugs. NEJM 340: 1888–1899
55. Lee A (1999) The effects of nonsteroidal anti-inflammatory drugs (NSAIDs) on postoperative renal function: a meta-analysis. Anaesth Intensive Care 27: 574–580
56. Mandell BF (1999) COX 2-selective NSAIDs: biology, promises, and concerns. Cleve Clin J Med 66: 285–292
57. Brookes P (1999) Interpreting the clinical significance of the differential inhibition of cyclooxygenase-1 and cyclooxygenase-2. Brit J Rheumatol 38: 779–787
58. Stichtenoth DO, Frolich JC (2003) The second generation of COX-2 inhibitors: what advantages do the newest offer? Drugs 63: 33–45
59. Kiefer W, Dannhardt G (2002) COX-2 inhibition and the control of pain. Curr Opin Investig Drugs 3: 1348–1358
60. Katz WA (2002) Cyclooxygenase-2-selective inhibitors in the management of acute and perioperative pain. Cleve Clin J Med 69: SI65–SI75
61. Cochrane DJ, Jarvis B, Keating GM (2002) Etoricoxib. Drugs 62: 2637–2651
62. Morrison BW (1999) Analgesic efficacy of the cyclooxygenase-2-specific inhibitor rofecoxib in post-dental durgery pain: a randomized, controlled trial. Clin Therapeutics 21: 943–953
63. Malmstrom K (1999) Comparison of rofecoxib and celecoxib, two

cyclooxygenase-2 inhibitors, in postoperative dental pain: a randomized, placebo and active comparitor controlled clinical trial. Clin Therapeutics 21: 1653–1663
64. Barden J, Edwards JE, McQuay HJ, Moore RA (2003) Single dose oral celecoxib for postoperative pain (Cochrane review). In: The Cochrane Library, Issue 2. Oxford
65. Ruoff G, Lema M (2003) Strategies in pain management: new and potential indications for COX-2 specific inhibitors. J Pain Symptom Management 25: S21–31
66. Pickering AE, Bridge HS, Nolan J, Stoddart PA (2002) Double-blind, placebo-controlled analgesic study of ibuprofen or rofecoxib in combination with paracetamol for tonsillectomy in children. Br J Anaesth 88: 72–77
68. Kart T, Christrup LL, Rassmussen M (1997) Recomanded use of morphine in neonates, infants and children based on literature review. 1. Pharmacokinetics. Paediatr Anaesth 7: 5–11
69. Pounder DR, Steward DJ (1992) Postoperative analgesia: opioids infusion in infants and children. Can J Anaesth 39: 969–974
70. Jin F, Chung F (2001) Multimodal analgesia for postoperative pain control. J Clin Anesth 13: 524–539
71. Morton NS (1993) Development of a monitoring protocol for the safe use of opioids in children. Paediatr Anaesthesia 3: 179–184
72. Kart T, Christrup LL, Rasmussen M (1997) Recommended use of morphine in neonates, infants and children based on a literature review: part 2. Clinical use. Paediatr Anaesth 7:93–101
73. William DG, Hatch DJ, Howard RF (2001) Codeine phosphate in paediatric medicine. Br J Anaesth 86: 413–421
74. Lehmann KA (1994) Tramadol for the management of acute pain. Drugs 47: 19–32
75. Viitanen H, Annila P (2001) Analgesic efficacy of tramadol 2 mg kg (−1) for paediatric day-case adenoidectomy. Br J Anaesth 86: 572–575
76. Finkel JC, Rose JB, Schmitz ML *et al* (2002) An evaluation of the efficacy and tolerability of oral tramadol hydrochloride tablets for the treatment of postsurgical pain in children. Anesth Analg 94: 1469–1473
77. Berde CB, Sethna NF (2002) Analgesics for the treatment of pain in children. New Engl J Med 347: 1094–1103
78. Holder KA, Dougherty TB, Porche VH, Chiang JS (1998) Postoperative pain management. Int Anesthesiol Clin 36:71–86
79. Mukherjee K, Esuvaranathan V, Streets C, Johnson A, Carr AS (2001) Adenotonsillectomy in children: a comparison of morphine and fentanyl for perioperative analgesia. Anaesthesia 56:1193–1197
80. Kiffer F, Joly A, Wodey E, Carre P, Ecoffey C (2001) The effect of preoperative epidural morphine on postoperative analgesia in children. Anesth Analg 93: 598–600
81. Tobias JD, Deshpande JK, Wetzel RC, Facker J, Maxwell LG, Solca M (1990) Postoperative analgesia. Use of intrathecal morphine in children. Clin Pediatr (Phila) 29: 44–48

82. Yee LY, Lopez JR (1992) Transdermal fentanyl. Ann Pharmacother 26: 1393–1399
83. Van Dijk M, Bouwmeester NJ, Duivenvoorden HJ, Koot HM *et al* (2002) Efficacy of continuous versus intermittent morphine administration after major surgery in 0–3-year-old infants; a double-blind randomized controlled trial. Pain 98: 305–313
84. Esmail Z, Montgomery C, Court C *et al* (1999) Efficacy and complications of morphine infusions in postoperative paediatric patients. Paediatr Anaesth 9: 321–327
85. Hendrickson M, Myre L (1990) Postoperative analgesia in children: a prospective study in intermittent intramuscolar injiection versus continuous intravenous infusion of morphine. J Pediatr Surg 25: 185–190
86. McDonald AJ, Cooper MG (2001) Patient-controlled analgesia: an appropriate method of pain control in children. Paediatr Drugs 3: 273–284
87. Ballantyne JC, Carr DB, Chalmers TC, Dear KBG *et al* (1993) Postoperative patient-controlled analgesia: meta-analyses of initial randomised controlled trials. J Clin Anesth 5: 182–183
88. Peters JWB, Bandell Hoekstra H, Huijer AS *et al* (1999) Patient controlled analgesia in children and adolescents. A randomised controlled trial. Paediatr Anaesth 9: 235–241
89. Walder B, Schafer M, Henzi I, Tramer MR (2001) Efficacy and safety of patient-controlled-opioid analgesia for acute postoperative pain. A qualitative systematic review. Acta Anaestesiol Scand 45: 795–804
90. Skues MA, Watson DM, O'Meara M (1993) Patient-controlled-analgesia in children. A comparison of two infusion techniques. Paediatr Anaesth 3: 223–228
91. Doyle E, Robinson D, Morton NS (1993) Comparison of patient-controlled-analgesia with or without a background infusion after lower abdominal surgery in children. Br J Anaesth 71: 670–676
92. Gunter JB (2002) Benefit and risks of local anesthetics in infants and children. Paediatr Drugs 4: 649–672
93. Gajraj NM, Pennant JH, Watcha MF (1994) Eutectic mixture of local anesthetics (EMLA) cream. Anesth Analg 78: 574–583

Author Index Volume 1–30

Advances and Technical Standards in Neurosurgery

Subject Index Vol. 1–30

Advances and Technical Standards in Neurosurgery

Springer-Verlag
and the Environment

WE AT SPRINGER-VERLAG FIRMLY BELIEVE THAT AN international science publisher has a special obligation to the environment, and our corporate policies consistently reflect this conviction.

WE ALSO EXPECT OUR BUSINESS PARTNERS – PRINTERS, paper mills, packaging manufacturers, etc. – to commit themselves to using environmentally friendly materials and production processes.

THE PAPER IN THIS BOOK IS MADE FROM NO-CHLORINE pulp and is acid free, in conformance with international standards for paper permanency.

MIX
Papier aus verantwortungsvollen Quellen
Paper from responsible sources
FSC® C105338

If you have any concerns about our products,
you can contact us on
ProductSafety@springernature.com

In case Publisher is established outside the EU,
the EU authorized representative is:
Springer Nature Customer Service Center GmbH
Europaplatz 3, 69115 Heidelberg, Germany

Printed by Libri Plureos GmbH
in Hamburg, Germany